STRATEGIC INTELLIGENCE MANAGEMENT

STRATEGIC INTELLIGENCE MANAGEMENT

National Security Imperatives and Information and Communications Technologies

Edited by
Babak Akhgar
Simeon Yates

ELSEVIER
AMSTERDAM • BOSTON • HEIDELBERG • LONDON
NEW YORK • OXFORD • PARIS • SAN DIEGO
SAN FRANCISCO • SINGAPORE • SYDNEY • TOKYO
Butterworth-Heinemann is an imprint of Elsevier
BH

Acquiring Editor: Pamela Chester
Editorial Project Manager: Amber Hodge
Project Manager: Punithavathy Govindaradjane
Designer: Mark Rogers

Butterworth-Heinemann is an imprint of Elsevier
225 Wyman Street, Waltham, MA 02451, USA
The Boulevard, Langford Lane, Kidlington, Oxford, OX5 1GB, UK

Notices

Knowledge and best practice in this field are constantly changing. As new research and experience broaden our understanding, changes in research methods or professional practices, may become necessary. Practitioners and researchers must always rely on their own experience and knowledge in evaluating and using any information or methods described herein. In using such information or methods they should be mindful of their own safety and the safety of others, including parties for whom they have a professional responsibility.

To the fullest extent of the law, neither the Publisher nor the authors, contributors, or editors, assume any liability for any injury and/or damage to persons or property as a matter of products liability, negligence or otherwise, or from any use or operation of any methods, products, instructions, or ideas contained in the material herein.

Library of Congress Cataloging-in-Publication Data
Application submitted

British Library Cataloguing-in-Publication Data
A catalogue record for this book is available from the British Library.

ISBN 978-0-12-407191-9

For information on all Butterworth–Heinemann publications
visit our website at store.elsevier.com

Printed and bound in China

13 14 15 16 17 10 9 8 7 6 5 4 3 2 1

TABLE OF CONTENTS

ACKNOWLEDGMENTS

We wish to thank to everyone who has contributed to this book. We would particularly like to acknowledge the following organizations and individuals.

ORGANIZATIONS

Athena Project Partners (European Union; EU)
Centre of Excellence in Terrorism, Resilience, Intelligence & Organised Crime Research (UK)
Critical Incident Analysis Group (U.S.)
Cyber Security Forum Initiative (U.S.)
European Commission (EU)
iSAR+ Project Partners (EU)
North East Counter Terrorism Unit (UK)
Raboud University (the Netherlands)
Sheffield Hallam University
Swedish Civil Contingencies Agency (Sweden)
Swedish National Defense College (Sweden)
TEKEVER (Portugal)
West Yorkshire Police (UK)

PEOPLE

Lord Carlile of Berriew CBE QC, Independent Reviewer of National Security Policy in Northern Ireland
Detective Chief Superintendent David Buxton, Head of North East Counter Terrorism Unit (NE CTU) of West Yorkshire Police and Association of Chief Police Officers (Terrorism and Allied Matters)
Meredydd John Hughes CBE QPM who served as Chief Constable of South Yorkshire Police from September 1, 2004 to 2011

ABOUT THE AUTHORS

Lord Carlile of Berriew CBE QC

Alex Carlile was called to the Bar by Gray's Inn in 1970. He is a Bencher of Gray's Inn. He sits as a Recorder of the Crown Court, as a Deputy High Court Judge, and as a Chairman of the Competition Appeal Tribunal. He was, until 2009, the Honorary Recorder of the City of Hereford. He was the Independent Reviewer of Terrorism Legislation 2001–2011 and the Independent Reviewer of the Government's new PREVENT policy. He remains the independent reviewer of National Security policy in Northern Ireland. He is the President of the Howard League for Penal Reform and of The Security Institute. He is a Fellow of King's College London and a Fellow of the Industry and Parliament Trust. He holds Honorary Doctorates of Laws in universities in Manchester, Wales, and Hungary. From 1983 to 1997 he was the Liberal then Liberal Democrat MP for Montgomeryshire in Mid Wales. During that time he served as spokesperson on a range of issues, including Home Affairs and the Law. He was Leader of the Welsh Liberal Democrats from 1992 to 1997. He was a leading proponent in the House of Commons of the War Crimes Act, and a founding officer of the All-Party War Crimes Group. He served on several Select Committees and Standing Committees of the House of Commons. He was appointed a Life Peer in 1999. Alex Carlile, until 2007, was Head of Chambers (Chairman) of one of the largest sets of barristers' chambers in London, 9-12 Bell Yard, and was awarded a CBE in 2012.

Babak Akhgar

Babak Akhgar is Professor of Informatics and Co-Director of the Centre of Excellence in Terrorism, Resilience, Intelligence and Organised Crime Research (CENTRIC) at Sheffield Hallam University and Fellow of British Computer Society. Professor Akhgar graduated from Sheffield Hallam University in Software Engineering. Afterwards he gained considerable commercial experience as a Strategy Analyst and Methodology Director for several companies. He consolidated this experience by obtaining a Masters degree (with distinction) in Information Systems in Management and a PhD in Information Systems. He has more than 100 referred publications in international journals and conferences. He is a member of editorial boards of three international journals and chair and program committee member of several international conferences. He has extensive and hands-on experience in development, management, and execution of large international knowledge management and security initiatives (e.g., combating terrorism and organized crime, cyber security, public order, and cross-cultural ideology polarization). Professor Akhgar has an established network of collaborators in various academic and law enforcement agencies locally, nationally, and internationally. The impact of his research on e-security, manifested in a multilingual portal for business crime reduction, and his research on combating organized crime and terrorism lead to an international research project with partners such as Europol and a number of law enforcement agencies (with project value of 3.2 M Euro). He has recently co-edited a book on intelligence management called *Intelligence Management: Knowledge Driven Frameworks for Combating Terrorism and Organized crime*.

Simon Andrews

Dr. Simon Andrews is a Senior Lecturer in Software Engineering within the Department of Computing at Sheffield Hallam University, a founding member of CENTRIC, and an international expert on Formal Concept Analysis (FCA). He is Editor in Chief of the *International Journal of Conceptual Structures and Smart Applications*

(IJCSSA). Originating from his PhD, awarded in 1996, his first interest was in writing and animating Z specifications. These interests lead him to FCA; applying advanced mathematical theory to contemporary industrial practice. He was a key participant at the FCA Algorithm Performance Competition at the 2009 International Conference of Conceptual Structures (ICCS) in which his FCA algorithm (In-Close) was the close runner up.

Simon Andrews is currently a Principle Investigator for a prestigious European Commission 7th Framework Program project, called CUBIST (www.cubist-project.eu/), where he is developing semantic tools and technologies for large-scale data analysis, which include knowledge discovery, visual analytics, and data scaling techniques. Working with public and private datasets, including those from local government, biomedical research, satellite telemetry, and human resources, he is developing algorithms, techniques, and software to elicit hidden meaning from data, including triples and linked data that are not accessible using traditional data analysis techniques.

Liz Bacon

Professor Liz Bacon BSc, PhD, CEng, CSci, FBCS, CITP, FHEA is Dean of the School of Computing and Mathematical Sciences at the University of Greenwich, Chair of the BCS Academy of Computing, a BCS (Chartered Institute for IT) Vice President and Trustee, and is a past Chair of the Council of Professors and Heads of Computing. She is/has been involved in many professional activities during her career such as working with e-skills UK, the Science Council, Parliamentary IT Committee (PITCOM), European Quality Assurance Network for Informatics Education (EQANIE), the National HE STEM Program, and was an ICT Thought Leader for the University of Cambridge International Examinations. Liz is a joint Director of the eCentre research group and has been involved in e-learning research for more than 10 years, which includes work on personalization, serious games, and enterprise development. She is an experienced systems designer and developer, with the bulk of her research and practice activity directly industry facing, through knowledge transfer and consultancy.

David Chadwick

David Chadwick is a senior lecturer in the Cyber-Security, Auditing, Forensics Education (C-SAFE) Centre at the University of Greenwich and specializes in teaching and researching systems auditing, cryptography, and steganography. Prior to lecturing David worked for many years as a systems project manager and systems auditor at Glaxo Smith-Kline and British Telecom. Before that, David worked at Lloyds of London (the insurance market). David is a member of the British Computer Society (Information Risk Management and Assurance Group) and also of the Information Systems Audit and Control Association (ISACA).

Mohammad Dastbaz

Professor Mohammad Dastbaz is the Dean of Faculty of Arts, Environment, and Technology at Leeds Metropolitan University. Mohammad's background is in design and development of digital media systems. His main research work over the recent years has been focused on the use of emerging technologies in learning and training and the development of "eGovernment." Professor Dastbaz, has been the Symposium Chair of Multimedia Systems in IEEE's Information Visualization (IV) conference since 2002 and has published widely, including numerous journal paper articles, conference papers, book chapters, and books on e-learning, eGovernment, and design and development of multimedia systems. He was the proposer of the EU's innovative PANDORA project that used emerging technologies to develop innovative virtual reality training packages to strategic decision makers dealing with national emergencies. Professor Dastbaz is a Fellow of the British Computer Society and UK's Higher Education Academy as well as the professional member of ACM and IEEE's computer society.

David Fortune

David had a 30 year career in policing where he undertook a variety of management and operational roles. During the latter years of his service he took a key role in innovation and strategic engagement with Europe for the police, focusing on organized crime and terrorism issues. He was a member of Europol's Expert group on property crime; and worked on the ODYSSEY project that looked at using new technology to address gun crime across Europe. He is currently a member of the European steering group of the FP7 EU funded COMPOSITE and DARIUS projects, which look at policing cultures across the EU and the police use of unmanned aerial vehicles, respectively. He is currently working with the DISASTER project looking at the interoperability of Emergency Management Systems across Europe and also works with Sheffield Hallam University at CENTRIC.

Dimitrios Frangiskatos

Dimitrios Frangiskatos is a senior lecturer in Computer Security and Audit at the University of Greenwich. He has a BEng and an MSc from the University of Greenwich and is teaching computer forensics, penetration testing, and computer security. Before joining the University of Greenwich in 2003, Dimitrios worked on secure network programming and system integration for Fluent Networks Ltd, then as an independent IT security consultant, and later joined NTA Monitor where he worked as part of their penetration testing team. The focus of Dimitrios' current research is on securing data in transit.

Peter Fussey

Dr. Peter Fussey is a Reader in Criminology in the Department of Sociology at the University of Essex, UK. The Department of Sociology at the University of Essex is renowned for its research excellence and is currently nationally ranked in joint first position for the quality of its research. Dr. Fussey's main research interests focus on surveillance, social control, and the city, particularly, in relation to crime and terrorism, and he has published widely in these areas. He is currently researching the form and impact of the 2012 Olympic security strategy on its wider urban setting and is also working on two large-scale ESRC and EPSRC funded research projects looking at counterterrorism in the UK's crowded spaces and at the future urban resilience until 2050. His other work focuses on organized crime in the EU with particular reference to the trafficking of children for criminal exploitation and is currently contracted by Routledge to write a monograph on the subject. Recent books include Securing and Sustaining the Olympic City (Ashgate) and Terrorism and the Olympics (Routledge).

Diane Gan

Dr. Diane Gan is co-founder and director of the C-SAFE Centre, which is a research and teaching group within the School of Computing and Mathematical Sciences at the University of Greenwich. She is involved in research on cyber security and digital forensics, particularly cyber malware. Diane has been instrumental in developing the university's Masters and short course programs with themes ranging from security and risk management to forensics and the law. She teaches communications systems, digital forensics, and cyber security at both undergraduate and postgraduate levels. Diane has strong links with industrial organizations related to penetration testing and has a PhD in Communications from the University of Greenwich.

Edward Halpin

Edward Halpin is a Professor in the Faculty of Arts, Environment, and Technology, and is co-chair of the Leeds Digital Research Centre, at Leeds Metropolitan University. Edward has a background in politics and social informatics/ information management, and describes himself as a political scientist with a particular interest in human rights, child rights, peace and conflict resolution, security, surveillance, and the use of information in the pursuit of these issues. He has worked as an expert for the European Parliament Scientific and Technical Options

(STOA) Unit, attended the European University in Florence as an Associate Schumann Fellow, and is a Fellow of the Royal Society for the Arts. Edward is Chair of the Geneva-based Human Rights Information and Documentation Systems International (HURIDOCS), an international nongovernmental organization that helps human rights organizations use information technologies and documentation methods to maximize the impact of their advocacy work. Professor Halpin has published widely including co-editing several books: *Cyberwar, Netwar, and the Revolution in Military Affairs* and *Human Rights and the Internet* (both Palgrave) and *Human Rights and Information Communication Technologies: Trends and Consequences of Use* (IGI).

Christopher P. Holstege

Christopher P. Holstege, MD, is an Associate Professor of Emergency Medicine and Pediatrics at the University of Virginia's School of Medicine and Chief of the University of Virginia's Division of Medical Toxicology. Dr. Holstege clinically provides care for poisoned patients and serves as a practicing member of the emergency medicine faculty. He has been integrally involved in a number of high-profile criminal poisonings. Dr. Holstege has over 100 publications in medical journals, periodicals, and books pertaining to emergency medicine and toxicology. He has edited or authored 10 books and is the lead editor of a book titled *Criminal Poisoning: Clinical and Forensic Perspectives*. He was the sole toxicologist on the Expert Behavioral Analysis Panel–Amerithrax Case, authorized by the Department of Justice. In appreciation of his work in both education and clinical service, Dr. Holstege received the 2003 University of Virginia School of Medicine Dean's Award for Clinical Excellence and the 2002 National Faculty Teaching Award from the American College of Emergency Physicians.

Hamid Jahankhani

Professor Hamid Jahankhani is the Associate Dean at the School of Architecture, Computing, and Engineering at the University of East London. He gained his PhD from Queen Mary College, University of London. In 2000 he moved to the University of East London to become the Professor of information security and cyber criminology. Professor Jahankhani's principal research area for a number of years has been in the field of information security and digital forensics. In partnership with the key industrial sectors, he has examined and established several innovative research projects that are of direct relevance to the needs of UK and European information security, digital forensics industries, critical national infrastructure, and law enforcement agencies. Most of his research work in the field has been manifested in a number of ways so that it contributed significantly to the measures governments must take to protect the security of information on the Internet, the implications of cybercrime in large corporations and individuals, resilience issues and threat assessment, risk analysis, and the formulation of security policies, vulnerability assessment, and forensics investigation of mobile devices. He is a respected international figure in the field of electronic security and digital forensics

Jan Kallberg

Dr. Jan Kallberg is Assistant Professor in Homeland Security at Arkansas Tech University and a Research Associate at the Cyber Security Research and Education Center (CySREC), Erik Jonsson School of Engineering and Computer Science, at The University of Texas at Dallas. Dr. Kallberg did his post doctorate work at CySREC. He has 20 years' experience in cyber security, and has recently been published in the *United States Air Force's Strategic Studies Quarterly* and *IEEE Security and Privacy*. Dr. Kallberg maintains his personal Web site under www.cyberdefense.com.

Mats Koraeus

Mats Koraeus is a PhD candidate at Radboud Universiteit Nijmegen and an analyst at the Swedish National Center for Crisis Management

Research and Training (CRISMART) at the Swedish National Defense College in Stockholm. Mr. Koraeus holds an MA in Political Science at Uppsala University. At CRISMART, he has been engaged in a 5 year applied research project for the Swedish Ministry of Defense, on the topic of the use (and abuse) of subject-matter experts and the application of highly detailed technical or expert knowledge in crises, as well as in numerous similar projects for the Swedish Civil Contingencies Agency. Other areas of special interest include cross-disciplinary research and methodology, such as applying literary theory or computer science in social-science research.

Kristian Krieger

Dr. Kristian Krieger is a Research Associate in the Department of War Studies at King's College London. His research focuses on disaster and risk governance by state and nonstate actors from a comparative interdisciplinary perspective. He is currently working on the EU-funded project Preparedness and Resilience against CBRN Terrorism using Integrated Concepts and Equipment (PRACTICE). Dr. Krieger holds a PhD in Human Geography from King's College London and was recently awarded the Third Giandomenico Majone Prize from the European Consortium for Political Research for his doctoral research.

Ivan Launders

Ivan Launders is a professional "Enterprise Architect" for British Telecommunications and software researcher. He has 23 years of software and telecommunication experience working across a wide range of products and solutions primarily in the UK, as well as specific projects in France, Germany, and Italy. His industrial experience has included implementing information system security on several government solutions and projects. Since joining the industry as a real-time application programmer, Ivan has progressed to an enterprise architect role leading technical sales and solution designs across a range of technologies and a wide range of market sectors that include heath, finance, central government, retail, and local government. In addition to his professional career at British Telecommunications, he is an associate lecturer in business computing within the Department of Computing at Sheffield Hallam University, UK. His research interests are in knowledge representation, particularly in capturing and modeling the exchange and use of knowledge in business transactions and business processes.

Glyn Lawson

Dr. Glyn Lawson MIEHF is a Lecturer within the Faculty of Engineering and member of the Human Factors Research Group at the University of Nottingham. His research expertise includes the human-centered development of new technologies, and he has held positions of responsibility on several EU- and nationally funded research projects. Glyn has particular expertise in the evaluation of methods for predicting behavior in emergency situations. He has also conducted research on deception detection and worked on requirements capture within the Security domain. Glyn is a registered member of the Institute of Ergonomics and Human Factors and sits on the Education and Training panel. In 2012, Glyn coedited a journal special issue on counterterrorism for "Applied Ergonomics: Detecting terrorist activities: Hostile intent and suspicious behaviors."

Peter Lehr

Dr. Peter Lehr is a lecturer in terrorism studies at the Centre for the Study of Terrorism and Political Violence (CSTPV), University of St. Andrews, Scotland/UK. As a regional specialist on the India Pacific, his teaching and research cover the following areas: maritime safety and security (including piracy and maritime terrorism), political violence and terrorism, and organized crime. He also works on critical infrastructure protection, with a focus on airport and seaport security. Dr. Lehr has published on political Islam in South/Southeast Asia and on Indian Ocean maritime security issues, including Somali piracy. He is the editor of *Violence at Sea: Piracy in the Age of Global Terrorism* (Routledge, 2007), and the co-editor (together with Rupert Herbert-Burns and Sam Bateman) of Lloyd's *MIU Handbook of Maritime Security* (Taylor & Francis, 2009). Currently, he is working on a book manuscript on piracy from ancient to modern times to be published by Yale University Press in Autumn 2013. He earned his PhD (Doctorate in Political Science) from the University of Heidelberg, Germany.

Eleanor Lockley

Dr. Eleanor Lockley is a researcher within the Cultural Communication and Computing Research Institute at Sheffield Hallam University. She predominantly works within the areas of communication and information studies. Over the last 4 years she has worked on a range of multidisciplinary projects that address aspects of human computer interaction from a sociological perspective.

George Loukas

Dr. George Loukas has a PhD in cyber security from Imperial College and is now a senior lecturer carrying out research and teaching on cyberphysical security technologies. Before joining the University of Greenwich, George was an advisor on technical R&D collaborations for defense and security companies and the UK government. He is currently on the editorial board of *BCS Computer Journal* and on the advisory board of three FP7 projects. George has participated in a number of high-profile international projects, funded by BAE Systems, Selex, BT, FP7, and EPSRC, on both cyber and physical security. He has published over 20 papers in prestigious journals and conferences in the areas of self-aware networks, denial of service attacks, disaster management, and robotic communications. George's current research focuses on the convergence of cyber and physical attacks in the context of terrorism.

Lachlan MacKinnon

Professor Lachlan MacKinnon BSc, PhD, FBCS, CITP, MIEEE, MACM, MAACE is a Professor of Computing Science (Strategic Development), Head of the Department of Smart Systems Technology, and Head of the Department of Creative Digital Technologies, in the School of Computing and Mathematical Sciences, University of Greenwich, UK. He is also Visiting Professor of Information and Knowledge Engineering at the University of Abertay Dundee, UK (where he was formerly Dean of the School of Computing and Creative Technologies), and Visiting Professor of Games and Multimedia Technology at Buskerud University College, Kongsberg, Norway. He is Chair of the Executive Committee of the British National Conference on Databases, and a member of the UK National Committee of the British Human Computer Interaction Group. Lachlan has a 20 year history in cyber security, starting from work in data security in heterogeneous, distributed systems, and while he was Dean of Computing at Abertay he introduced the first undergraduate and postgraduate degrees in ethical hacking in the UK. Currently his research is focusing on crisis management training, designing secure and usable systems and protocols, and countermeasures for cyber terrorism, combining research work in the eCentre, of which he is a joint Director, with work in C-SAFE.

Bárbara Manso

Mrs. Bárbara Guerra Manso is the Head of Image and Communication at TEKEVER and has been involved in strategic market research activities in the defense and security market since 1999. She was responsible for the social network analysis in experimentation research projects on network-enabled environments and new C2 approaches subcontracted by the U.S. Naval Postgraduate School held in partnership with the Portuguese Military Academy. She is graduated in International Relations by the Lisbon Technical University and is postgraduated in Information Warfare/Competitive Intelligence at the Portuguese Military Academy.

Marco Manso

Mr. Marco Manso is the Head of the Communication Systems Department at TEKEVER. Since 2004, he worked in R&D projects and activities involving national and international consortia, including the coordination of an R&D department. Between 2005 and 2012, he was the principal investigator of experimentation research projects on network-enabled environments and new C2 approaches and C2 agility subcontracted by the U.S. Naval Postgraduate School held in partnership with the Portuguese Military Academy. He is a coordinator member of the European IMG-S (www.imgs-eu.org), a group that involves more than 100 organizations from 24 EU countries. He participated in several NATO SAS and NIAG studies (SG-102/103: Soldier Systems Interoperability, SG-115: NEC Infrastructure for dissemination of ISR data, SAS-065: NATO NEC C2 Maturity Model, and SAS-085: C2 Agility). He is Graduated in Electronics and Computer Sciences by the Technical University of Lisbon, postgraduated and Masters Student in Information Warfare/Competitive Intelligence by the Portuguese Military Academy, and PhD Student in Earth and Space Sciences in the University of Évora in Portugal.

Franck Mignet

Franck Mignet, MSc is a researcher at Thales Research and Technology Nederland. He studied is MSc in Wave Propagation Modeling for Geophysics at Institut Francais du Petrole in France. He started his career as a research engineer for TotalFina Elf based at University of California at Santa Cruz. He then moved toward the software development industry holding positions as a tester and test manager for various applications including telecommunication products. His current research interest includes Bayesian networks, data fusion, and data mining.

Seyed Mohammad Reza Nasserzadeh

Seyed Mohammad Reza Nasserzadeh got his Master's degree in Information Technology Management from University of Tehran, the most prominent university in Iran. He is now a PhD candidate in System Management in the aforementioned university and his research interests are online behavior modeling, phenomenology of online consumer behavior, and simulating systems behavior using intelligent models and specifically cognitive maps. He has refereed publications in line with his research interests in *IEEE Computer Society* and *International Journal of Business Information Systems*.

Troy Nold

Troy Nold is a Captain in the U.S. Army and serves as an advisor to CIAG at the University of Virginia and as a Project Manager for Research Strategies Network (RSN). He graduated from Virginia Military Institute in 2007 with a BA in International Studies and Modern Languages and Cultures. After commissioning, he was assigned to the 4th Brigade, 10th Mountain Division in Fort Polk, Louisiana, where he served as an Infantry Platoon Leader, Executive Officer, and Battalion Rear Detachment Commander, earning the Army Commendation Medal in 2011. Captain Nold currently consults on forensic issues relating to military medicine and national security.

Patrick de Oude

Dr. Patrick de Oude received his Bachelors and Masters degree in Artificial Intelligence at the University of Amsterdam, The Netherlands, and graduated in 2004 and 2006, respectively. He continued with his PhD in Artificial Intelligence at the same university and received his PhD degree in 2010. From 2010 he worked as a researcher at the Thales Research and Technology Netherlands in Delft, The Netherlands. His interests and work are centered on (distributed) probabilistic reasoning and modeling, information fusion, graphical models, causality, multi-agent systems, data mining, and machine learning. At Thales Research and Technology his research focuses on environmental crisis management, cyber security, threat detection, and credit card fraud detection. In addition, he maintains and is responsible for the development of dynamic process integration framework, which is a multi-agent system that supports distributed hybrid information fusion involving both automatic reasoning processes and human experts.

Gregor Pavlin

Dr. Gregor Pavlin is senior researcher and project manager at Thales Nederland B.V. He received his PhD in computer science from the Graz University of Technology, Austria, in 2001. His current research interests are sound processing techniques and architectures for distributed information fusion and collaborative reasoning with humans in the loop, algorithms and modeling techniques supporting robust decentralized information fusion, self-organizing modular information fusion systems, and techniques for efficient implementation of complex service-oriented processing systems. Gregor is also the visiting researcher at the Intelligent Autonomous Systems group at the University of Amsterdam.

Julia M. Pearce

Dr. Julia Pearce is a Research Associate in the Department of War Studies at King's College London, where she also convenes the MA module "Social Dimensions of Terrorism." Her recent research has focused on public information needs following CBRN terrorist events (through the EU funded PIRATE research project), risk and crisis communication following a chemical incident emergency (through the EU-DG SANCO funded CIE Toolkit project), and stakeholder perceptions of the resilience of the UK's future energy and transport infrastructure to natural and malicious threats (through the EPSRC/ESRC funded Resilient Futures project). Dr. Pearce is a social psychologist by training and her research interests include risk and crisis communication, social and national identity, social representations, and moral panic.

Simon Polovina

Dr. Simon Polovina is a Senior Lecturer in Business Computing within the Department of Computing at Sheffield Hallam University, UK. He has 8 years of industrial experience in accounting and information and communication technologies (ICT) and has published widely with over 90 learned publications to date. His interests are in the use of smart applications and how they can detect novel or unusual transactions that would otherwise remain as lost business opportunities or represent illicit business or criminal activity. An underlying theme of this work is conceptual structures (CS), which harmonize the human conceptual approach to problem solving with the formal structures that computer applications need to bring their productivity to bear. Simon has technical expertise in SAP technologies, service-oriented architecture, business process management, object-oriented analysis and design, conceptual modeling of organizations as multi-agent systems, and in interaction design. He is a Principal Investigator for the prestigious European Commission 7th Framework Program project CUBIST, where he is applying CS and co-managing the project. CUBIST is centered on applying smart technologies (namely CS and the Semantic Web) to business intelligence.

Thomas Quillinan

Dr. Thomas Quillinan BEng, MSc, PhD, CISSP is a researcher at Thales Research and Technology Nederland. He received a PhD in the area of Security for Distributed Systems, and an MSc in Computer Science from University College Cork in Ireland. He previously worked as part of the IIDS group in the Vrije Universiteit Amsterdam, working in the areas of Distributed Systems and Crisis Management. He originally studied at the University of Limerick in Ireland, where he received an undergraduate degree in computer engineering. He is also a Certified Information Systems Security Professional. His research interests include the security of systems and networks.

Kellyn Rein

Kellyn Rein is a research associate in Command and Control Systems at Fraunhofer FKIE in Germany. There she is part of the C2LG development team and is currently working on a model for evaluation of uncertainty in information fusion. She is also a member of the Affiliate Faculty in the Center of Excellence in Command, Control, Communication, Computing, and Intelligence in the Volgenau School of Engineering at George Mason University in Fairfax, Virginia, in the United States. She holds a Bachelors degree in German and Mathematics; two Master's degrees, one in Management and a second in Computer Information Systems; and is currently finishing doctoral work in Communication Science at the University of Bonn with a focus on quantification of uncertainty in natural language-based information for information fusion.

M. Brooke Rogers

Dr. Brooke Rogers is a Senior Lecturer in Risk and Terror in the Department of War Studies, and a Co-director of the MA in Terrorism, Security, and Society at King's College London. Her current research follows two key themes: responding to terrorism and violent radicalization. The majority of her projects include cross-disciplinary investigations of public and practitioners' responses to CBRN terrorist incidents (i.e., Home Office, PIRATE, CIE Toolkit, PRACTICE), protecting crowded places (Safer Spaces; Designing with Intent: Influencing Behaviour in Transitional Spaces), and the resilience of the critical national infrastructure (Resilient Futures). Dr. Rogers is a social psychologist by training. She holds honorary associations with a number of organizations and has membership on the Community Resilience Programme Steering Group for the Cabinet Office (CCS), the Community Resilience Guiding Principles Advisory Board (Cabinet Office UK), the Health Protection Agency (HPA) Psychosocial and Behavioural Issues Expert Advisory Sub-group of the ERDG, and the Royal Society Advisory Committee on Scientific Aspects of International Security (SAIS). Dr. Rogers has been involved in teaching and training for the Home Office, MOD, Metropolitan Police, and Police National CBRN Centre. She lectures on NATO courses in five countries, and has held advisory positions with a number of government organizations on a national and international level. She also enjoys membership on the editorial board for the journal, *Mental Health, Religion and Culture*.

Gregory B. Saathoff

Gregory B. Saathoff, MD is Associate Professor of Research in Public Health Sciences, and Associate Professor of Emergency Medicine at the University of Virginia's School of Medicine. He also serves as Executive Director of the University of Virginia's Critical Incident Analysis Group (CIAG). In this capacity, he directs the operation of the group, which operates as a "ThinkNet" that provides multidisciplinary expertise in developing strategies that can prevent or mitigate the effects of critical incidents. In 1996 he was appointed to a U.S. Department of Justice Commission charged with developing a methodology to enable the FBI to better access nongovernmental expertise during times of crisis. Dr. Saathoff has testified to the Senate Committee on Homeland Security, House Committees, and the U.S. Commission on Civil Rights on the issue of prison radicalization. In addition to scientific articles and book chapters, his books include the *Crisis Guide to Psychotropic Drugs and Poisons*, and co-editorship of *Criminal Poisoning: Clinical and Forensic Perspectives*. Dr. Saathoff served as the Chairman for the Expert Behavioral Analysis Panel of the Amerithrax Case. In 2009 this independent panel of behavioral experts was authorized by the Chief Federal Judge Royce Lamberth to examine the mental health issues of Dr. Bruce Ivins and the lessons that can be learned, which may be useful in preventing future bioterrorism attacks.

Rose Saikayasit

Dr. Rose Saikayasit is a Research Fellow within the Human Factors Research Group in the

Faculty of Engineering at the University of Nottingham. Rose's current research is focused on understanding and eliciting user requirements of frontline security personnel and stakeholders to develop requirement specifications for further applications in security and counterterrorism, as part of the EPSRC "Shades of Grey" project. Her research interests include understanding and supporting collaboration in colocated and virtual engineering and design teams through the use of user-centered design of novel collaborative technologies. She is experienced in user requirements elicitation, scenario development, and human factors evaluation of technologies in a variety of security and nonsecurity sectors.

Paul de Souza

Paul de Souza is the Founder/President/Director of the Cyber Security Forum Initiative (CSFI) and its divisions Cyber Warfare Division (CSFI-CWD) and Law and Policy Division (CSFI-LPD). He is a Federal Director of Training and Education for Norman Data Defense Systems, and he also teaches PSSL 6247 Cyber Defense Strategies at George Washington University. de Souza has over 13 years of cyber security experience and has worked as a Chief Security Engineer for AT&T, where he designed and approved secure networks for MSS. He also worked for CSC and U.S. Robotics as a security engineer. He has consulted for several governments, military organizations, and private institutions on best network security practices.

Andrew Staniforth

Andrew Staniforth is a serving police counterterrorism Detective Inspector and former Special Branch detective. He has extensive operational experience across multiple counterterrorism disciplines, now specializing in security-themed research at the North East Counter Terrorism Unit, one of four such units across the UK. As a professionally qualified teacher, Andrew has designed national counterterrorism exercise programs and continues to deliver counterterrorism training to senior police commanders from across the world as part of the International Commanders Programme at the National Police College, Bramshill, England. Andrew is the author of the *Blackstone's Counter-Terrorism Handbook* (Oxford University Press 2009, 2010, 2013), and *Blackstone's Handbook of Ports & Borders Security* (Oxford University Press, 2013), both used by operational practitioners across the national security domain in the UK. Andrew is also the author of the *Routledge Companion to UK Counter-Terrorism* (Routledge, 2012), the forthcoming *Blackstone's Practical Policing: Preventing Terrorism & Violent Extremism* (Oxford University Press, 2013), and has over 150 counterterrorism-related articles and papers published in policing and security journals. Andrew is a Senior Research Fellow at CENTRIC; Visiting Researcher at C3Ri at Sheffield Hallam University; and Senior Research Fellow, Criminal Justice Studies, School of Law, University of Leeds.

Alex W. Stedmon

Dr. Alex W. Stedmon, FIEHF, CPsychol, CSci, FRSA is a Chartered Psychologist, Chartered Scientist, Fellow of the Institute of Ergonomics and Human Factors, and Fellow of the Royal Society for the encouragement of Arts, Manufactures, and Commerce. He is a Reader in Human Factors within the Culture, Communication, and Computing Research Institute at Sheffield Hallam University. He worked for the Ministry of Defence before moving into academia and explores human factors issues of technology use in security applications as well as contextual methods for investigating suspicious behaviors in various domains. Alex is one of the technical leads for a strategic security consortium (EPSRC project: Shades of Grey) and received Centre for Defence Enterprise funding for projects on automated CCTV issues, identifying human pheromones associated with dishonest behaviors, and issues of collaborative intelligence information gathering. Alex is also part of the GEOCORE consortium working in the development of methods and techniques to underpin and improve the handling and understanding of GEOINF/GEOINT within the

Ministry of Defence. More recently Alex has edited a journal special issue on counterterrorism for *Applied Ergonomics* and has been commissioned to write a book on counter terrorism and hostile intent: human factors theory and application.

Eric Stern

Dr. Eric Stern is Visiting Professor at the University of Virginia and Professor of Political Science/Crisis Management at the Swedish National Defense College in Stockholm, where he served as Director of the Swedish National Center for Crisis Management Research and Training (CRISMART) from 2004 to 2011. Dr. Stern holds a PhD from Stockholm University and a BA from Dartmouth College. He has published extensively in the fields of crisis and emergency management, security studies, executive leadership, foreign policy analysis, and political psychology. Among his monographs and edited volumes are *The Politics of Crisis Management: Leadership Under Pressure* (Cambridge University Press, 2005), winner of the American Political Science Association's 2007 Herbert Simon Award, and *Beyond Groupthink: Political Group Dynamics and Foreign Policymaking* (University of Michigan Press, 1997). Particular areas of expertise include post-crisis evaluation and learning, interactive education, and case research/teaching methodologies. In addition to his scholarly work, Professor Stern has collaborated closely with many government agencies and international organizations on a wide range of consulting, educational (including training and exercise development), and applied research projects.

Fahimeh Tabatabaei

FahimehTabatabayi received her Master's degree in Information Technology Management from MehrAlborz University following her Bachelor's degree in Applied Mathematics from Shiraz University in Dubai. Her thesis concentrated on simulating online behavior, which was nominated for the best thesis and won the contest across the university. Her research interests also include online behavior modeling using artificial intelligence techniques. She has a number of publications in artificial intelligence techniques.

Bhavani Thuraisingham

Dr. Bhavani Thuraisingham is the Louis A. Beecherl, Jr. Distinguished Professor and Director of CySREC at The University of Texas at Dallas. She is an elected Fellow of IEEE, the AAAS, the British Computer Society, the Society for Design and Process Science, and the Society of Information Reuse and Integration (subcommittee of IEEE Systems, Man and Cybernetics Society). She has unique experience working in the commercial industry, research laboratory, U.S. government, and academia, and her 30+ year career includes research and development, technology transfer, product development, program management, and consulting for the federal government. Her work has resulted in 100+ journal articles, 200+ conference papers, 3 U.S. patents, 12 books, and numerous awards. Under her leadership, CySREC has generated over $20 million in research and education funding from AFOSR, NSF, IARPA, NGA, NASA, ONR, NIH, DARPA, Raytheon, Tektronix, and others.

Steve Wright

Dr. Steve Wright is a Reader and lecturer in the School of Applied Global Ethics at Leeds Metropolitan University, teaching a variety of courses on peace, conflict, state security, and advanced surveillance. Steve lectures worldwide on these issues and apart from interfaith dialog, his recent publications include work on climate change and border control, sublethal weapons, cities under siege, and global telecommunications surveillance. Steve studied at Manchester University, earning a BSc in Liberal Studies in Science. He went on to work for the Council, first as the Head of Manchester's controversial Police Monitoring Unit, then later in the Chief Executive's Department working on such diverse issues as emergency planning, nuclear war, and accessible public transport. He co-founded the Omega Foundation in 1989, which is still based in the city today, working on tracking the supply of military, security, and police equipment to repressive regimes. Today, he retains his link with the city through being a member of the board of the Mines Advisory Group, which removes explosive remnants of war from battlefields, cities, and villages all over the world.

Simeon Yates

Professor Simeon Yates is the Director of C3RI at Sheffield Hallam University in the UK. The C3RI includes the Art and Design and Communication and Computing Research Centers. Simeon is also a Chair at CENTRIC. His research focus is on the social, political, and cultural impacts of digital media. This includes a longstanding focus on digital media and interpersonal interaction, issues of digital exclusion, and more recently projects that address the use of digital technologies in the context of security and crises. Over the last 3 years Simeon was one of the leads on a major interdisciplinary program at Sheffield Hallam and has undertaken research on best practice in supporting cross-disciplinary working.

FOREWORD

It is a real privilege to be invited to write the Foreword for this edited collection of works from leading practitioners and academics concerned in the field of national security intelligence management. As Independent Reviewer of Terrorism Legislation in the United Kingdom for 9 years and Independent Reviewer of National Security policy in Northern Ireland, I have spent considerable time with operational practitioners who manage intelligence in support of delivering security to the nation. This is, and remains, a complex task. Much of this work is conducted quietly and patiently by many extremely able practitioners, the unsung guardians and sometimes heroes of national security and public safety. Our streets, and especially public places, would be much less safe were it not for these men and women.

To protect a nation and to remove or mitigate the diverse range of security threats necessitates a preventative and intelligence-led approach. In the post 9/11 era the appetite for gathering and recording intelligence has been relentless. While the drive and determination to gather new intelligence by the security apparatus of the state has served to overcome many security challenges it has also created new ones. The ability to capture, assess, store, and effectively mine information from ever-increasing intelligence databases fragmented across government agencies and departments is one such challenge. When combined with the phenomenon of data created by social network media and new smart mobile technology adds greater complexity for the effective management of intelligence.

Of course, in the field of national security attention is always focused (and rightly) on terrorism. There are few terrorists, and happily many of them are not especially competent in bringing their aspirations to fruition. However, the really dangerous terrorist is difficult to detect, often very security conscious, and a huge danger to society as a whole. International terrorist events continue to provide a worrying and disturbing geopolitical context. The reality that violent Jihadists all over the world are working together against the established order, and that with rare exceptions there are links of some kind to weave all the terrorist cells into an international destructive tapestry, is now well argued. When added to the aspiration of terrorist groups to acquire the capacity and capability to mount chemical, biological, radioactive, or nuclear attacks provides deeper concerns for citizen's safety and for national security.

Yet on a domestic level, we have come to learn more over recent years about the home-grown al-Qaeda inspired terrorist threat, and most important, that the issues, which give rise to the development of extremist perspectives and to terrorism, begin at the most local level. Thus, national security increasingly depends upon neighborhood security and good local community-based intelligence to prevent terrorism at its source has come to the fore.

Like terrorism, cyber attack is not simply a risk for the future. Government, the private sector, and citizens are under sustained cyber attack today from both hostile states and criminals. They are stealing our intellectual property, sensitive commercial and government information, and even our identities in order to defraud individuals, organizations, and the government. Cyber security has been assessed as one of the highest priority national security risks to the UK and rightly so as cyberspace, via the Internet and social network media, is already woven into the very fabric of our society. Therefore, activity in cyberspace will continue to evolve as a direct national security and economic threat, as it is refined as a means of espionage and crime, and continues to grow as a terrorist enabler.

In addressing these contemporary challenges for national security this edited collection achieves the rare combination of fusing practitioner's real-life experience and true academic rigor with private sector expertise. Those of us with a daily interest in the field of national security will find this new work to be an insightful reference and one which will be required reading for policy makers, practitioners and academics across the landscape of national security. This timely volume, which fills a lacuna in intelligence management literature, has been driven by the Centre of Excellence for Terrorism, Resilience, Intelligence and Organised Crime Research (CENTRIC), a multidisciplinary and end-user centric research body, located within the Cultural, Communication and Computing Research Institute (C3ri) at Sheffield Hallam University. Their mission, to provide a platform for researchers across the security domain, is evident throughout the very fabric of this volume, the reading of which will serve to ensure that we are better informed today to meet the national security challenges of tomorrow.

Lord Carlile of Berriew CBE QC

CHAPTER

1

Introduction: Strategy Formation in a Globalized and Networked Age—A Review of the Concept and its Definition

Babak Akhgar, Simeon Yates, Eleanor Lockley

INTRODUCTION

Threats to nations and their citizens—be they human made or natural—are often moments of crisis and can have huge impacts on individuals, communities, businesses, and societies. Preventing, mitigating, or supporting resilience in the face of such crises remains a major role for nation states. Yet we inhabit a world where nations may be replaced by networks and where threats are global in reach. How do the governmental and policy functions of nation states address such changes? How can they develop strategies to deal with threats to themselves and/or their neighbors? Importantly, how can the technologies that have created this globalized and networked world—information and communication technologies (ICT) as well as global travel and trade—provide tools to support such strategies? Or are these networks opening us up to new threats and dangers?

In this book we are concerned with the links between ICT and national security strategies. Overall the concern is with "national security"; however, that may be conceptualized. This book addresses contemporary understanding of national security and threats to national security, predominantly from those tasked with providing this security, or those providing academic and research support to these practitioners. It also addresses issues of technological development and design, information, and communication technology use as well as the threats that such new technologies bring to nation states.

It is interesting to note that journalist and writer Simon Winchester argued that the Internet was born in 1883 with the eruption of Krakatoa. The international telegraph network spread the news of this disaster around the world almost as fast as the sound waves from the explosion. This disaster also demonstrated the fragility of this fledgling global ICT network as the resultant earthquakes, tsunamis, and eruption cut undersea cables. We now live in a networked world and this reality needs to be part of our security planning and strategies.

NATIONAL STRATEGY AND STRATEGY FORMULATION PROCESS

To discuss the notion of National Security strategy formulation process, it is essential to first define strategy, particularly at the national level. The concept of strategy originates from the military. The word derives from the Greek *strategia*, the office of the military council. It is commonly used in many domains to imply an overall plan or approach. Since the widespread adaptation of the concept of strategy by academics and practitioners in the 1950s, particularly in business and military schools, there has been a proliferation of research on "strategy." Despite this, a general consensus about what strategy is has not been reached. Across a broad range of disciplines a variety of conceptual frameworks and methodologies have been developed for the formulation and implementation of "strategies." All of these are based on different interpretations of the meaning of strategy. Some key examples include: Andrews (1971), Mintzberg (1976), Hofer and Schendel (1984), Porter (1985), Rainer (1989), Johnson and Scholes (1993), Stacey (1993), Thompson (1993), Levy (1994), Lynch (1997), Hussey (1998), and Wickham (2000). In this chapter we are not aiming for a critical evaluation of strategy as a concept, and we are not introducing various schools of thought on strategy or strategy formulation process. Instead, we are using a concept of strategy that is, in our view, more closely aligned to the notion of *national security*.

Nickols (2000) stated:

> *Strategy is at once the course we chart, the journey we imagine and, at the same time, it is the course we steer, the trip we actually make. Even when we are embarking on a voyage of discovery, with no particular destination in mind, the voyage has a purpose, an outcome, an end to be kept in view.*

Based on this view, and that of Mintzberg (1976), we put forward the proposition that strategy at the national level can consist of the epistemological combination of proposition, perspective, position, results of an oriented long-term plan, and pattern. In essence national strategy is the bridge between governmental policy or high-order goals for maintaining and preserving national interests on the one hand and concrete actions on the other. In short, national strategy is a term that reflects an evaluable framework that provides specific guidance for specific actions in pursuit of a national interest by utilizing resources within dynamic local and global settings. This setting and the strategy that is cognizant to some degree of the complex web of thoughts, ideologies, visions, doctrines, ideas, insights, knowledge, legal and constitutional frameworks, experience, goals, expertise, values, perceptions, and expectations of those making, implementing, or impacted by the strategy. Importantly the strategy will reflect the collective mental constructs of those individuals in positions of power or responsibility with regard to the strategy. Nickols (2000) stated that strategy "has no existence apart from the ends sought." This implies that the necessary precondition for formulating strategy is a clear and widespread understanding of the ends to be obtained. Without these ends in view, action is superficial and likely to lead to "strategic failure." Thus national strategy formulation also can be defined as a pragmatic, action-oriented, and goal-driven process of transforming current national status (AS IS) to the desired status (TO BE) based on mental constructs (e.g., vision, values, and motivation) of individuals with governing and policy-making responsibilities. This has to take place within the constraints of relevant material, social, cultural, constitutional, and legal frameworks.

When considering national strategy, the following properties should also be considered:

- The time frame for implementation is usually medium to long term.
- The evaluation of the value of the impact of a national strategy should be based on the return on investment and return on capital employed; however, these may be consistently measured.

- Overall national strategy usually combines more functional strategies (e.g., defense strategy, financial strategy, and social and economic strategy), which also decompose to operational strategies (e.g., naval strategy, banking sector strategy, policing strategy).
- National strategy should support/shape a national competitive advantage or reduce its rivals' competitive advantage. National strategy is not limited to geographical and territorial boundaries of a nation; therefore, the notion of competitive advantage should be considered at a global level.

Johnson and Scholes (1993) provided a framework (see Figure 1.1) for summarizing the key stages of strategic management. In the next section this view of strategy is adapted and applied in the context of a national strategy.

At the heart of Johnson and Scholes' (1993) framework lie three interconnected components:

1. Strategic analysis for understanding the strategic position of a nation
2. Strategic choice for generation of strategic options, evaluation of strategic options, and selection of strategy
3. Strategy implementation for the translation of strategy into action by considering national resources, legal and constitutional frameworks, and the management of deployment/implementation

The glue that provides the linkage between the three components is the *strategic intelligence* of the nation. This consists of all of the critical data, information, and knowledge needed for realization of the national strategy. Therefore the systems, platforms, and infrastructure that support or create such intelligence are considered one strategic national asset needed for realization of national strategy.

The realization of the above model requires a set of analytical tools that can aid the strategy formulation process. There are various tools and techniques that have been used in the strategy formulation process in military and business planning, and their application can be extended and adapted for a national strategy formulation process. A simple taxonomical classification can place their application into three categories:

FIGURE 1.1 Elements of national strategy life cycle. *(Johnson and Scholes, 1993).*

1. Those who aid the analysis of the environment in which strategy is formulated
2. Those who identify choices, options, and alternative scenarios
3. Those tools and techniques that facilitate realization/implementation and evaluation of strategy.

Table 1.1 summarizes some of the most important (widely used) techniques, models, frameworks, and tools. Here we have to re-emphasize that the tools and models stated in Table 1.1 are only a sample representation, and in-depth evaluation of every tool and model stated in this table is outside the scope of this book. We have used a control vocabulary to adapt the application of these techniques, tools, and framework, for national strategy.

TABLE 1.1 Strategy Techniques, Models, Frameworks, and Tools

Tools	Purpose/Rational	Developer	Application in National Strategy Life Cycle
SWOT (strengths, weaknesses, opportunities, and threats)	To formulate a generic strategy by analyzing a nation's environment and the internal resources of a country	Andrews (1971)	Analysis
PEST (political, economic, sociocultural, and technology)	Used as a checklist to consider prompt analysis of different environmental influences	Used by a number of researchers and practitioners with no specific reference	Analysis
SFP model	Formulates strategy based on environmental and resource analysis	Andrews (1971); Hofer and Schendel (1984)	Implementation
Five forces analysis	Enables nations to examine their position within a competitive environment within a global setting	Porter (1985)	Analysis and Choice
Value chain analysis	Examines governmental offices' internal effectiveness and their competitiveness; also can be used for capability assessment	Porter (1985)	Analysis and Implementation
Force field analysis	Used to identify the influential forces involved in complex situations such as crisis management	Lewin (1951)	Analysis
Porter's diamond	Used to identify the national competitiveness	Porter (1985)	Analysis
Cost/benefit analysis	Used to place financial values on all costs and benefits of a particular strategic option	Mirsham (1980)	Analysis and Choice
Critical success factors analysis	Provides analysis and assessment of a nation's most important area of concern when failing to draw attention to it can result in catastrophic failure	Denning (2000, 2012)	Analysis and Implementation
Stakeholders analysis	Used to identify any state or international actors, group, or individual who can affect the deployment of a particular strategy.	Ansoff (1965); Argenti (1968)	Analysis

NATIONAL SECURITY

The term national security has no universally accepted definition and concepts linked to it are often ambiguous with an emphasis on freedom from military threat. A common understanding of national security focuses on the protection of society and citizens against threat or risk by government or nation states. It is traditionally associated with military defense and the security services, and is often focused on a local level. The provenance of the concept of national security appears to be the U.S. National Security Act of 1947 that was in response to WW II. It is therefore unsurprising that much of the literature around national security focuses upon the American perspective. The National Security Act (1947) was considered legally ambiguous in part, although this was allowed for broad interpretation whenever issues were deemed to threaten the interest of the state (Romm, 1993).

Although the National Security Act did include a realization that it was more than just military security, this was understated from the beginning (Paleri, 2008). The Clinton administration's National Security Strategy for a New Century (1999) said:

Nearly 55 years ago, in his final inaugural' address, President Franklin Delano Roosevelt reflected on the lessons of the first half of the 20th Century. 'We have learned,' he said, 'that we cannot live alone at peace. We have learned that our own well-being is dependent on the well-being of other nations far away. We have learned to be citizens of the world, members of the human community.

Once again though this strategy similarly focuses on strong military might. The terrorist events of recent years (New York twin towers attack of 2001, Madrid bombing 2004, London 2005 metro bombing, Mumbai 2008, etc.) have forced governments to think about national security in a broader context than just military operations and issues of competitive advantage in armed forces. For example, the current UK national security strategy readily admits that the country faces a different and more complex range of threats and includes terrorism, cyber warfare, and natural disasters as examples. Current national security strategies have therefore had to expand to consider sophisticated antiterrorism strategy. This is of course especially true as domestic attacks have been made on civilians and there has been a greater diversity of potential threats across the globe (see UK National Security Strategy; HM Government, 2011).

Therefore, today's national security strategy does not focus just on military- and defense-related responses; it also emphasizes a broader range of issues such as the economic, energy, environmental, social, and cultural security of a nation (Romm, 1993). In particular, both the United States and UK refer to the concept of "national interest." According to Knight (2002, p. 2):

a national security strategy is a place to spell out national interests, threats to those interests, and the organization and allocation of national resources to pursue and defend those interests.

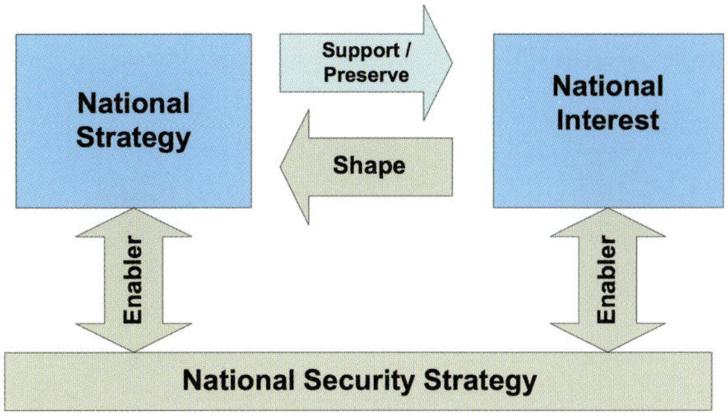

FIGURE 1.2 Inter-relationship between national strategy, national interest, and national security strategy.

Hence national security becomes a derivative of national interest. There are many methodological approaches to national security strategy development (e.g., U.S. Army War College Methodology, French National Security Strategy, and HM Government National Security Strategy).

In our view the ultimate objective of national strategy should result in a set of actions for the preservation of long-term national interest. Figure 1.2 provides a representation of the inter-relationship between national strategy, national interest, and national security strategy.

STRATEGIC INTELLIGENCE

It should be noted that a lack of understanding in the above inter-relationships can often lead to failure of the state (or government) to preserve its interest in the long term. For example, a state may provide military, diplomatic, financial, and tactical training support for a specific militant group, a social movement, or another state. This may be done under the pretext of its "national values" to partially gain a short-term national security objective (i.e., a containment solution for an international conflict). However, in the long term the same group/state might create a fundamental threat to the security and interests of the nation. In addition, the lack of "strategic intelligence" about the "situation of concern," the local and global context, can also lead to consequential threat to the national interest of a nation. Such lack of intelligence might cover the worldview, ideological stances, geopolitical relations, international and local legal frameworks, and the social and cultural understandings of all parties.

We argue then that maintaining a robust body of strategic intelligence available quickly to policy makers when developing strategy is key. Below is a canonical set of requirements needed for establishing strategic intelligence as an effective instrument for national security strategy. These requirements represent a framework that we believe is needed to link national strategy, national interest, and national security.

- **Assessment:** Provides the tasking and coordination mechanisms across a national security architecture with an accurate picture of the situation in its area of responsibility, how that picture is changing now, and how it may change in the future. The assessment component in essence should address global environmental scanning, which includes national strengths, weaknesses, opportunities, and threats (both current and future).
- **Knowledge:** Ensures that all parts of national security intelligence gathering machinery is focused upon relevant threats and risks to provide key decision makers the knowledge and learning processes they require to effectively frame strategic policies, priorities, and resource use. Additionally, this canonical component emphasizes knowledge-based decision-making processes and the need for utilizing data and information within national data/information repositories (Akhgar and Yates, 2011).
- **Holistic:** Ensures that intelligence is gathered, assessed, disseminated, and actioned from sources pertaining to the full range of risks and threats, which includes both internal (domestic) and external (international). An "all hazards" approach to strategic intelligence is now required in this increasingly interdependent and globalized world.
- **Goal driven:** Provides all national security intelligence apparatus operating across the full landscape of government to adopt a collaborative approach, each bringing their own specific area of expertise to bear in concert with one another to tackle a specific hazard in the national interest.
- **Adaptive:** Provides a flexible and dynamic approach to protect national security interests ensuring that the national security apparatus and strategic focus has the ability to swiftly respond to all new and emerging threats: whether from opposing states, international or domestic groups, lone actors, or from cyber space.
- **Result oriented:** Ensures that specific and measurable targets are set via a performance-driven approach that is embedded into the very culture of national security strategic intelligence.

The above set of requirements represents a framework that we believe is needed to link national strategy, national interest, and national security.

INTERCONNECTED WORLD

According to Buzan (2009), globalization brings with it specific new challenges for national security. The increase in the range, scale, and intensity of human interactions goes hand in hand with increased interdependence. Buzan argued that interdependence increases the importance of the system level (the global economy, the cyber sphere, the global environment, international society) over the unit level (states) and this in turn increases the number of shared interests and shared fates (global epidemics, economic crises, cyber war, environmental change). Buzan (2009) stated that:

> *National security thus ceases to be largely self-referential, instead becoming embedded in threats, possible fates and courses of action shared with many other states and humankind generally.*

Through globalization, national interest becomes relevant on a local and global level and, as previously suggested, national interest directly impacts upon national security.

Linked to globalization, and a key consideration for national security, is the notion of the knowledge society and its enabling technological factors such as the Internet, cloud computing, and big data solutions. This constellation of technologies, networks, and global linkages has been variously described as the "Networked Society." These are sociological not technology based models of contemporary society. Like Buzan (2009) the key authors in this domain (Hiltz and Turoff, 1978; van Dijk, 1991; Castells, 2000; Rainie and Wellman, 2012) focus on the social, economic, and political interdependencies that globalization and new digital media bring. Importantly, they note the centrality of "networks" of people or systems to contemporary societies. A key feature of their ideas is the mapping of flows of information with flows of power, influence, and economic exchange. Although they see these as often supranational networks and flows, they also note that the hub locations for networks, both of people and systems, are often cities where major social, economic, and political elites are based.

This networked society provides one of the key contexts for contemporary national security strategy. On one level it challenges the idea of "nations" and the anthropomorphic view of nations as individual actors. As Buzan (2009) noted it creates the need to look at the links between systems, institutions, and networks within, between, and among nation states. It also implies that the physical network infrastructure of the Internet, digital media, and transnational travel is the major potential target for threats to both national and global security. Last, but just as important, is the implication that the networked society is nearly ubiquitous; reaching into even the most socially deprived or isolated communities and nations. As a result it may also be a key resource in developing security strategies.

As many recent natural and human disasters such as the Japanese tsunami, the Haitian earthquake, 7/7, and 9/11 have all shown, contemporary societies are heavily reliant upon integrated networks. Sometimes their failure adds to the disaster, such as the loss of mobile connectivity during 7/7. Sometimes the technologies, hub places, and links of networks are the target of security threats, as in 7/7 and 9/11. On the other side of the coin our networks provide for resilience after and solutions to security threats. Examples here might include the use of Facebook and Google maps after the Haitian earthquake to support the relief effort. As Castells, Wellman and others have noted, contemporary networks also open up information and knowledge flows about and during crisis events. This might include such things as citizen journalism with mobile phones during 7/7; or protests, blogging, and micro-blogging by those caught up in crises; or SMS warnings to others about dangerous locations during the Mumbai attacks. Importantly, future security threats are unlikely to occur and be managed solely within one nation state. An attack on any network infrastructure, whether a communications network, financial network, or political network event, will have global consequences. Major security threats such as "doomsday scenarios" of nuclear, biological, or chemical terrorism are also not likely to remain isolated within one nation, either in their effects or the communication of information and knowledge about it.

NATIONAL SECURITY, ICT, AND STRATEGY

The focus of this book is the development of national security strategy in the context of these new global networks. In this introduction we have set out three key issues:

- The development of strategy in the context of national security driven by national interest
- The importance of high-quality information and knowledge in the development of strategy—strategic intelligence
- The nature of globalized networks based on ICT systems and people that are concurrently locations for threats to national and global security, tools to defend or to threaten national security, and sources of information for both nations and citizens about threats and events

This book is therefore arranged in four sections. Section 1 addresses a range of current issues around national security. This includes the formation of strategy (Chapter 2), internal threats to national security (Chapter 3), and the specific issue of maintaining elements of global (Chapter 4) and national (Chapter 5) security. Section 2 addresses the challenges faced by both governments and citizens in the context of a globalized and networked world. These challenges include using networked media to communicate risks and appropriate behaviors (Chapter 6), the risks associated with the use of such media (Chapter 8), or their use to support citizens in crisis situations (Chapter 9). Other challenges include resilience in the face of major security threats (Chapter 7) or the challenges to citizen rights in the face of expanding networked security infrastructure (Chapter 10). Section 3 details in depth the opportunities and concerns that networked media bring to the job of addressing national and global security. These include processes for thinking through user-focused designs for security solutions (Chapter 11) and the role of knowledge management in crises (Chapter 12). Chapters 13–16 address the specifics of employing state-of-the-art ICT solutions in the context of security. Section 4 looks to the future and addresses the potential threats to our networked society and especially to our ICT infrastructure by addressing issues of cyber security and defense. The section reviews current thinking on cyber security and defense strategies (Chapters 17 and 18) and then explores both cyber terrorism and state-sponsored cyber operations (Chapters 19 and 20). The section concludes by examining strategic approaches to limiting the impacts of cyber crime on citizens.

SECTION 1

National Security Strategies and Issues

CHAPTER

2

Securing the State: Strategic Responses for an Interdependent World

Andrew Staniforth

A CATALYST FOR CHANGE

For several generations terrorist groups pursuing their political, religious, and ideological beliefs have planned and executed attacks on the free and democratic communities of the Western world. The UK in particular has a long history of tackling terrorists and extremists who have had cause to challenge and violently oppose British values of tolerance, human rights, and the rule of law. And yet the largest loss of life of British citizens during a single terrorist attack did not occur in Britain at all but in the United States. Among the 2,977 people murdered on September 11, 2001, 67 were British. Acts of terrorism—in its widest definition—have been as endemic to human history as the ideological complexities that belie them, but the attacks of September 11 were very different. Seen from the West, there was something particularly horrific about the suddenness and sheer magnitude of the events that became known by their date. In lower Manhattan, on a field in Pennsylvania, and along the banks of the Potomac, the United States suffered its largest loss of life from an enemy attack on its own soil (The 9/11 Commission Report, 2004). Within just 102 minutes, four commercial jets would be simultaneously hijacked and used as weapons of mass destruction to kill ordinary citizens as part of a coordinated attack that would shape the first decade of a new century of counterterrorism and prove to be a catalyst for change in national security policy.

To regard 9/11 as merely another episode in the long history of terrorism would, from a national security perspective, be folly. The nature of the tactics used and the devastation caused during the attacks literally and metaphorically marked the end of one epoch and the dawn of another. As news channels beamed their now infamous live images around the world of aircraft crashing into the World Trade Center's "Twin Towers," viewers could hardly comprehend the events they were witnessing. Part of the stunning horror of the attacks in New York was the very spectacle of immense physical structures, monoliths of Western industrialized success, invested with the spirit of the age suddenly not being there. In some ways the Twin Towers' absence from what is arguably the world's most famous skyline has become an eerie and enduring anti-monument to terrorism. Despite the increasing complexity and proliferation of international terrorism, it is still rare that for the post-World War II West to have to confront mass destruction usually associated with interstate warfare.

Everyone who saw the footage of 9/11, at the time and in every broadcast since, has witnessed the pernicious potential of terrorism: to deliver fatal strikes at the very center of the world's one true super power. Variously condemned as crimes against humanity and acts of war by world leaders, governments quickly called for retaliatory action and sought to swiftly review existing national security strategies.

Amplifying the sense of vulnerability felt by the global audience of 9/11 was another vital feature of the tactics: the lack of a visible and identifiable enemy. These attacks were not state sponsored, nor were they part of conventional or even recognizable warfare. Stunned by the attacks, not only the "lay" audience but also the global intelligence community had to come to terms with the emerging fact that it had been wrong-footed by a tiny band of terrorists dispatched by al-Qaeda, a loosely constructed organization based in one of the poorest, most remote, and least industrialized countries on earth (Staniforth, 2011). The al-Qaeda transnational terrorist network had demonstrated with chilling effect its expertise at recruiting in one location, training in a second, attack planning in a third, and delivering mass murder in a fourth. The uncomfortable truth seems to be that a single nation at the turn of the century could not, in reality, comprehend the size and scale of the threat, nor could they meet the challenges of multiple determined terrorists bent on killing themselves and others. The indiscriminate killing of thousands of people by al-Qaeda on 9/11 introduced a new form of relationship between national governments and those who threaten them—a relationship that would evoke a new type of national security response.

Beyond the thousands of people who perished during the 9/11 attacks many more were seriously injured, both physically and psychologically. Members of Manhattan's many diverse communities joined forces with the emergency services and other professional organizations calling upon their natural instincts as much as their training, professionalism, and resourcefulness to save the lives of others and mitigate the consequences of attacks in a way more reminiscent of wartime Europe. The series of coordinated suicide bombings created a new genre of terrorist conflict and people caught up in the horror of its consequences had to face a reality formerly known only by war-ravaged nations for the very first time. In the years that followed the 9/11 attacks we would come to learn in forensic detail about the attacks and the people who perpetrated them. We would also learn a great deal about the ideology and methodology behind the 102 minutes of terrorism that would frame the threat and shape the response to the internal security of nations across the world.

LESSONS LEARNED

On November 27, 2002, the United States Congress and President George W. Bush established the National Commission on Terrorist Attacks Upon the United States. Known as the 9/11 Commission, this independent, bipartisan panel was directed to examine the facts and circumstances surrounding the September 11 attacks. Its aim was to identify lessons learned and provide recommendations to safeguard against future acts of terrorism. In pursuing their mandate the 9/11 Commission reviewed more than 2.5 million pages of documents, interviewed more than 1,200 individuals in 10 countries, and conducted hearings over 19 days taking public testimony from 160 witnesses. The 9/11 Commission report was critical of the United States administration in a number of key areas. It questioned how 19 suicide terrorists could be imported into America without examination and how were they able to live undetected within local communities for several months while seeking English language classes and flight instruction. Furthermore, the 9/11 Commission sought to identify why the al-Qaeda operatives were not identified by United States agencies working overseas in the first place. Most important, however, the 9/11 Commission focused upon how the United States, having learned from its experiences, could develop their response to tackle this new threat and protect its citizens in the future.

The 9/11 Commission identified that the United States had committed enormous resources to national security and specifically to counterterrorism. Between the fiscal year 2001, the last budget adopted before 9/11, and the fiscal year 2004, total federal spending on defense, (which included expenditures on both Iraq and Afghanistan campaigns), homeland security, and international affairs rose more than 50% from $354 to 547 billion (The 9/11 Commission Report, 2004). In comparison, the United States had not experienced such a rapid surge in national security spending since the Korean War, reflecting the economic impact upon federal finances from the al-Qaeda terrorist threat. The 9/11 Commission recognized that this pattern of attack and response had occurred before in America's history. The United States understood that some programs developed in direct response to a crisis, alongside new agencies that had been created, may be discarded while others may be invented or redesigned. The 9/11 Commission recognized that during the period of its publication in 2004 that it was a time for reflection and evaluation as they considered their next steps, the first of which was to understand the full extent of the threat they were facing.

It is difficult to comprehend the United States authority's failure to recognize the signs that a major terrorist plot was developing within their communities. The sheer size and scale of the United States security machine did not correctly analyze, assess, and prioritize intelligence on a national level. A large number of agencies that held critical data did not share its information. These organizations were working in isolation each with their own "need to know" principles and limited "need to share" protocols. A full national picture of the emerging threat was not put together (Staniforth, 2011). Like a giant jigsaw puzzle many of the smaller pieces were missing that would have provided United States authorities with a greater opportunity to identify and disrupt the plot. Despite collating intelligence to develop a picture of unfolding events, the United States authorities were behind the activities of the terrorist cell and as the al-Qaeda planes operation drew into its final phases in September 2001, time simply ran out. The 9/11 Commission concluded that the United States

> *domestic agencies were not mobilized in response to the threat. They did not have direction, and they did not have a plan to institute. The borders were not hardened. Transportation systems were not fortified. Electronic surveillance was not targeted against a domestic threat. State and local law enforcement were not marshaled to augment the Federal Bureau of Investigations efforts. The public was not warned.*
>
> ***PNLD and Staniforth, 2009***

The message was clear: to effectively manage the new and emerging threats from al-Qaeda would require a new type of national security response.

CONTESTING TERROR

The Canadian Government responded quickly to support its neighbor during the 9/11 attacks. As the United States closed its airspace to incoming flights Canadian traffic control accepted 224 diverted flights within 45 minutes of the attacks at the World Trade Center (PNLD and Staniforth, 2009). The diverted flights, which contained 33,000 passengers, were provided with landing slots at airports across Canada. Remarkably, small communities like Gander, Newfoundland, and Labrador accommodated 12,000 of these passengers despite their population of only 10,000 (PNLD and Staniforth, 2009). Canada's response is now widely recognized as an important part of preparing for the wider consequences of international terrorism—consequences that need to form part of national security contingency planning. It is important to remember that the Canadian Government accepted these diverted flights with no absolute guarantee that they did not contain further suicide terrorists. Then Canadian Prime Minister, Jean Chrétien, urgently established an *ad hoc* cabinet committee on Public Safety and Anti-Terrorism, which began to review policies, legislation, regulations, and programs across the Canadian Government. Their aim was to strengthen all aspects of their approach to fight terrorism and ensure public safety. From this committee they developed a new counterterrorism strategy known as the Anti-Terrorism Plan, which continues to be used to this day.

The government of the United Kingdom also responded swiftly to the catastrophic attacks on the United States. While Stephen Lander, Director General of the Security Service (MI5), remained in London to brief Prime Minister Tony Blair, Deputy Director Eliza Manningham-Buller flew to Washington on September 12 with Chief of the Secret Intelligence Service (MI6), Sir Richard Dearlove and the Director of Government Communication Headquarters. At a time when U.S. airspace remained officially closed, this UK delegation was escorted to Andrews Air Force Base in the United States by eight United States Air Force F-16 fighter planes (Andrews, 2011). Upon arrival they were whisked away to dinner and prolonged crisis talks at the Central Intelligence Agency's Langley headquarters with George Tenet, the Director of the National Security Agency, and the Deputy Director of the Federal Bureau of Investigation. Stephen Lander informed colleagues at the Security Service that, "It was clear that the American side, who were exhausted and angry, deeply appreciated seeing friends from the UK, coming to offer support and help" (Andrews, 2011).

The security agencies of the UK were no strangers to responding to terrorist attacks. At the time of the devastating delivery of bin Laden's most ambitious attack, the Planes Operation of 9/11, the UK government, like many other countries in the developed world, had no sophisticated or coherent cross-departmental strategy to counter international terrorism. In short, the UK had no plan to institute of any rigor that would have been able to effectively respond to a major al-Qaeda indiscriminate attack. Of course, the UK security apparatus had memories of the long counterterrorist campaign in Northern Ireland to draw on, and the foundations that had been laid down in terms of a corpus of emergency terrorism legislation on the statute book. Throughout the history of counterterrorism practice in the UK, collaboration between government departments had been key to the success of many operations and the intelligence community had learned the value of close co-operation with the police service. Nevertheless, the characteristics of jihadist terrorism with its vaulting ambitions, strident ideology, and disregard for civilian casualties—indeed for all human life, with adherents prepared to give their lives in their attacks—represented very new challenges for Parliament and public, government, and law enforcement alike.

In the immediate aftermath of 9/11, Sir David Omand GCB, was appointed in the new position of UK Security and Intelligence Coordinator in the Cabinet Office in London. He initiated work on developing a comprehensive national counterterrorism strategy. The threat from jihadist terrorism had been assessed as severe and numerous plots affecting British interests overseas and at home were of great concern to the security authorities. Sir David Omand launched work in November 2002 on a UK counterterrorism strategy (called CONTEST: CouNter-Terrorism STrategy). The strategy was later presented to the Cabinet and adopted in 2003, but the details remained confidential and were not published by the government until 2006. An updated version, CONTEST 2, was published in 2009, and a third generation, CONTEST 3, was published in 2011 by the coalition government. Building the CONTEST strategy commenced with assessing the strategic risks to the UK and the first steps were to establish the "ends, ways, and means" of al-Qaeda's terrorist intentions. In summary, Sir David Omand identified that the ends of al-Qaeda were to "liberate and unite the global community of Islamic believers in a Caliphate covering all 'Muslim' lands" (Omand, 2010). Its ways of achieving this were to "weaken the 'Zionist and Crusader' enemy by building strength through global alliances with existing terrorist organisations." It was assessed that al-Qaeda would complete this by means of "extreme and dramatic terrorism to expose weakness and decadence in the West; to create confidence that resistance to US influence was possible; and to inspire an 'awakening' of revolutionary consciousnesses" (Omand, 2010).

From identifying these key issues countermeasures could then be constructed, but such a strategy could not be bold enough to state that it could stop terrorist attacks. An important and implicit assumption was therefore contained within CONTEST from the outset that there was no complete defense against contemporary determined terrorists, especially as they continued to develop new ways in which to deliver death and destruction on an unimaginable scale. Even during its early stages of development, CONTEST indicated that there were no frameworks that could guarantee peace. Sir David Omand progressed the work to finalize a new counterterrorism strategy for the UK believing that "The aim has to be to take sensible steps to reduce the risk to the public at home and to our interests overseas, on the principle known as ALARP, to a level 'as low as reasonably practicable'" (Omand, 2010). CONTEST provided measures to counter the identified ends, means, and ways of al-Qaeda's terrorist strategy. Therefore, to mitigate the ends of al-Qaeda's intentions, CONTEST must "protect the UK and its interests from security risks including terrorism in all of its forms." The ways in which to do this were to "work systematically to reduce terrorism risk through concerted action by government." From these were mapped what became known as the "4P" campaigns, which included:

> *Pursue, to stop terrorists attacks; Prevent, to stop people becoming terrorists, or supporting violent extremists; Protect, to strengthen protection against terrorist attack; and Prepare, where an attack cannot be stopped, to mitigate its impact.*
>
> ***Omand, 2010***

The means to achieve this was to be "seen to uphold human rights, the rule of law, legitimate and accountable government, and the core values of justice, and freedom." The strategy that emerged from this work had a clear strategic aim: "to make it possible for society to maintain conditions of normality so that people could go about their normal business, freely and with confidence, even in the face of suicidal terrorist attacks" (Staniforth, 2012). The conditions, freely and with confidence, were an important reminder to seek security in ways that upheld British values such as liberty and freedom under the law.

In developing the Prevent, Pursue, Protect, and Prepare structure of CONTEST, Sir David Omand believed that the strategy was

> *easily understood as a logical narrative, translated into specific programmes of action across government, the private sector and the voluntary sector, and as has been shown, capable of being updated and extended in response to developments in the threat and in UK security forces technologies for countering it.*
>
> ***Staniforth, 2012***

It was important that the complexities of such a wide-ranging strategy were simplified and focused as successful delivery would depend upon a joined approach across government, the strength of partnerships, and professional relationships. The simplicity and effectiveness of the CONTEST strategy was soon recognized by the European Commission as a potential tool in tackling security threats across Europe. During December 2005 the Justice and Home Affairs Council adopted the European Union Counter-Terrorism Strategy. The Strategy, divided into the four pillars of Prevent, Protect, Pursue, and Respond, had been directly borrowed from CONTEST. The strategy was welcomed by all European Heads of States and Governments, which committed the European Union to "combat terrorism globally while respecting human rights, and to make Europe safer, allowing its citizens to live in an area of freedom, security and justice." This new pan-European strategy progressed the European Action Plan on Terrorism, a program of work established by the European

Commission at the 2004 European Council. This Council session was conducted in the wake of the al-Qaeda Madrid bombings on March 11, 2004, which resulted in the death of 191 innocent civilians and leaving more than 1,584 injured. This coordinated attack, carried out exactly 911 days after 9/11, was the first atrocity to mark the date of the al-Qaeda planes operation in the United States and was a planned attack that remains Europe's deadliest terrorist incident so far this century.

The UK CONTEST strategy had provided an effective framework to coordinate, direct, and shape not only the UK government's response to the threat from international terrorism, but all 27 Member States across Europe. A decade since its first presentation to the Cabinet in 2003, it continues to provide clear and focused direction. Over time CONTEST has grown in size, scale, and scope and is now the world's most sophisticated counterterrorism strategy. The latest version, published in July 2011 (CONTEST 3), is unprecedented in its size. It now stands at a combined total of 236 pages—62 pages longer than its 2009 iteration, and 203 pages longer than that of 2006. This expansion reflects the increasing breadth and depth of the UK's counterterrorism policy, but its expansion is only part of the story. While it is true that CONTEST 3 covers a greater diversity of threats and recommendations than its predecessors, the reason for this growth has been a calculated move by successive governments to bring counterterrorism strategy out into the open. Whereas the first iteration of CONTEST, released in 2003, remained classified, its updated 2006 version did not. The 2009 CONTEST took this transparency one stage further, offering an unparalleled insight into the UK's strategic approach to countering terrorism. CONTEST 3 has continued in this vein. It offers a candid assessment of the current threat landscape, situating UK counterterrorism policy against a changing geopolitical backdrop in the Middle East and in light of shifting modes of terrorist attack. The development of CONTEST reveals an important element of counterterrorism practice in the UK, that the public, very often the victim of terrorist attacks, can help prevent them, but in order to do so they need to be informed and kept updated so that citizens and their protectors can all work together toward achieving the shared values of freedom, tolerance, democracy, and human rights. In the aftermath of 9/11, and the virulent threat from those intending to attack the UK's free and democratic society, it was clear that CONTEST on its own could not protect the public; other tools were required to support its strategic aim of ensuring that people could go about their lives freely and in confidence. So the British government set about consolidating and coordinating all of its national security activities into a single strategy.

NATIONAL SECURITY FRAMEWORKS

The UK's CONTEST strategy had served to prove that Britain was stronger, safer, and more resilient when all of its government departments brought their expertise to bear in concert with one another toward achieving an identifiable common goal in the nation's interest. It was also quickly apparent to Ministers that the threat from international terrorism was interconnected with other threats and risks to the security of its citizens. CONTEST therefore not only provided the foundations upon which to construct a wider security framework for the whole nation, it also revealed strong connections between the threats and risks posed to UK national security. To maximize the potential of the CONTEST experience, during 2007 the British government established the Commission on National Security in the 21st Century. Under the auspices of the Institute for Public Policy Research (IPPR), the UK's leading think tank that since 1998 has been at the forefront of progressive debate and policy making in the UK, the all-party Commission was to prepare the ground for an independent national security strategy for the UK. Co-chaired by Lord Paddy Ashdown, former leader of the Liberal Democratic Party and Lord George Robertson, former Secretary of State for Defense and Secretary General of NATO, the Commission had 14 members, which included the creator of CONTEST, Sir David Omand GCB, as well as Sir Chris Fox, former police Chief Constable of Northamptonshire and President of the Association of Chief Police Officers. The membership of the Commission brought together leading experts from the fields of security, policing, defense, intelligence, academia, diplomacy, science, and politics. Based on extensive research and broad deliberations over a two-year period, the Commissions' final report, published on the June 30, 2009, Shared Responsibilities: A National Security Strategy for the United Kingdom, set out a wide range of proposals designed to make the UK, its citizens, businesses, and communities more secure.

This report highlighted that globalization was diffusing power among many different actors in the international system, which was fueling a massive redistribution of economic and political influence across the world. It also revealed that while globalization was impacting traditional bases of power, at the same time the global population was growing rapidly with an estimated world population of 9.2 billion by 2050, of which only 1.25 billion would live in developed countries. The report worryingly suggested that

> *This shift in population means the end of the West as the pivotal region in world affairs, intense pressure on natural resources, an increasing marginalized global majority, and increased migration flows from poor to rich states.*

The report held no punches, especially concerning the impact of climate change revealing that it was set to transform the security environment and was likely to reduce and shift the availability of habitable land, food, and water. As a direct result, the report went on to claim that the change in climate may exacerbate interstate tensions generating forced movements of people identifying that weak and unstable states now outnumber strong and stable ones by more than two to one, and state failure and collapse would be a highly visible feature of the international security landscape for decades to come (IPPR, 2012).

While the Commission was tasked solely to provide the foundations for a UK national security strategy, the impact from its findings for all societies in the Western world was clear to see. The UK were also rightly concerned with its close links with Member States across Europe, especially the impact upon policing the UK, and none more so than increasing threats from transnational crime networks who were assessed as capable of taking full advantage of failing states and mass migration in the future. Today we already see that organized criminals have expanded their trafficking operations in drugs, arms, and people and are undermining and corrupting state governance arrangements in many countries, facilitating and profiting from violent conflict in the process. The threat from international terrorism was also assessed by the Commission as developing and despite this specific threat to national security that will most likely come from the use of conventional weapons and improvised explosive devices, the report identified that

> *The threat of technologically sophisticated chemical, biological, radiological, or nuclear terrorism is real and no longer comes only from organized groups like al Qa'ida and its imitators, but also from lone individuals with relevant expertise and access to the necessary technological infrastructure. Insufficiently secure government laborites around the world remain a particular worry in relation to bioterrorism.*

The Commission also identified that the rapid advances in information technology and biotechnologies would create new vulnerabilities for national and international security and that cyber crime and cyber terrorism were already realities, warning that "new discoveries in biotechnology put to deadly purposes would have terrifying implications."

In concluding its observations of threats to national security, the Commission recommended that the UK could not be isolated from any of the security developments. It explained, quite rightly, that although the UK, like many countries across the world, benefited enormously from their participation in an open world economy and society, they were also reliant on world energy markets and were increasingly vulnerable to their instability. The report concluded by stating that "More generally, the country will clearly be less secure if the wider international security environment deteriorates and the UK has a clear stake in ensuring that it does not happen." The Commission made a series of recommendations in its final report providing the foundations upon which to construct a new national security strategy including the following:

- A new approach to the situation in Afghanistan and Pakistan
- Proposals to improve the UK's energy security
- Measures to address radicalization and the threat of terrorism in the UK
- The call for a transformation in our approach to defense policy, and measures to strengthen both NATO and the European pillar of the transatlantic alliance
- Measures to strengthen and improve the institutions handling security at the center of government
- Proposals for improved global governance
- A call to strengthen and deepen the legitimacy of the security strategy we pursue

On publication of the final report, Lord Ashdown, Co-Chair of IPPR's Commission, stated that

> *We cannot carry on as we are. We need to change the way we think and change not just what we do but also how we organize ourselves to do it. In a world where power is no longer the sole preserve of nation states, and where security is no longer only about defense, we need new joined-up machinery in Whitehall, a truly integrated strategy that links all of our policy instruments together, and a much greater focus on how we link the UK effort to the efforts of others around the world. The most important part of what we do today is what we do with others.*
>
> ***IPPR, 2012***

STRATEGIC RESPONSES

During March 2008, Prime Minister Gordon Brown and Lord West of Spithead, the Permanent Undersecretary for Counter Terrorism and National Security, introduced a new National Security Strategy for the UK—"Security in an Interdependent World." For the very first time the Cabinet Office had published a single, overarching strategy that brought together the objectives and plans of all departments, agencies, and forces involved in protecting UK national security. The Cabinet Office stated that its publication was a significant step and the latest in a series of reforms bringing greater focus and integration to the UK's national security approach. The findings and recommendations of the Commission were woven through the very fabric of the strategy and its aim—to protect the UK and its interests, enabling its people to go about their daily lives freely and with confidence, in a more secure, stable, just, and prosperous world—bore the hallmarks of CONTEST and the influence of its creator, Sir David Omand GCB. The strategy had little time to embed itself into the culture of Whitehall before a new coalition government was elected during May 2010, but an important precedent had been set and the national security advisors to incoming Prime Minister David Cameron knew the value of a cross-party and cross-government collaborative approach to national security.

To coordinate the UK's security response under the new coalition British government, a review and rewrite of the national security strategy was swiftly undertaken and a revised strategy—"A Strong Britain in an Age of Uncertainty: The National Security Strategy"—was presented to Parliament by the Prime Minister by Command of Her Majesty during October 2010. This new strategy, designed to span the full extent of this parliamentary term until 2015, now provides a cohesive, pan-government response to the variety of threats the UK must counter. The strategy reveals that new threats can emanate from states, but also from nonstate actors: terrorists, home-grown or overseas; insurgents; or criminals, as well as highlighting that the security of UK energy supplies increasingly depends on fossil fuels located in some of the most unstable parts of the planet. Nuclear proliferation is also identified as a growing danger as well as the UK's security vulnerability to the effects of climate change and its impact on food and water supply. But most important, the new national security strategy revealed that Britain today was both more secure and more vulnerable than in its long history. More secure, in the sense that it did not currently face, as it has so often in the past, a conventional threat of attack on its territory by a hostile power, but more vulnerable, because it remains one of the most open societies, in a world that is more networked than ever before. In the long history of preserving the UK's security, previous governments had grappled with the brutal certainties of the Cold War: with an existential danger that was clear and present, with Soviet armies arrayed across half of Europe, and the constant threat of nuclear confrontation between the superpowers. Contemporary Britain faces a different and more complex range of threats from a myriad of sources—terrorism; cyber attack; unconventional attacks using chemical, nuclear, or biological weapons; and large-scale accidents or natural hazards—any one of which could do grave damage to British citizens and its interests at home and overseas.

For the UK, and other countries of the Western world, the concept of national security today is very different that it was 10 or 20, let alone 50 or 100 years ago. Geographically Britain is an island, but economically and politically it is a vital link in the global network. That openness brings great opportunities, but also great vulnerabilities. The British government knew, for example, that terrorist groups like al-Qaeda were determined to exploit its openness to attack its cities and infrastructure by plotting to kill as many UK citizens as possible or to inflict a crushing blow to the UK economy. This specific threat is the most pressing threat that the UK faces today. Any strategy to build a strong Britain and to protect the UK's national security must begin with the role that Britain wants to play in the modern world. In a world that is changing at an astonishing pace, Britain's interests remain surprisingly constant. The UK is an open, outward-facing nation that depends on trade and has people living all over the world. In fact, 1 in 10 British citizens now live permanently overseas. It is a country whose political, economic, and cultural authority far exceeds its size. The global force of its language, the ingenuity of its people, and the intercontinental reach of its time zone, allowing the UK to trade with Asia in the morning and with the Americas in the evening, means it has huge advantages. To build a strong Britain the coalition government believed that a radical transformation was required in relation to the way it considers national security, as well as how it was to organize itself to protect it. This is particularly important to the UK and other Western democracies as they enter an age of uncertainty, especially economic uncertainty where an effective national security strategy can weigh on all threats a state faces, and prepare its government to counter them effectively.

The UK's ability to meet current and future threats depends crucially on tackling the budget deficit. UK national security depends on its economic security and vice versa. An economic deficit is also a security deficit. So at the heart of protecting the UK's security were some tough choices to bring the defense and security apparatus budget back into balance. These choices are informed by the risks, analysis, and prioritization now set out in the National Security Strategy. The coalition government believes that the UK's financial position in the world is the largest single challenge facing the government, which affects both national security and all other areas of public policy. The most urgent task of the British government is therefore to return its nation's finances to a sustainable footing, as it cannot have effective foreign policy or strong defense without a sound economy and a sound fiscal position to support them. In addition, all government departments, including those contributing to national security, will be required to play their part and programs of austerity are already impacting those government agencies that supply port and border security. Finding new and efficient ways of working and maximizing the potential of current assets while seeking innovative collaborative approaches to an increasingly integrated border security system are all key components to protect our nation's security while rebuilding our finances in this age of economic uncertainty.

NATIONAL SECURITY MACHINERY

The British government believes that the UK is well placed to benefit from the world of the future, but in its National Security Strategy it sets out three key steps for a secure and prosperous Britain. These steps include using all of its national capabilities to build Britain's prosperity, to extend its national influence in the world, and to strengthen its security (see Figure 2.1).

To drive the UK national security cycle and to embed the new approach into the very culture of British governance, two essential components were built into its national security machinery: the creation of a new National Security Council (NSC) and second, a National Security Risk Assessment (NSRA) process. One of the new coalition government's first acts in leading the UK in 2010 was to create the NSC. This development, which marked a key change in the security machinery of government, brought together all the senior ministers concerned under the chairmanship of the Prime Minister. The NSC ensures a strategic and tightly coordinated approach across the entire government to the risks and opportunities the country faces. It is an essential part of the national security cycle ensuring that key decisions from across the full landscape of government are made to strengthen the UK. The British government understood all too well that a truly strategic approach to national security was required to go further than just simply assessing domestic civil emergencies. The new national security approach, as well as looking at short-term domestic risks, also had to consider for the first time all aspects of national security. Therefore, an "all hazards" strategy was created, underpinned by the first ever NSRA, which continues to assess and prioritize all major areas of national security risk, both domestic and overseas. To develop the NSRA, subject-matter experts, analysts, and intelligence specialists were asked to identify the full range of existing and potential risks to UK national security that might materialize over a 5- and 20-year horizon. All potential risks of sufficient scale or impact requiring action from government and/or with an ideological, international, or political dimension were assessed based on their relative likelihood and relative impact. The impact was assessed based on the potential direct harm a risk would cause to the UK's people, territories, economy, key institutions, and infrastructure. From the rigorous risk analysis process the results of the first NSRA suggest that, over the next 20 years, the UK could face risks from an increasing range of sources, and that the means available to its adversaries are increasing in number, variety, and reach. In our increasingly networked world, which creates great opportunities but also new vulnerabilities, the UK must, in particular, protect virtual assets and networks on which its economy and way of life now depend. This has quickly become just as important as directly protecting physical assets and presents a seismic shift in national security planning. It

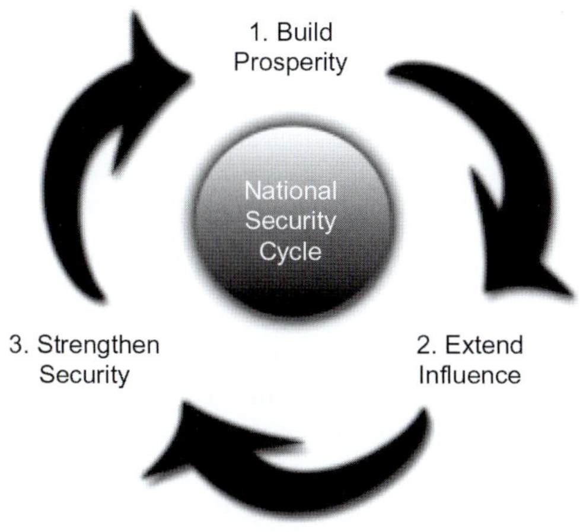

FIGURE 2.1 The national security cycle.

is, however, most important to stress that the NSRA serves to inform strategic judgments—it is not a forecast. The British government, indeed no government no matter how powerful, cannot predict with total accuracy the nature or source of the next major national security incident it will face, but a process such as the NSRA significantly helps key decision makers along the corridors of power and influence government to make informed choices. In particular, it assists the British government in prioritizing the risks that represent the most pressing security concerns in order to identify the actions and resources needed to deliver responses and countermeasures to mitigate those risks.

SECURITY CONTEXT TODAY

Many countries in the Western world continue to endure a real and pressing threat from international terrorism, particularly that inspired by al-Qaeda and its affiliates. The International Security Assistance Force (ISAF) is engaged in conflict in Afghanistan because of this threat. ISAF personnel are assisting the Government of Afghanistan to prevent Afghan territory from again being used by al-Qaeda as a secure base from which to plan attacks on the West and its interests overseas. Terrorists can also exploit instability in countries like Somalia or Yemen. This instability can spread from one country to another as was observed in the Balkans at the end of the last century. Lawless regions provide a haven for terrorist groups and organized criminal networks alike. Western democracies want an Afghanistan that is not a threat to their security or that of the wider international community. To achieve this they are supporting an Afghan-led process to develop the Afghan security forces and build a more effective Afghan state that can control its own security and, ultimately, achieve a lasting political settlement. While there has been significant cost, the Western governments believe that they are making progress identifying that the Afghan security forces are now 260,000 strong, increasingly showing that they can provide their own security. The transition of security responsibility to the Afghans has begun and the joint Afghan and international operations across the country are putting pressure on the insurgency. The Afghan economy is growing rapidly and the Afghan government's ability to deliver key services such as health and education has significantly improved. Western governments state that they will continue to work with the Afghans to secure further progress is made on corruption, regional engagement, and political and economic reform, but they must not be complacent. The insurgency remains strong and adaptable and continued resolve and commitment is required to ensure success.

While the struggle against terrorism in theaters of conflict overseas continues the principal threat from international terrorism to Western democracies will continue to come from al-Qaeda, its affiliates, and terrorists inspired by its ideology. The core of al-Qaeda in the mountainous border regions of Pakistan and Afghanistan is under increasing pressure. Military action has weakened al-Qaeda and other terrorists there, but has not destroyed them entirely. This increased pressure has forced al-Qaeda to adapt, and this threat is already becoming more diverse and this trend is likely to continue. Al-Qaeda has affiliates located in Somalia, Yemen, and Iraq, through which it can exert its influence on others. Fragile, failing, and failed states around the world also provide the environment for terrorists to operate as they look to exploit ungoverned or ill-governed space. Those who have combat experience in theaters of conflict overseas may return home to their Western communities with practical, operational, and technical know-how to conduct attacks. This insider threat, also categorized as home-grown or neighbor terrorism, remains a clear and present danger to Western communities. The current al-Qaeda-aligned insurgency in Somalia provides another node of potential risks as this environment may provide a similar training ground for individuals with terrorist ambitions. When all is considered, state protectors must be prepared to tackle different types of terrorist attacks and keep their intelligence antenna alive to weak signals, which are early signs and indicators of any potential new and emerging threats. Al-Qaeda still aspires to mass-casualty attacks, but the increased pressure it is under and the success of the security forces in disrupting attacks has forced its members to explore other methods. For example, al-Qaeda and other groups have stated an aspiration to develop unconventional (chemical, biological, radiological, or nuclear) capabilities. Al-Qaeda has a long-held desire to maximize the impact of its attacks through the use of such weapons. As far as Western governments are aware or prepared to disclose publicly at this time, al-Qaeda has yet to develop such capability, but all indications are that it will continue to seek all the means to do so.

The dangers to Western communities are further amplified by the view held by the global intelligence community that senior al-Qaeda figures have urged Muslims in the West to conduct attacks without training or direction from established groups. Such lone terrorists are inherently unpredictable and their plots are difficult to detect. Al-Qaeda may consider smaller scale attacks against softer targets, which would nonetheless attract considerable media attention. Although Western governments continue to work to prevent terrorism at its source and pursue terrorist plotters, they must not underestimate this challenge. Any degree of complacency based upon the absence of major terrorist events remains both misplaced and unwise. These remain challenging times. Although intelligence and law enforcement agencies across the West have had success in disrupting the great majority of planned attacks, international terrorism can greatly affect their interests at home or overseas. It is easier to disrupt terrorist capability than to remove terrorists' underlying motivation, but all engaged in this struggle must still work to stop people from becoming terrorists in the first place and all national security strategies of Western democracies expect international terrorism to continue to pose a significant threat in terms of both likelihood and potential impact for the foreseeable future.

FROM THREAT TO THREAT

The contemporary security context remains diverse and contemporary national security strategies must recognize that there are numerous hazards to overcome alongside the immediate threat from international terrorism and the continued struggle against terrorism and violent extremism in conflicts overseas. Such hazards include transnational threats, organized crime, espionage, and domestic terrorism. There are a number of significant transnational threats that require the attention of Western governments, many of which directly impact the security of ports and borders. The UK and the United States, alongside other nations, are at a crucial stage in international efforts to prevent nuclear proliferation in the Middle East. If Iran acquires nuclear weapons technology, there is a strong possibility that other states in the region would follow. A Middle East with several nuclear weapons states would lead to high instability, precarious energy security, and would have a severely damaging effect on the Middle East peace process.

While the proliferation of nuclear capabilities attracts the attention of national security professionals, other risks also require a dedicated and determined response. Organized crime, for example, affects national interests and the lives of citizens at home and overseas. At present there are approximately 38,000 individuals involved in organized crime just in the UK, costing its economy and society between £20 and 40 billion per annum. Organized crime is a global phenomenon, shaped and sustained by a number of strategic factors. These factors include conflict and instability where the rule of law is poor and where corruption associated with organized crime is prevalent. This is often coupled with poverty. It is facilitated by a more interconnected world aided by new technologies, and driven by a desire for personal gain and community tolerance, or ignorance, of organized criminality. Serious and organized crime is a major problem that affects real people. Its impact is seen on the streets and is felt in our communities every single day. The drug dealing on street corners, the gangs terrorizing communities, the burglary and muggings by addicts, the trafficking of vulnerable young woman into prostitution, and the card cloning and credit card fraud that robs so many are all fundamentally driven by organized criminals. As new opportunities arise, such as computer-enabled crime, organized criminals will be quick to take advantage, and unless governments develop a flexible and effective response it will be the ordinary citizens who will pay the price damaging the economic well-being at a time of financial uncertainty in world markets and rising national levels of debt.

Other key threats, which must be considered in the planning of contemporary national security strategies, include traditional espionage. This continues to threaten Western democracy interests, with their commercial sector under threat alongside diplomatic and defense interests. The revolution in global communications and increased movement of people, goods, and ideas has also enabled the use of cyberspace as a means of espionage. This enables operation from a safe distance and makes the attribution of attacks more difficult, thus reducing the political risk associated with spying. Like espionage, cyber attack is not simply a risk for the future. Governments, the private sector, and citizens are under sustained cyber attack today from both hostile states and criminals. They are stealing intellectual property, sensitive commercial and government information, and even citizens' identities in order to defraud individuals, organizations and governments. In the future, unless the governments take positive action, this threat could very well increase. For this reason, cyber security has been assessed as one of the highest priority national security risks to the many Western democracies. Cyberspace is already woven into the fabric and contemporary culture of Western society. It is integral to the economy and security and access to the Internet, the largest component of cyberspace, is already viewed by many citizens as the "fourth utility," and is a right rather than a privilege. In less than 15 years, the number of global Web users has exploded from 16 million in 1995 to more than 1.7 billion today (Staniforth, 2012).

While cyberspace provides countries with massive opportunities, the risks emanating from their growing dependence on it are huge. By 2015, there will be more interconnected devices on the planet than humans; everything from mobile phones, cars, and fridges will be networked across homes, offices, and classrooms across the globe. Activity in cyberspace will continue to evolve as a direct national security and economic threat, as it is refined as a means of espionage and crime, and continues to grow as a terrorist enabler, as well as a military weapon for use by states and possibly others. But creating a counter-cyber security posture across the full spectrum of activities is also a great opportunity for democracies to capitalize on their national economic and security comparative advantages (Staniforth 2012).

CHALLENGES AHEAD

Throughout the evolution of national security practice in the Western world, shocking and dramatic incidents at home and overseas have only served to deepen the resolve of governments to develop and strengthen their response. New events will doubtlessly provide the genesis for the development of a stronger and more sophisticated approach to national security designed to meet the future challenges of an unpredictable world. Yet beyond resolute determination, such events must instill a re-dedication to preparedness so that new ways of working can be identified; so that practitioners push the boundaries of national security practice beyond the current state of the art; and most important, so that those in authority work to embed progressive developments to ensure that the primary driver for change in national security practice is not simply the next successful attack, disaster, or crisis. Effective national security provision is no longer the solitary endeavor of a single state.

Collaboration is key to safeguarding security for all citizens, but upon the initial creation of a highly sophisticated all-hazards national security strategy, it must be understood by those in authority that despite such a plan there are no guarantees of a risk-free life and bad things still happen. It is therefore essential that developments to protect a nation will have to be guided by the state's relationship with its citizenry, but in a world of startling change and developing hazards the first duty of any government remains the security of its country and its citizens.

CHAPTER

3

We Have Met the Enemy and They Are Us: Insider Threat and Its Challenge to National Security

Gregory B. Saathoff, Troy Nold, Christopher P. Holstege

INTRODUCTION

When he wrote "We have met the enemy, and he is us" in 1970, the cartoonist Walt Kelly paraphrased United States Navy Commodore Oliver Hazard Perry's 1813 naval victory statement by simply replacing the word "ours" with "us." In a new millennium of national security challenge, we suggest yet another change of pronouns, from "he" to "they" (Oppel and Bowley, 2012). Indeed, attacks by people with privileged access to an organization's systems and resources are a serious threat faced by all industries and sectors today. The consequences of insider threat include not only financial loss, but also the disruption of business operations, a decrease in competitive advantage, damage to reputation, and harm to individuals. In the realm of national security, the negative impact of an insider attack may be far more catastrophic. While insider threat is not new, it has evolved dramatically in response to changing technology, social, business, and cultural factors (Colwill, 2009). The majority of current literature on the subject comes from the fields of information technology and cyber security. It is therefore not surprising that many of the solutions available to government and industry for identifying vulnerable insiders focus almost exclusively on technology. However, technical solutions alone are not sufficient. Equally important, insider risk management should directly address the underlying behavioral component of insider threat by implementing security controls with a strong focus on behavior and environment.

Insider threat is a conundrum: it is so damaging because it is frequently ignored. In some instances, ignorance may be rooted in wholesale denial of the existence of the threat itself. This chapter begins by defining the insider threat, the characteristics and motives of the insider, the types of attacks, and their consequences. It examines the scope of the problem in the context of both technology and insider threat behaviors, with an emphasis on the latter, which is all too often overshadowed by a focus on the former. Because of its impact in shaping our understanding of insider threat, the Amerithrax case is examined in detail for the homicidal, suicidal, and substance-dependence factors that influenced the perpetrator of the attacks.

The behavioral impact of insider threat is best conveyed through case studies. Some of the notorious examples of insiders who have impacted national security include two Cold War spies, Aldrich Ames and Robert Hanssen. Both Bradley Manning, a U.S. Army intelligence analyst who was arrested in 2010 for disclosing classified material to the whistle-blower Web site WikiLeaks, and Dr. Bruce Ivins the perpetrator of the 2001 anthrax attacks, albeit less traditional in their security breaches, demonstrate the protean and insidious nature of the emerging insider threat.

The Amerithrax case is particularly revealing because it portrays today's insider threat in a field other than information technology. It demonstrates the ubiquitous nature of the problem, and it is an example of an attack of sophisticated means that would not have been detected by computer forensics, the most offered solution. Confronting the insider threat requires a multifaceted approach of technical methods and organizational policy. Lessons from the Amerithrax case and others recommend that behavioral tools are essential for promoting a culture of security and responsibility, for example, employee screening that includes relevant toxicology testing for substances of abuse, compliance with policies and procedures, leadership presence, security awareness training, and a focus on occupational health and wellness in the workplace. Although most organizations have explicit policies for ensuring personnel security, there are often problems in application and execution of relevant policies and procedures. For example, during the transition from NATO to Afghan local police forces, the killings of NATO personnel by Afghan trainees have been noted to occur due to the failures to implement policies and procedures designed to thwart insider threat (Oppel and Bowely, 2012).

DEFINING THE INSIDER THREAT

Of note, there is no universally accepted definition of "insider threat." The variety of definitions exemplifies both the complexity of the insider threat and thus far siloed approaches to risk management among the most affected industries. Our current understanding of insider threat comes primarily from information technology, the nuclear industry, and biosecurity. Quite frequently, the insider problem is compartmentalized within a single industry and/or a well-defined aspect of an organization's security framework. Many organizations have taken steps to identify insider threat prior to entry into a system. This perimeter-based approach, however, does not sufficiently address the most serious insider threats to national security. Ames, Hanssen, Manning, and Ivins did not presumably enter the work force with intent to do harm. Of the four, only Ivins' background checks, had they been thorough, would have disqualified him from initial employment by the United States Government.

From these various definitions, one can infer the underlying characteristics that make the insider threat a discrete security challenge. Blackwell (2009, p. 9) defined an "insider" as "one who has legitimate access to an organization, its systems, information or other resources," and the "insider threat" more precisely as "a risk that an insider can use access or knowledge to cause harm to the organization." Paradoxically, an employee is an organization's most valuable asset and greatest risk. The inherent problem of insider threat arises when an "insider," or employee, compromises the trust and access he is given. Gelles et al. (2012, p. 2) also provided a useful definition:

> *Insider threat exists within every organization where employees (insiders) comprise the core of an organization's operational plan and are the key drivers of its mission execution. As a result (threat) of some perceived injustice, retaliation, sense of entitlement, or unwitting need for attention and/or validation, the employee takes some action as part of a contrived solution that results in negative consequences for the organization. Asset loss is the end result of actions taken by an employee or insider who has access to sensitive, classified, or proprietary information that when disclosed causes damage to an organization's interests.*

The concept of an "insider" is somewhat ambiguous. "Employee," on the other hand, has a visceral meaning because it implies the threat is "one of us." In addition, Gelles articulated insider threat as a process, such as an employee, triggered by internal and external factors, makes a decision, which leads to a chain of events that cause harm to an organization. Blackwell's definition underscored the vulnerability of giving an employee privileged access to a system, the only definitive aspect of insider threat. Gelles elucidated those characteristics that are indeterminate such as, predisposing risk factors, situational precursors, motivations, and the methods by which insiders perpetrate an attack. Knowing how the insider threat differs from external threats and understanding the granularity of the threat is essential for creating an overarching risk mitigation strategy.

How the Insider Threat is Different from the External Threat

Disproportionate attention focuses on mitigating external threats. Most organizations worry about people breaching existing external boundaries and firewalls. Therefore security tools are more attuned to threats from the outside and averse to threats from within the organization. The insider threat is unique because being an insider implies a relationship of trust and privilege, an assumption that can be misleading. According to Cole and Ring (2005), insider threat has received limited attention for three reasons:

1. Organizations do not know it is happening.
2. It is easy to be in denial.
3. Organizations fear bad publicity.

"Insider threat is occurring all the time, but since it is happening within a company, it is a private attack. Public attacks like defacing a Web site are hard for a company to deny. Private attacks are much easier to conceal" (Cole and Ring, 2005, pp. 6, 17). One of the limitations in defining the scale of the insider threat can be attributed to a lack of credible data and information sharing, which tacitly supports Cole's argument. Pfleeger and Stolfo (2009, p. 10) suggested two reasons for the paucity of data in the information technology (IT) sector:

> *First, many organizations are loathe to reveal the nature and magnitude of the cyber incidents they've experienced for fear of reputational harm. Second, most cyber surveys are convenience surveys; it's impossible to know what population the results represent.*

Their statement reflects the ongoing culture of denial. The proclivity for trust in the workplace is perhaps the greatest contributor to institutional denial—it is painful and embarrassing to be betrayed by those with whom we are close. The opacity of insider threat prevents many individuals and organizations from acting. Unfortunately, it does not make the consequences any less real.

It is important to distinguish between outsider versus insider threat because it effects how we approach the problem. One of the dominant paradigms in cyber security is the notion of perimeter defense, which is sometimes called the Maginot Line. The analogy implies that an organization protects what is inside the system from an attacker that is outside the system. The Maginot Line was a massive series of defense fortifications built before World War II that was designed to protect the eastern border of France. The Germans outflanked it within days, rendering its purpose strategically irrelevant. In fact, the diversion of valuable resources to the Maginot Line was blamed for preventing the modernization of the French military. Threats change and they require constant reassessment. Although a perimeter defense is a suitable tool for mitigating external threats, it is far from a comprehensive security strategy. Used alone, it gives only an illusion of security (Waterman et al., 2008).

Insiders, by default, are already inside the perimeter, therefore many of the countermeasures that companies deploy to prevent external threats are ineffective against the insider threat. Cole (2005, p. 7) argued that because insiders already have access, "the amount of damage they can cause is much greater than an external attacker, and the chances of getting caught are much lower."

Enhanced Capabilities of the Insider

Insiders have information and capabilities unavailable to outsiders. One way to assess the insider threat is by using the law enforcement CMO model, which proposes that the inside attacker has capability, motivation, and opportunity. Probst and Hunker (2011, p. 6) summarized four key attributes of the insider. They also serve as broad indicators of an insider's capability.

1. Access to the system
2. Ability to represent the organization to outsiders
3. Knowledge
4. Trust by the organization

Insiders are potentially more dangerous to an organization because it is easier for them to carry out an attack. They have legitimate and privileged access to an organization, its facilities, and information. Controlling access is the first line of defense against insider threat. Employers make their first security determination by either granting or denying access to its organization as a result of pre-employment screening. A study by the CERT Insider Threat Center, Common Sense Guide to Prevention and Detection of Insider Threats, suggests that employers do not take this responsibility seriously enough. CERT reported that 30% of insiders who committed IT sabotage had been arrested previously, including arrests for violent offenses. Of those insiders with criminal records, over half had multiple prior arrests and convictions (Cappelli et al., 2009, p. 43).

Both the number of people with access to information and the amount of data held by a single individual are significant criteria for assessing overall risk. There are a number of organizational tools to control these two variables. The "principle of least privilege" is used to determine the appropriate level of access. This approach implies that you give employees the least amount of access needed to do their job. When it's not possible to fully limit access, methods such as the "two-person rule" and "separation of duties" can be implemented to provide additional supervision to critical processes (Cole and Ring, 2005). For example, the two-person rule is frequently enforced in high-containment labs where scientists are working with biological select agents that pose the greatest risk to national security. A disciplined approach to controlling and monitoring access minimizes the potential for insider threat.

An insider is vulnerable to exploitation by outsiders.

> *It is usually far more cost-effective and quicker for an external threat source to place, or subvert, an insider to exploit vulnerabilities to steal information rather than launch an attack through multiple layers of protection. Insiders have time and are given capability and opportunity.*
>
> *Colwill, 2009, p. 2*

Examples of intrusion by outside entities include blackmail and espionage.

Two prominent cases of espionage include Aldrich Ames and Robert Hanssen. Ames, one of the most infamous traitors in U.S. history, was a Soviet counterintelligence expert in the Central Intelligence Agency (CIA). In April 1985, he arranged a meeting with a diplomat at the Russian Embassy, Sergey Chuvakhin, and he disclosed the names of two KGB agents spying for the United States in exchange for $50,000. In June 1985, Ames subsequently provided documents that compromised all human intelligence operations in Moscow. He exposed over 10 top-level CIA and Federal Bureau of Investigation (FBI) sources spying for the United States, many of whom were executed. CIA officials have stated that this was the "largest amount of sensitive documents and critical information, that we know anyway, that has ever been passed to the KGB in one particular meeting" (Senate Select Committee on Intelligence, 1994, p. 7) Ames continued to sell secrets to the Soviets during his follow-on assignments in Rome, the Soviet--East European Division at CIA Headquarters, and the Counterintelligence Center until his arrest in

1994. The compromise of these identities and as many as 30 operations led to the virtual collapse of the CIA's Soviet operations. According to the 1994 Select Committee, Ames has caused more damage to the national security of the United States than any spy in the history of the CIA.

Seven years after Ames was arrested, Robert Hanssen, a former FBI agent, plead guilty to charges of espionage by revealing highly classified counterintelligence and military information to the KGB during intermittent periods from 1979 to 2001. He compromised over 6,000 pages of known documents, and he disclosed numerous human sources, three of whom were ultimately executed (Office of the Inspector General, 2003). During his assignment at the New York FBI field office, Hanssen was put in charge of installing the FBI's automated counterintelligence database. Ironically, he used his tradecraft to exploit the information system he helped establish and thereby avoid detection (Herbig and Wiskoff, 2002).

Knowledge of the organization, its systems, and the location of critical assets make it easier for an insider to carry out an attack. An inside attacker has a high probability of success because he knows when and where to attack. It is common for insiders to exploit a known vulnerability or loophole. In contrast, outsiders are not privileged to access the same information, and an attack requires a greater investment of time and effort. The 2012 CERT Insider Threat Study on fraud in the U.S. financial services sector concluded that the aggregate insiders' means of carrying out an attack were not technically sophisticated.

> *Of the 80 fraud cases coded, 57 (71 percent) cases relied on some form of authorized use or non-technical bypass of authorized processes. Of the 57 cases, 52 involved subjects using some form of previously authorized access to carry out the fraud. Finally, in 5 of the 57 cases, the subject used some form of non-technical method to bypass authorized processes and commit the fraud... while the insiders' methods were largely non-technical, the insiders themselves also held non-technical positions.*
>
> ***Cappelli et al., 2012; p. 16***

This finding illustrates how a confluence of access and knowledge contribute to the capabilities of an inside attacker. The Insider Threat Study recommended that organizations pursue "commonly accepted security principles" in order to minimize the threat such as, "access control, least privilege, and separation of duties" (Cappelli et al., 2012, p. 19).

Organizations inherently want to trust their employees. Shaw and colleagues (1998, p. 3) stated,

> *As vetted members of the organization, employees are in a position of trust and are expected to have a vested interest in the productivity and success of the group. Considered "members of the family," they are often above suspicion—the last to be considered when systems malfunction or fail.*

An environment of trust makes it difficult to distinguish malicious and benign intent in the workplace. The CERT Insider Threat Study on cyber activity in the banking and finance sector discovered that 83% of insider threat cases involved attacks that took place physically from within the insider's organization. In 70% of the cases, the incidents took place during normal working hours (Randazzo et al., 2004, p. 20). Failure to recognize and report concerning behaviors in the workplace is sometimes attributed to legal concerns. It is controversial to report individuals for breaking the rules or acting suspiciously in fear of being accused of defamation.

The assumption of trust also leads to a culture of denial, which was evident in the Amerithrax Case. The U.S. Department of Justice has determined that Bruce Ivins was solely responsible for the anthrax mailing attacks of September and October 2001. There were multiple Department of Defense (DoD) procedures and policies in place. Specifically, Dr. Ivins had signed multiple waivers that would have allowed DoD background investigators to review his psychiatric records prior to his initial hiring in 1980. These records documented serious criminal behaviors that would have automatically disqualified Dr. Ivins from receiving employment in his sensitive position, which also required a security clearance. Other less serious criminal behaviors, like vandalism directed against scientific colleagues, demonstrated his capacity for both stealthy behavior and rage (Saathoff et al., 2011; Willman, 2011). Had existing policies and procedures been utilized over the years, they would have prevented Bruce Ivins from having access to weapons-grade anthrax. The culture of denial, that is, "blindness" to the possibility that it could have been "one of us" diminished the effectiveness of those policies.

Categories of Insider Threat

Insider threat can be divided into two broad categories according to either malicious or negligent intent. Whereas many authors describe threat as the "intent to cause harm or danger," Blackwell's definition leaves open the possibility that insider threat can be unintentional, meaning that misuse of one's privileged access can also expose an organization to risk. An employee who poses a threat to an organization due to substance intoxication may be unintentional, although attempts to evade drug screen detection are intentional by their very nature. The distinction between malicious and non-malicious insiders is an important factor to consider for determining an appropriate policy response.

The two categories of insider threat are manifested in recent security breaches from within the intelligence community. As a case in point, In May 2010, U.S. Army intelligence analyst Bradley Manning was arrested for disclosing the largest set of classified information ever leaked to the public including more than 250,000 diplomatic cables, more than 90,000 intelligence reports on the War in Iraq and Afghanistan, and a video of a helicopter attack in Baghdad thought to have depicted indiscriminate killing of civilians. Reportedly, Private Manning exploited a security loophole to copy thousands of files onto compact discs over a six-month period. Private Manning was charged with "aiding the enemy," a capital offense (*U.S. v. Bradley Manning*, U.S. House of Representatives, 2006). The dissemination of documents through WikiLeaks represents the potential magnitude of the threat of an insider attack.

In contrast, there are also instances where a person who is entrusted with classified information makes an unintentional disclosure, known as "spillage," which results when classified or sensitive information is transferred to an unauthorized system. Spillage is caused by improperly handling media and/or media releases, improper data transfers, mislabeling data, or contaminated data received by an outside source, for example, malware (Committee on National Security Systems, 2007). Ignorance of a leak is dangerous because it prevents the individual or system from responding to the problem. Whether intentional or negligent, these two routes illustrate the perils of insider threat when information is compromised.

To the extent that we can understand the discrete risks posed by different types of insiders, are we able to implement effective security control measures? After the 9/11 attacks, the DoD expanded overall access to its Secret Internet Protocol Router Network (SIPRNet) to more than 3 million users. SIPRNet is an encrypted server used by DoD to globally transmit material up to and including "secret," the government's second highest level of classified information. In an effort to prevent future attacks, the network was linked with embassies and consulates to enable U.S. Government agencies to share classified information more easily. Ironically, increased access led to yet another national security risk, giving the WikiLeaks informant access to confidential diplomatic messages (Greenemeier and Choi, 2010). In response to the WikiLeaks incident the government restricted access, and implemented a host-based security system (HBSS), a technology formerly used by credit card companies to detect suspicious and anomalous behavior. The

HBSS is an automated system that includes a firewall, intrusion detection system, and other controls to monitor and detect threats to IT systems. The government previously relied on "perimeter-based solutions," such as access controls and encryption to protect its network. The DoD's response to WikiLeaks shows how it adapted security mechanisms to prevent an insider attack. The Cybersecurity Director to the Department of Homeland Security, Amit Yoran (Greenemeier and Choi, 2010), advocated a risk-response approach for addressing insider threat in response to WikiLeaks. His remarks also highlight two unmistakable characteristics of the threat: access and trust.

> *Once you have access to these classified systems and are inside their tough perimeter, they have historically been very trusting. And when you have a trusted insider who is interested in causing harm or inappropriately accessing information, that sort of architecture with strong perimeters is quite flawed.*

Insiders can be further classified according to their level of access. According to Blackwell there are three different types of insiders: (1) trusted insider, (2) regular insider, and (3) partial insider. He stated that a trusted insider includes management, administrators, financial, and technical staff. They require the greatest amount of access to complete their work. A regular insider, or insider, is an employee that is given access for his work. Insiders may use legitimate access and/or obtain unauthorized access to conduct an attack. A partial insider has limited access and includes people like security guards, maintenance, and cleaning staff (Blackwell, 2009). Other authors present more detailed classification schemes; however, Blackwell's model is concise, illuminating the two most important elements of an insider—intent and level of access—that can easily be applied to a range of different scenarios. Within this framework, an insider may be an employee, contractor, vendor, business partner, or visitor who is given internal access privileges (Blackwell, 2009).

Carrying Out the Attack

The previous examples have illustrated that some of the most significant risks to national security have been perpetrated by employees on the inside and not the result of attacks conducted by outsiders. In order to prevent and detect insider threat it is vitally important to understand the characteristics of the attacker, employee risk factors, motivations, the types of attacks, and how they are impacted by technology. Some of the most comprehensive studies on insider threat have been conducted by the CERT Insider Threat Center, the DoD Personnel Security Research Center, and the RAND Corporation. From these studies one can make two important assumptions. First, individual actors, their motives for committing crime, and the type of attacks they commit are highly diverse. These variables make profiling the insider threat a futile endeavor. Second, the studies consistently acknowledge concerning behaviors, situational stressors, and organizational culture as having the most direct implications on insider risks. Insider threat is manifested when human behaviors deviate from normal standards of conduct and established policies.

The CERT Insider Threat Center has conducted the only longitudinal study of insider threat. Since 2001, CERT has collaborated with the U.S. Secret Service to examine relevant cases and recommend best practices for the prevention and detection of insider threat. CERT has collected over 700 cases, which serve as the foundation of their work. The Insider Threat Study focuses particularly on the threat of the individual. This study has resulted in four case study reports that analyze the discrete nature of insider threat in the following sectors: (1) banking and finance, (2) critical infrastructure, (3) IT and telecommunications, and (4) government. Their recent work has culminated in the publication of The CERT Guide to Insider Threats: How to Prevent, Detect, and Respond to Information Technology Crimes (Theft, Sabotage, Fraud), 2012.

According to The CERT Guide to Insider Threats, there are three major types of insider cyber crimes including: (1) IT sabotage, (2) intellectual property theft, and (3) fraud. For each crime profile, CERT describes how the insider threat has evolved over time, as well as the motivations, attack methods, organizational issues, and precursors that may have helped the organization prevent or detect the attack. One of CERT's major findings was that insiders did not share a common demographic profile, rather their characteristics varied according to the type of crime and sector. For example, they determined technical users were predisposed to insider IT sabotage. They were primarily motivated by revenge and provoked by situational stressors in the workplace. Crimes were often planned while the individual was employed, but executed following termination. In contrast, insider theft of intellectual property was frequently perpetrated by technical scientists, engineers, and programmers seeking business advantage. Fraud, on the other hand, was committed by lower level employees who were motivated by financial gain. They were determined most likely to be recruited by outsiders to steal information (Moore et al., 2012). The case studies reveal the diverse topography of insider threat, albeit limited to the IT sector. The challenge of mitigating insider threat is onerous because of the various ways insiders cause harm to an organization.

Insider attacks are low-frequency high-impact events. The limited number of real-world datasets makes it difficult to accurately model the insider threat by most technical standards, and therefore analysis often leads to a heuristics-based approach. As a case in point, Shaw and others (Shaw et al., 1998, p. 5) categorized a group of insiders labeled Critical Information Technology Insiders (CITIs), which they feel are particularly vulnerable to insider threat.

> *Case studies and survey research indicate that there is a subset of information technology specialists who are especially vulnerable to emotional distress, disappointment, disgruntlement and consequent failures of judgment which can lead to an increased risk of damaging acts or vulnerability to recruitment or manipulation.*

He stated that the majority of CITIs possess the following risk factors: (1) introversion, (2) social and personal frustrations, (3) computer dependency, (4) ethical flexibility, (5) reduced loyalty, (6) entitlement, and (7) lack of empathy (Shaw et al., 1998).

Contrary to this approach, the National Infrastructure Advisory Council concluded that oversimplification of the insider threat could cause organizations to focus on the wrong people. The report stated, "Discussions on insider threat betrayals often describe insiders as disgruntled, and while this is a convenient and somewhat accurate stereotype, it is limiting in the sense that the insider psychology is more complex" (National Infrastructure Advisor Council, 2008, p. 16). The study found no direct correlation between disgruntled workers and insider threats. This example illustrates the problem with profiling employees as a means of preventing insider threat. Schouten and Saathoff (2010, p. 228) provided the following rationale:

> *The quest for a screening tool that relies on personality and other characteristics to detect employees at risk of posing an insider threat is understandable, but remains unfulfilled. Like other low incidence phenomena, such as workplace violence and suicide, inside attacker profiles suffer from the problem that even a highly sensitive test will yield an unacceptably high level of false positives. As a result, many individuals will be identified falsely as posing a risk.*

Not having enough factual data and the prevalence of risk factors in the general population makes it nearly impossible to detect insider threat using technical screening methods. In contrast, prior concerning behaviors, substance abuse, and employment history are considered to be more reliable indicators of insider risk.

Understanding the behavioral factors that shape insider threat is critically important for preventing an attack. Gelles et al. (2012) described insider threat as a process where an individual engages in a pattern of behavior to achieve a desired solution. Gelles suggested that some of the following risk indicators and characteristics can be used as a guide to identify potential insider threats:

- Behavior that is consciously pursued over an extended period of time
- End results of a complex set of problems, conflicts, and disputes, generally reflected in an individual's personal life
- Individuals feeling the organization was unresponsive to their needs
- Individuals seeking validation of their self-aggrandized view of their abilities and achievements
- Self-centered, entitled, and undervalued persons
- Individuals seeking immediate gratification and satisfaction
- Individuals that, if their needs are not met, act in ways that are rebellious, passive aggressive, or destructive
- Individuals who seek out others who will meet their needs or undermine the efforts of those they feel have neglected them, or who did not recognize their potential
- Intolerance of criticism, inability to assume responsibility for their actions, blaming others and minimizing their mistakes or faults (Gelles et al., 2012, p. 5)

In September 2009, Chief Judge Royce C. Lamberth of the U.S. District Court for the District of Columbia issued an Order authorizing an independent panel of experts, the Expert Behavioral Analysis Panel (EBAP) to examine the mental health issues of Dr. Bruce Ivins and provide analysis of lessons learned. The report directly addressed the individual and organizational behavioral aspects of insider threat, and it revealed that had existing policies and procedures been followed, Dr. Ivins' documented criminal behaviors would have come to the attention of government officials and would have disqualified him from employment. Three of the most apparent behavioral risks presented by Bruce Ivins were substance dependence, homicidality, and suicidality.

THE AMERITHRAX CASE

Dr. Bruce Ivins was identified as the individual who was solely responsible for the mailing of the anthrax letters to the media and U.S. Congress in September and October of 2001. Killing five individuals, infecting many others, and causing incalculable damage to the national security research infrastructure relating to biological select agents and toxins, the case is significant in that Dr. Ivins constituted a "triple threat." This threat involved some of the most significant behavioral factors that can motivate and influence the actions of an insider: substance dependence, homicidality and suicidality.

Using the Amerithrax case as an example of insider threat, we will review the three factors followed by a discussion of the emergence of substance abuse of drugs that are difficult to detect and its potential impact upon national security.

Substance Dependence

Dr. Ivins' case exemplified the dangers associated with personnel who engage in substance abuse. Dr. Ivins abused prescription drugs (Saathoff et al., 2011; Willman, 2011). He reportedly started taking diazepam due to a back injury early in his career and then began taking sleeping pills when he later experienced insomnia. At one point in his career, his reported average daily intake was a staggering four eszopiclone tablets, six zolpidem tablets, three diazepam tablets, and diphenhydramine, all taken on top of 12 ounces of vodka, an absolutely extraordinary quantity of sedating substances. He admitted that those medications would independently cause him to experience "blackouts." He reported that during those blackouts, he would send e-mails that he did not remember sending, reporting that at times they would be very embarrassing. He also admitted to having thoughts about killing himself with alcohol and pills at that time. His prescription drug abuse is consistent with changes in substance abuse within the United States. In 2007, prescription drug abuse visits to the country's emergency departments surpassed illicit substance abuse, and those numbers continue to climb (Substance Abuse and Mental Health Services Administration, 2010).

Homicidality

Throughout his life Dr. Ivins told mental health professionals of his plans to poison various people (Saathoff et al., 2011; Willman, 2011). Although he never consummated any of these plans, he reported that he had not only carefully designed them but also had carried them through to the point of poisoning substances intended for the victims' ingestion.

Dr. Ivins acknowledged his plans to poison others during many psychiatric sessions. The first reported incident occurred during a psychiatric session in 1978. At that time, he plotted to disfigure a coworker with a scalpel, kill with a bat, or poison. He elaborated by stating that poisoning would have been easy and explained how he would put tea mixed with poison into the jar from which the coworker typically drank. He also admitted that he had some thoughts about harming various people in 2000 and how he obtained certain substances, including ammonium nitrate to cause an explosion and cyanide to poison. He admitted to thinking about these malicious things only when he was in his "quiet rage" states and planned, in minute detail, how he would carry out these plans. During a psychiatric session in 2000, Dr. Ivins confided that he had visited a previous coworker, who was in New York, and had brought poison in a bottle to kill her, but changed his mind. As late as March of 2008, Dr. Ivins discussed with his therapists "how to murder someone and not make a mess."

In addition to Dr. Ivins' demonstrated interest in using poisons and explosives to harm others, he also acquired firearms and bulletproof vests in planning for an assault against a list of people who he felt had wronged him. Even though his statements caused a subordinate to express fear for her personal safety, Dr. Ivins' supervisor's only advice was for the subordinate to "hide" in the laboratory that Dr. Ivins was no longer authorized to access. Two days later, therapists at a group therapy session heard him make similar homicidal threats and as a result arranged for him to be assessed in an emergency department of a hospital where two separate mental health professionals assessed him and agreed to have him involuntarily hospitalized for both homicidality and suicidality.

Suicidality

Dr. Ivins also often acknowledged that he had detailed plans to commit suicide with various substances. His own suicide exemplifies how he was able to meticulously plan a successful poisoning. For example, during a psychiatric session in March of 2008, Dr. Ivins pulled a handwritten plan out of his wallet and let his therapist read it. The plan was extremely detailed and involved his drinking Manischewitz wine in the exact amount, to the ounce, necessary, and taking a specific number of pills of zolpidem and benzodiazepines. Dr. Ivins conducted Internet research to determine exactly how much alcohol and pills he needed, according to his height and body weight, to kill himself.

During its investigation and review of his mental health records, the EBAP also learned that Dr. Ivins repeatedly threatened to kill himself with anthrax. That a scientist with a security clearance and access to biological select agents and toxins such as anthrax would contemplate suicide with one of these agents is significant. This expression of suicidality indicated the microbiologist's comfort with the nefarious use of a dangerous agent that had been entrusted to him by virtue of his insider status and professional expertise.

Substance Abuse

Drug testing is exceedingly limited in detecting substances abused in today's society. Most drug testing programs employ antibody identification of drug metabolites, with many specifically looking for only a limited number such as marijuana, cocaine, amphetamines, and opiates. The positive identification of drug metabolites is influenced by chronicity of ingestion, fat solubility, and co-ingestions. Many illicit drugs of abuse are not detected on most urine drug screens, including such substances as gamma-hydroxybutyrate, khat, and lysergic acid diethylamide. The recent increase in a plethora of Internet-acquirable drugs, such as synthetic cannabinoids (e.g., "Spice" and "K2") and "bath salts" (synthetics, such as mephedrone and methylenedioxypyrovalerone; MDPV), are not detected on any drug screens. Further confounding the use of drug screens has been the explosion of prescription drug abuse (e.g., oxycodone, methylphenidate, zolpidem), which also do not make typical drug screens positive. The utility of urine drug screens is fraught with significant testing limitations, including false-positive and false-negative results.

Certainly substance abuse has played a role within the context of insider threat. Bruce Ivins' case exemplifies this issue. Substance abuse can also lead to other individuals becoming easy targets for espionage. As an example, Sgt. Roderick Ramsay was arrested in 1990 for spying for Hungary and sentenced to 36 years in prison (Fischer, 1997). He was a clerk responsible for accounting for all classified documents in the G3 Plans Section of the 8th Infantry Division in Germany. From 1983 to 1986 he passed secret and top secret documents containing military operational plans for the defense of Western Europe to Hungary. He reportedly was paid $20,000 for the transfer of the information. Sgt. Ramsay used drugs regularly while in the military and recruited other American soldiers to provide classified documents. He used drug use as a selection criterion when recruiting other soldiers to spy and determined that those willing to take the risk of regularly using drugs had already shown a propensity to violate military regulations.

There is also a rapidly emerging change in the prevalence of substances abused. Prescription drug abuse is currently a major issue worldwide and has been growing in prevalence over the past decades. However, designer synthetic product availability is a growing trend due to easy access via the Internet. For example, substituted cathinone abuse occurred primarily in Europe after aggressive Internet marketing of nonhousehold, "legal high" products containing mephedrone. During 2009, the UK Forensic Science Service noted a rapid rise in designer stimulant seizures, the majority of which contained mephedrone. Surveys conducted in the UK at that time and prior to legal restriction found lifetime self-reported mephedrone use ranging from 20% in secondary and university students to over 40% in dance club attendees (Winstock et al., 2011; Wood et al., 2011). In contrast to the UK's mephedrone expansion, Finland experienced growing use of MDPV, which in the fall of 2009 prompted initiation of systematic forensic testing of all drivers suspected of being under the influence of drugs (Kriikku et al., 2011). Over the next year MDPV was detected in 5.7% of 259 cases.

The U.S. rise in designer stimulant use lagged a couple years behind Europe. The Drug Enforcement Agency's National Forensic Laboratory Information System identified substituted cathinones in 34 forensic samples from 8 States in 2009. By 2010, 628 samples from 27 States contained substituted cathinones (Drug Enforcement Agency, 2011). Although the findings suggest increased availability throughout the United States, regional variation in access and consumption may exist, given that most samples were obtained from the South (57%), followed by the Midwest (25%), Northeast (16%), and West (2%). The U.S. forensic data are paralleled by a concerning increase in calls to U.S. poison control centers, which provides further evidence that the prevalence of designer-substituted cathinone use is widespread and growing. After receiving no calls in 2009, the American Association of Poison Control Centers received 304 calls about bath salts exposure in 2010 and 6,138 calls in 2011, a 20-fold increase from the previous year (American Association of Poison Control Centers, 2012). None of the substances can be detected on typical urine drug screens and all have been associated with psychotic and irrational behavior.

SUMMARY

While the current focus on insider threat is based on lessons learned within the cyber and technology arenas, supervisors and investigators should not lose sight of the need for vigilance in identification and reporting of insider threat behaviors.

The case of Dr. Bruce Ivins, who has been determined by the U.S. Department of Justice as being solely responsible for the mailing of the anthrax letters in 2001, is instructive. In addition to the five people who died as a result of the mailed anthrax, many more were subject to inhalational and cutaneous anthrax infections. The cost to society in terms of expense and social disruption remains incalculable.

As an individual, Dr. Ivins represented a threat that was both malicious and negligent. These threats included his long-standing homicidality, suicidality, and substance dependence. The fact that his criminal behaviors were unrecognized by most of his colleagues speaks to both his compartmentalized, stealthy behavior and the trust and denial that existed within his laboratory culture.

CHAPTER

4

An Age of Asymmetric Challenges—4th Generation Warfare at Sea

Peter Lehr

INTRODUCTION

The last couple of years saw a vivid and controversial discussion centering on two different military-related concepts: revolution in military affairs (RMA), and asymmetric warfare (AW). While RMA dominated discussions in the decade following the first Gulf War of 1991, the events of September 11 and the ensuing global war on terror (GWoT) plus the re-emergence of piracy firmly tilted the balance in favor of the latter concept, as illustrated by the number of conferences and workshops covering this topic, and by the proliferation of new ideas and assumptions related to AW. In the realm of maritime security, and under the general and widely believed impression that an "age of asymmetric challenges at sea" (Kaplan, 2011) is upon us, new concepts related to (naval) AW emerged, such as the Royal Navy's Maritime Irregular Activities (MIA), or the U.S. Navy's Confronting Irregular Challenges (CIC). MIA focuses on all naval activities a regular navy can conduct short of war, and short of all other conventional usages of navies as specified by Geoffrey Till, for example. In a sense, both MIA and CIC seem to be the naval equivalent of the general "nexus issues" debate, which tends to lump together all kinds of threats, whether they are political or criminal in nature. Furthermore, and apart from arguably also being a defensive measure to make navies appear more relevant in the era of severe (defense) budget cuts, this also implies an unspoken acceptance of the notion popular with many Western scholars that conventional war, and, by extension (or rather reduction), conventional naval war is a thing of the past, and that we have moved from third generation warfare (3GW) to a fourth generation warfare (4GW). As laudable as this postmodernist stance may be, the Georgian–Russian crisis and the South Ossetian War of August 2008 should serve as a reminder that such a notion may well be a bit too hasty, and yet another example of misunderstanding current and near-future threats–with potentially fatal consequences.

This contribution takes a look at the Indo-Pacific to discuss nontraditional security issues such as piracy, maritime terrorism, and maritime AW as well as the security problems posed by weak and failed coastal states (including coastal states undergoing regime or policy change), the future of conventional energy resources and energy security, and finally traditional security issues such as the enfolding arms race between India and China. On the basis of these findings, and drawing on the 4GW studies of scholars such as William S. Lind and practitioners such as Thomas X. Hammes, the nature of current and near-future maritime conflicts and the effects on Western national security will be analyzed. In this context, a critical look at the very idea of a transition from a conventional war (3GW) to a form of warfare in which the state has lost its monopoly on violence, finding itself fighting against violent nonstate actors, and in which the boundaries between combatants and noncombatants become blurred (4GW; Lind, 2004) is essential. This is not just for academic reasons, but for the way we construct "our" threat, which, quite logically, affects the formation of national security policies and strategies, the defense budgets, and, last but not least, the acquisition of future weapons systems. The gist of this chapter is that the current preoccupation with the asymmetric and/or irregular threat posed by al-Qaeda or Somali high sea piracy might lead to long-term defense decisions, which may prove fatal under different circumstances.

DEFINITIONS: FROM NAVAL AW TO MIAs

First let's turn to the longer established concepts of RMA, symmetric/asymmetric warfare, and regular/irregular warfare in order to provide some historical depth and theoretical background to the current debates around the "reductionist" concepts MIA/CIC on one hand and the "expansionist" 4GW on the other. Although the concepts can be traced back to the 1980s, conferences and workshops focusing on RMA appeared on the agenda mainly under the impression of the First Gulf War of 1990–1991. Attempts were made to define the concept and to assess its implication for the mass-based military of emerging nations aspiring to great power status, especially the People's Republic of China (PRC) and India. There are many definitions for RMA, but here, we shall use Clifford Roger's (2000, p. 22) definition:

> *[A] paradigm shift in the nature and conduct of military operations which renders obsolete or irrelevant one or more core competencies of a dominant player.*

The gist of most discussions was finding ways to transform mass-based armed forces of the past into "leaner and meaner," for example, high-tech and highly mobile, fighting forces of the twenty-first century. Ironically, the definition given above for this anticipated high-tech revolution could also be used to describe the concept of AW, which gained currency in the context of the global alliance against terror: basically, AW also aims at rendering irrelevant core competencies of a dominant player. It does so—or at least tries to do so—not by reaching for the high ground of (expensive and sophisticated) high-tech solutions, but by looking for comparatively cheap, unsophisticated, and unconventional low-tech approaches. Since AW arguably is an even broader concept than RMA, these approaches may include, for example, cyber or "hacker" attacks, information warfare, guerrilla tactics, or acts of terrorism directed against armed forces or any other vulnerable assets of the enemy. This indicates that asymmetries can also include asymmetries of purpose on top of the usual asymmetries of weapons and tactics. Tellingly, one United States definition of the concept draws our attention to the act that the adherents of AW basically deny (mostly Western) high-tech powers what this author would like to call a "level battlefield":

> *Put simply, asymmetric threats or techniques are a version of "not fighting fair," which can include the use of surprise in all its operational and strategic dimensions and the use of weapons in ways unplanned by the United States.*
>
> ***National Strategical Studies:1998 Strategic Assessment, 2012***

In itself, this "not fighting fair" is nothing new. Probably the most famous as well as earliest example is the contest between David and Goliath mentioned in the Bible: the use of a projectile weapon against an unsuspecting warrior prepared for a decisive single combat based on certain rules of conduct is not fighting fair indeed, but it undoubtedly achieved the objective of "rendering obsolete a core competency of a dominant player" (Kriwaczek, 2012, p. 93).

Just as the more general concept of AW, the narrower notion of either naval or maritime asymmetric warfare is not new. The bulk of the mostly Western literature on this kind of war focuses on more "conventional" and decidedly naval forms of asymmetric warfare, such as the Battle of the Atlantic during the WW II, where wolf packs of German submarines attacked convoys protected by surface warships ranging from corvettes to battleships to carriers, and on the present use of missile crafts and diesel-powered submarines by smaller navies against superior forces in a littoral scenario. The doctrines of the naval branch of the Iranian Revolutionary Guard and the Iranian Navy proper would be a case in point. It is also quite interesting to note that al-Qaeda was quick to draw the attention of its followers to forms of maritime AW more suitable to their means (see Ulph and Heffelfinger, 2004). But we do not intend to go into history here. Suffice it to say that state actors as well as substate and nonstate actors are aware of the merits, and possibly also the demerits, of naval AW. This also explains why, in its modern avatar, the concept of AW re-emerged in the context of the global war against terrorism, prompted mainly by the USS Cole attack: How can a modern power possessing overwhelming technical superiority protect itself against low-tech threats?

However, even a short and superficial glance at maritime issues that keep navies occupied at the moment demonstrates that a focus on naval AW or on naval irregular warfare does not provide us with a full picture of the current threats at sea. For example, the activities of Somali pirate action groups (PAGs) in the Somali Basin may well contain aspects of asymmetry and irregularity, but it is problematic to define them as "warfare" without using a lot of qualifications and caveats. Hence, the Royal Navy prefers the "reductionist" term maritime irregular activities instead of naval asymmetric warfare, as used by the U.S. Navy. MIA can be defined as "behaviour [at sea] that attempts to effect or prevent change through the illegal use, or threat, of violence, conducted by ideologically or criminally motivated non-regular forces, groups or individuals, as a challenge to authority." In support of MIA, a Royal Navy discussion paper argues that "[contemporary] experience confirms that today's irregular threats—criminality, disorder, insurgency and terrorism, are more complicated, less geographically constrained and less containable than before." Likewise, the U.S. Navy moved away from the terms asymmetric warfare or irregular warfare toward confronting irregular challenges, defined as situations "wherein states and nonstate actors leverage uncontrolled or ungoverned space to employ informational, economic, technological, and kinetic methods against civilian populations and targets to achieve their objectives." Of both concepts, CIC is the broader one since it explicitly includes state actors, while MIA only covers nonregular forces illegally using, or threatening, violence, thus more in league with the older concept of violent gray area phenomena (GAP). Rather interestingly, a Congressional Research Service (CRS) Report for Congress acknowledges the new term, only to then discard it in favor of the older term irregular warfare for "purposes of convenience" (O'Rourke, 2012, p. 3). This shows how difficult it can be for new terms to gain currency, especially if they do not seem to offer more academic rigor.

CASE STUDY: THE INDIAN-PACIFIC AND SEA LINES OF COMMUNICATION SECURITY

For our own purpose of convenience, we shall follow CRS for the moment to take a close look at maritime activities, irrespective of whether they are asymmetric, irregular, or something entirely different, and return to our concepts later. Also for the purpose of convenience, and to avoid any tedious cataloging of maritime conflicts or "activities" anywhere in the (maritime) world, we will restrict ourselves to the Indo-Pacific, as it is arguably the most important maritime strategic arena of the twenty-first century. Our oil and gas supplies have to travel through this area, negotiating several maritime choke points such as the Strait of Hormuz, the Bab el-Mandeb, and the Straits of Malacca, and this is where we have faced acts of maritime terrorism (launched by al-Qaeda cells) as well as acts of maritime guerrilla warfare (conducted by the Iranian Republic Guard Corps Navy; IRGCN). In this area we currently face a wave of high-seas piracy, which has resulted in a wide range of extraregional navies. Hence, this maritime region is home to potentially all sorts of MIA or CIC as defined above.

We already mentioned the core component of both RMA and AW, that is, "standing up to and/or rendering irrelevant" the overwhelming technological superiority of modern blue-water navies, especially the U.S. Navy. This is quite a tall order even for the blue-water navies of the rapidly emerging new great powers such as India and China, and even more so for substate and nonstate actors. In the latter's special case, that is, guerrilla groups, terrorist networks and/or organized crime networks, achieving sea control is an impossibility, but a more modest sea denial in littoral waters or near maritime choke points may not. Again, what they would need, just like a "real" navy, are secure land bases from which to launch their attacks, and again, preferably situated near restricted waters. From a geostrategic point of view, the Indo-Pacific features several locations where such attacks could possibly be launched, even by nonstate actors: the Red Sea, Bab el-Mandeb, and Gulf of Aden; the Persian Gulf and the Strait of Hormuz; the Palk Strait between Sri Lanka and India; the Straits of Malacca plus the Sunda, Lombok, and Ombai-Wetar straits; and, at least in theory, the Cape of Good Hope, from which movements of ships can easily be monitored. It is thus far from coincidental that the relatively few acts of naval asymmetric/irregular warfare, MIA or CIC, on record did actually occur near these maritime choke points, as the following examples demonstrate:

- During a phase of the Iran–Iraq War—the so-called "tanker war" starting in 1984—the IRGCN conducted wolf-pack style attacks based on fast attack crafts on Kuwaiti tankers in the vicinity of the Strait of Hormuz, "swarming" the tankers and their escorts in an attempt to overwhelm their defenses.
- Likewise, until their defeat in April/May 2009, the Liberation Tigers of Tamil Eelam (LTTE) conducted such swarming attacks on vessels of the Sri Lankan Navy (SLN) in the Palk Strait and the Gulf of Mannar, destroying or at least seriously damaging many SLN warships as well as capturing commercial vessels.
- Various al-Qaeda cells conducted suicide attacks on Western vessels as well as offshore oil installations in the Gulf of Aden (USS Cole attack, October 2000; M/V Limburg attack, October 2002), the Persian Gulf (al-Basra and Khor al-Amaya oil terminal attacks, April 2004), and the Strait of Hormuz (M/V M-Star attack, July 2010).

Somali organized pirate gangs operated in the Gulf of Aden and adjacent waters for about a decade before being pushed to the high seas, preying on international commercial shipping hundreds of nautical miles off their shores.

This also indicates that, basically, three categories of actors need to be dealt with in our context: state actors, substate actors, and nonstate actors, with the latter subdivided into criminal and political/ideological/religious actors. For our purposes, state actors are defined as sovereign states and their respective navies, coast guards, or other maritime agencies, including paramilitary entities with maritime capabilities. Substate actors are, in our understanding, insurgent or guerrilla groups, that is, groups or movements in control of a certain territory from which to launch strikes. Nonstate actors in our context should be understood either as "pure" terrorist groups not in control of territory, or as organized crime groups or networks. To avoid an otherwise lengthy examination and tedious cataloging of all possible actors along the shorelines of the Indian Ocean, we will only consider navies whose order of battle include forces tailor-made for asymmetric naval warfare, and substate and nonstate actors that possess a proven capability to conduct seaborne operations.

Criminal Nonstate Actors: Smugglers, Traffickers, and Pirates

On the lowest level of our possible actors, we find an amorphous network of criminal nonstate actors in the shape of smuggling rings and pirates. The bulk of the former simply conduct a continuation of legitimate trade by other means, so to speak, transporting all sorts of commodities such as ghee, TV sets, refrigerators, or cars from places like Dubai to the Iranian coast, usually to return with illicit commodities in demand in the emirates, including recreational drugs and (oftentimes underage) female prostitutes of South Asian origin. Other smugglers transport weapons from the Arabian shores of the Gulf of Aden to the coasts of Somalia, while some also ferry al-Qaeda operatives from locations along the shores of Pakistan into the gulf or from Yemen to Somalia. Prospective Jihadists from the Arabian Peninsula also oftentimes travel by sea either to Afghanistan via Pakistan or to Somalia across the Gulf of Aden. Like the opportunistic Persian Gulf pirates preying on local inshore fishing vessels as long as they are not sailing in the wake of patrolling Western warships, and the organized Somali pirate groups preying on vessels on the high seas, the smuggling rings are usually replenished from the ranks of fishermen who find it more and more difficult to earn a decent living against the backdrop of nearly exhausted fish stocks. Of particular note for those who have already encountered the highly sophisticated "narco-trafficantes" active in the Caribbean and their fleets of "go-fast" or "cigarette boats," semi-submersibles, and even submersibles is the fact that smuggling rings active in the waters of the Indo-Pacific do not (yet) field similar high-tech equipment. This is because of a still rather embryonic level of maritime surveillance and, thus, a near absence of what we now know as maritime domain awareness; making use of inconspicuous local wooden crafts (dhows) and small freighters to ply their illegal trade still works perfectly fine for them.

Apart from fishermen moonlighting as opportunistic pirates when the circumstances are right (or the "catch of the day" proved elusive again), most of the sea lines of the Indo-Pacific are also infested by organized pirate gangs preying on vessels on international voyage in the vicinity of maritime choke points or, in the case of the Somali PAGs, on the high seas. Many articles and even book-length studies have been written about modern day pirates, especially Somali and Southeast Asian pirates, covering basically everything from historical roots to modus operandi to countermeasures at sea and on land. Hence, this author assumes that the scope of this maritime crime is well known. Suffice it to say then that the presence of PAGs at two of the world's most important maritime choke points, such as the Straits of Malacca and the Bab el-Mandeb, and their attacks on oil tankers and container ships, made the industrialized nations of East Asia, Western Europe, and North America quite nervous. In theory at least, pirate fleets could interrupt the flow of maritime traffic, thus disrupting our "just-in-time, just enough" economies. For this reason, piracy as an essential criminal activity or, in its most sophisticated form, seaborne organized crime, was effectively securitized, reframed as a "clear and present danger" to our respective country's national security, and somehow lumped together with our ill-defined global wars on terrorism and drugs (Thorup, 2009).

From an East Asian perspective (Japan, South Korea, Taiwan, and the PRC), the two most neuralgic potential maritime trouble spots with regard to their seaborne energy supplies are the Persian Gulf/Strait of Hormuz and the Strait of Malacca/South China Sea maritime security complexes. In the case of China, the very possibility of a disruption of Chinese crude oil and gas transports led to the so-called Malacca Dilemma that we will discuss later. All East Asian industrial powers thus have a keen interest in the security of the straits and so do the rim states of this all-important waterway. Despite their mutual animosities, Indonesia, Malaysia, and Singapore pooled their resources to conduct combined anti-piracy patrols known as MALSINDO Malacca Straits Coordinated Patrols. Also, a government-to-government agreement named Regional Cooperation Agreement on Combating Piracy and Armed Robbery Against Ships in Asia (ReCAAP) was launched to facilitate operations, with a ReCAAP Information Sharing Center as its most important result so far.

From a Western (mainly European Union; EU) perspective, it is the complex of the Gulf of Aden/Bab el-Mandeb/Red Sea complex that is of second-most importance after the Persian Gulf/Strait of Hormuz security complex. The bulk of Europe-bound oil and gas tankers takes the route through the Bab el-Mandeb, the Red Sea, and the Suez Canal to the Mediterranean. This means that, just like in the case of the Straits of Malacca, an interruption of traffic through this maritime choke point would have serious consequences for the national security of EU countries. Hence, it is not surprising that the emergence of Somali high-seas piracy also led to a swift securitization of this essentially criminal problem, and the first naval out-of-area operation of a combined EU fleet in the shape of EUNAVFOR Operation Atalanta, as well as a NATO operation named Operation Ocean Shield. The reason behind the comparatively rapidly decided and organized Western naval action is the fact that all of the Gulf of Aden rim states are either failed (Somalia) or failing states (Yemen), or at least weak states without any meaningful maritime capabilities. Unfortunately, discussions between Saudi Arabia, Egypt, and Oman to launch a regional antipiracy patrol patterned after the MALSINDO never materialized, even before the onset of the Arab Spring. On the other hand, the EU and NATO found support from some unexpected sides: Russia China, Japan, and South Korea deployed warships, and even Thailand dispatched some of their ships, such as the HTMS Similan. The deployment of Chinese warships was not necessarily greeted with enthusiasm by India as the biggest power of the Indian Ocean, for reasons that will be discussed below. In any case, we can state that antipiracy operations in the Indo-Pacific are keeping many navies quite busy at the moment.

Political Nonstate Actors: (Maritime) Terrorists

The more serious danger for the national security of states depending on maritime traffic—and the reason behind the presence of the Combined Task Forces (CTF) 150 (Gulf of Aden), CTF 152 (Persian Gulf), and CTF 158 (Northern Arabian Gulf)—is posed by politically/religiously driven nonstate actors in the shape of al-Qaeda offshoots such as al-Qaeda in the Arabian Peninsula (AQAP), and al-Qaeda-affiliated groups in Southeast Asia such as Jemaah Islamiyah or the Abu Sayyaf Group (ASG). Al-Qaeda-affiliated cells already demonstrated their ability to strike at sea several times, for example, in October 2000 when they launched a suicide boat (WBIED) attack on the destroyer USS Cole, in October 2002 when they damaged the tanker M/V Limburg, or in April 2004 when they attacked the al-Basra and Khor al-Amaya oil terminals. In July 2010, and ending a six-year hiatus, another al-Qaeda cell also chose the now well-established suicide boat tactic to attack the Japanese tanker M/V M Star. Although the damage was limited probably due to a suboptimal rigging of the explosives load, the attack itself was a rather sophisticated variation of the previous WBIED attacks: it was the first incident in which a vessel underway was targeted at night, and not a stationary one in broad daylight. This implies that at least some al-Qaeda operators possess the expertise to plan and execute maritime terrorist attacks with a certain chance of success.

Interestingly, evidence collected in two unrelated events in May 2011 indicates that maritime terrorism could still play a role in al-Qaeda's strategic and operational future plans: the May 2, 2011, attack on Osama bin Laden's hiding place in the city of Abbottabad, Pakistan, and the arrest of Maqsood Lodin, an Austrian of Pakistani origin, on May 16, 2011, by German police in Berlin. The documents retrieved in both incidents, especially one titled Future Works, which was found in Lodin's possession, leave no doubt that al-Qaeda still has an interest in maritime terrorism, with a special focus on attacking oil tankers at maritime choke points. A noticeable increase in e-Jihadist forum chatter related to maritime terrorism in addition to recently reported plans of AQAP to target tankers sailing through the Bab el-Mandeb also point at a renewed interest in operations at the maritime front, after a lacunae of a couple of years. One of the plans featured the hijack of a cruise liner with the intent to put the hostages into orange boiler suits like those worn by the inmates of the detention center at Guantanamo Bay, and to shoot them one by one until the terrorists' demands (not specified in the plan) would have been fulfilled. Whether those plans ever see the light of the day, however, is quite uncertain: enthusiasm is not exactly the same as capability. Furthermore, compared with land terrorism, maritime terrorism is far more complex with only poor results so far. while on the other hand, ships seem to be somewhat less iconic as targets than airliners.

Substate Actors: From al-Shabaab (Somalia) to Moro Islamic Liberation Front (Philippines)

The next level of terrorists consists of substate actors in control of territory, that is, guerrilla movements. With regard to the Arabian Peninsula, we can state that since the ousting of the Shiite Mahdi Army from Basra in March/April 2008, no such actor with access to the shorelines exists in the Persian Gulf, although AQAP might actually be able to exert a modicum of control over certain remote corners of Saudi Arabia due to clan loyalties. The waters of the Red Sea in the vicinity of the Bab el-Mandeb and of the Gulf of Aden, however, are an entirely different matter: on the Arabian side, we find Yemen as a weak or even failing state embroiled in a civil war of government forces against clan militias, although it is unclear how much of the coast is affected by it. Here, it should be noted that the two most successful maritime terrorist attacks—against the USS Cole in October 2000 and the tanker M/V Limburg—occurred in Yemeni sovereign waters. On the African side of the Gulf of Aden, the failed state of Somalia proper plus the two (semi-) autonomous breakaway provinces of Somaliland and Puntland as arguably weak entities form the coastline—a coastline of about 3,300 km in length, including the Horn of Africa, which juts into the Arabian Sea like a dagger. Since the collapse of the Siad Barre government in early 1991, Somalia is mired in a bitter civil war between government plus African Union forces, clan militias, and the Islamist Al Shabaab, fighting with and against each other in ever shifting alliances. al-Shabaab, which controls parts of the southern Somali coast, and shore-based clans and their militias in the northern regions of Somalia including Puntland, are ideally placed to take their guerrilla warfare to the sea, targeting Western warships or hijacking merchant vessels in acts of "political piracy." Here this is defined as piracy not for private gains, but for political objectives, including funding their costly wars.

With regard to al-Shabaab, there is no circumstantial evidence that they are interested in developing maritime capabilities, or at least in "buying in" those of the pirate groups active in regions under al-Shabaab control. It is true that spokespersons of al-Shabaab like Muqtar Ali Robow declared the pirates "mujahideen at sea" strikes against "the crusader fleet," but this rhetoric was not followed with any noticeable action. Rather, al-Shabaab seems to be content to get its cut from the pirates' booty. There is more evidence of the direct involvement of shore-based clans and their militias in piracy, and the existence of a network of facilitators and other land-based connections offering support and protection to the pirates. Acts of major criminal hijack (MCHJ) such as hijacking large vessels including super tankers for ransom, and anchoring them in certain Somali harbors, would not be possible otherwise. Although it is unfair to claim that all coastal villages are involved, at least not in the sense of "actively involved," it seems that an "enabling environment" exists that could be compared to the famous English Cinque Ports in the medieval ages, or the ports of New York, Philadelphia, and Charleston in the times of Captain Kidd—nothing more and nothing less. Hence, we could argue that organized piracy may be on the brink of moving up the ladder from being a nonstate actor to being a substate one, and parts of the Somali coast might end up in the hands of a loose federation of pirate republics (some observers would probably argue that Puntland should already be seen as such). So, potentially, and if the lawlessness in Somalia persists, its shores would form an ideal launch pad for acts of political piracy or even for long-distance strikes against Western shipping, including warships in sustained acts of maritime terrorism. This could also be the future of the coasts of Yemen if the government loses its control of the region adjacent to the Gulf of Aden. A guerrilla-style naval AW, however, will stay out of the reach of local actors, be they non-state or substate, because for this, they are too weak and too unorganized.

With regard to MCHJ and political piracy, that is, hijacking or kidnapping for ransom not necessarily for private gains but rather for political objectives, the activities of ASG on the other end of the Indo-Pacific are the most famous of such cases due to the involvement of Western vacationers in one such incident. However, from a Philippine context, their activities are hardly new. First, "piracy has been an endemic national scourge almost endowed with a cultural/religious element," dating back to "as early as the 15th century…" (Guingona, 2002). Second, kidnapping for ransom is a thriving business not only in the Southern parts of the Philippines, but also on Luzon. ASG simply was the first criminal outfit venturing into Malaysian waters in modern times and kidnapping Western hostages on Sipadan Island in April 2000, which resulted in a celebrity status within the circles of al-Qaeda-affiliated global Jihadism and international fame through large-scale sensationalist media coverage. Since ASG can be seen as a "hybrid group" vacillating between mainly fighting for political objectives at one time (thus being a guerrilla group) and mostly fighting for private gains at another (thus being an organized crime group), it is difficult to establish with any degree of precision whether the majority of their acts actually constitute "political" piracy or whether they should be categorized as "ordinary" piracy without any political connotations. Since the early 1990s, the Islamist Gerakan Aceh Merdeka (GAM; Free Aceh Movement) based on the northern tip of Sumatra also conducted maritime kidnapping for ransom, usually aimed at Malaysian trawlers fishing in the Malacca straits (Mak, 2011). While the case of ASG remains ambiguous, it can be argued with far more confidence that GAM operatives were embarking on MCHJ not for private gain but as an additional source of income needed to finance their costly guerrilla war against Indonesian government forces, as we argued above. Since GAM is still observing the 2005 peace agreement with the Indonesian government, this leaves only ASG as a possible actor in this respect.

Compared with the clan-based militias of Somalia, GAM, or ASG, the LTTE Sea Tigers were playing in an entirely different league until the LTTE's defeat in April/May 2009. The Sea Tigers basically were the *de facto* "navy" of the LTTE, protecting what they saw as the sovereign waters of a free "Tamil Eelam," a homeland for Sri Lanka's Tamils, situated in the northern parts of the island. Just like navies of sovereign states, the Sea Tigers had uniforms, ranks, insignia, flags, and a Commander-in-Chief, named Soosai. Unlike navies of sovereign states, however, they also fielded a naval suicide squad, the Black Sea Tigers. So far, no other substate actor attained a similar grade of sophistication. This astonishing capability was mirrored by a wide range of naval actions the Sea Tigers and the Black Sea Tigers conducted over the years, setting them apart from the run-of-the-mill maritime terrorists and bringing them into the orbit of some smaller conventional navies, or at least the IRGCN, which will be discussed below. Apart from suicide attacks typical for acts of maritime terrorism, the Sea Tigers launched attacks on several of Sri Lankan Navy's naval bases, such as the one on Trincomalee in October 2000 or the attack on the Dakshina Naval Base at Galle in November 2006, using heavily armed gunboats and speed boats. The Sea Tigers also conducted amphibious operations in support of land-based operations, such as the large-scale landing during the battle for Elephant pass in 2000, when they landed no less than 1,500 guerrillas. Sea Tiger commandos allegedly even conducted scuba-diving attacks on ships moored in several Sri Lankan naval bases. Furthermore, they also introduced a certain amount of "improvised stealth features" to some of their attack vessels to increase their survivability; they experimented with home-made torpedoes and mines; and they even constructed their own mini-submarines built at a secret base on Phuket, Thailand, as well as in areas under their control (Davis, 2000). Given this wide range of activities, it is not very surprising that Velupillai Prabhakaran once even asked "for a de facto naval status to the Sea Tigers," claiming that "two thirds of Sri Lanka's coastline [would] come under Tiger control" (Suryanarayan, 2004).

State Actors: Iran and the IRGCN

If one widens the concept of maritime guerrilla warfare as conducted by the Sea Tigers to "asymmetric naval warfare" proper, a state actor even more capable than the Sea Tigers appears on the radar screen: the Iranian Navy, especially the IRGCN. To start with the regular Iranian Navy, the acquisition of three Kilo-class submarines of the export type 877EKM is of particular interest in our context. For the Iranian Navy, these three boats represent a quantum leap in regard of naval warfare, opening possibilities previously out of reach. For example, mining the Saudi Arabian Red Sea ports of Yanbu and Jeddah would now be possible, at least in theory. Originally, the three submarines were ordered in the mid-1980s to balance the expected fleet of six Italian frigates, which were never delivered because of the war. Lacking other assets, the Iranian Navy then redefined the role of the Kilo-class subs as deep water interdiction and sea denial (Bruce, 1995). The 877EKM version of the Kilo-class are 2,500-tons, diesel-electric submarines that are very quiet and very capable. In naval terminology, such boats are called hunter–killers. Naval experts assume that the Kilo-class boats will operate in the deep-water zones of the Gulf of Oman outside the Persian Gulf, the Arabian Sea and, at least potentially, in the Red Sea, but not in the Persian Gulf. The reasons for that is the shallowness of the Gulf combined with the relatively high temperatures of the waters, thus making it impossible for a submarine to hide beneath a layer of water with a higher degree of salinity to deflect sonar systems. The use of these submarines would be one form of naval asymmetric warfare, although a rather conventional one. Iran's navy could also embark on another form of asymmetric naval warfare, based on small high-speed vessels. However, the IRGCN is another and possibly more formidable as well as unpredictable actor when it comes not only to asymmetric, but also to irregular warfare.

It is largely forgotten that during the Iran–Iraq war in the 1980s, the IRGCN developed the concept of a guerrilla war at sea in the Persian Gulf and the Strait of Hormuz. From their bases along the Iranian coast, they repeatedly attacked Western tankers and warships in wolf-pack style fashion, using lightly armed high-speed crafts. In a future military confrontation caused by Iran's alleged interest in acquiring nuclear weapons, the Revolutionary Guard's flotillas of stealthy high-speed boats might pose more serious problems to Western commercial shipping and Western warships than the three Kilo-class submarines of the regular navy. In the littoral environment, shallow seas and narrow straits, their "relatively short range and light armament ... can be offset by stealthy approaches, geographical concealment and employment en-mass" (Jia, 2004, p. 19). It also goes without saying that these vessels are quite cheap when compared to a modern warship, which means (1) that they are also perfect weapons for suicide missions and (2) that they are also affordable for nonstate actors. Also part of the Iranian Navy's order of battle are four mini-submarines of North Korean provenance. These boats would be used for commando operations of Revolutionary Guard Special Forces against port installations, oil terminals, oil rigs, or against ships at anchor. Together with the Kilo-class boats and the fast attack crafts, they form an Iranian "maritime triad," as some observers have put it. It is interesting to note that the frequency, duration, and complexity of Iranian naval maneuvers increased measurably from the 1990s onward. The usual scenario of these combined maneuvers in cooperation with army and air force units focuses on amphibian operations and commando-style actions against oil rigs and oil terminals.

DISCUSSION: FROM ASYMMETRIES AND IRREGULARITIES TO 4GW

The analysis given above shows that, after the end of the LTTE, at the present time and in the near future, only one actor remains in the Indo-Pacific capable of waging a full-range guerrilla-style naval asymmetric war: the Iranian Navy, especially the IRGCN, as a state actor. This capability is based on a naval triad consisting of three Kilo-class submarines, a couple of mini-submarines of North Korean provenance, and a fleet of fast attack crafts of various sizes, ranging from Houdong-class attack crafts, also from North Korean origin, down to Zodiac-style rigid raiders. With respect to the capabilities of other substate actors such as the Somali militias and ASG, and particularly of nonstate actors, these are still very limited: nearly all attacks are launched using small speed boats, either fiberglass or rigid raiders, either for boarding and piracy (Somalis, ASG), or for suicide terrorist attacks (al-Qaeda cells). All other modes of spectacular attacks usually discussed by some maritime or terrorism experts are highly speculative: we simply do not know whether any of these groups is really interested in scuba-diving attacks or in subsurface attacks using mini-submarines, or in (trying to) detonating liquefied natural gas tankers. Why should these groups bother anyway? Obviously, suicide attacks seem to work just fine for them. Hence, labeling their operations either as "asymmetric warfare" or as "irregular warfare" indeed is not appropriate—calling them maritime irregular activities or challenges is the better choice for them.

The examples above indicate yet another shortcoming of the concept of (naval) asymmetric warfare: it usually focuses on the asymmetry of means only, but it does not address other possible asymmetries, for example, asymmetries of actors and purposes. In the examples provided above, actors range from state actors (IRGCN) to substate actors (Sea Tigers) to politically motivated nonstate actors (al-Qaeda) to criminally motivated nonstate actors (Somali pirate gangs). Hence, if one prefers to keep the focus on (naval) "warfare," then William S. Lind's "expansionist" concept of 4GW is quite useful, and, although still under discussion at the moment, potentially better equipped to describe and define the realities of twenty-first century warfare. Retired U.S. Marine Corps colonel T.X. Hammes offers the following descriptive definition of the concept:

> *Fourth generation warfare uses all available networks—political, economic, social, and military—to convince the enemy's political decision makers that their strategic goals are either unachievable or too costly for the perceived benefit. It is an evolved form of insurgency. The key concept in this definition is that 4GW opponents will attempt to directly attack the minds of enemy decision makers. The only medium that can change a person's mind is information. Therefore, information is the key element of any 4GW strategy. Effective insurgents build their plans around a strategic communications campaign designed to shift their enemy's view of the world.*
>
> ***Lind, 2003***

On the surface, this newly evolved form of insurgency resonates with the Vietnamese concept of dau tranh (struggle), that is, "the sustained application of total military and nonmilitary force over long periods of time in pursuit of an objective" in an armed dau tranh and a political dau tranh that also "includes activities directed against the foreign enemy in his home country," and also contains elements of the U.S. Navy's CIC concept. Important conceptual differences lurk beneath the surface, however. First of all, Hammes argued that these newly evolved insurgencies tend not to be hierarchical but networked, or even "networked coalitions of the willing." Secondly, these newly evolved insurgencies seemingly bring together different kinds of actors, which means "[it] is essential for us to understand that, even within a single country, the highly diverse armed groups that make up a modern insurgency have widely differing motivations." The following quote (although referring to asymmetry of means) illustrates this quite well:

> *[Asymmetry of means] evokes images of heavily armed and weighed down U.S. soldiers succumbing to the ragged, nearly naked, AK-47-armed Somali—what do we call them? Not soldiers, not civilians. Guerillas? Fighters? Gangsters? We envision those movie scenes of little boys with cell phones calling back to warn the warlords of the departure of American helicopters.*
>
> ***Perla et al., 2006***

And on a more theoretical level, an al-Qaeda member named Ubed al-Qurashi allegedly discussed the benefits of the 4GW concept for the "theoretically weaker side" (i.e., al-Qaeda) on the basis of two publications contained in the *Marine Corps Gazette* (see Hammes, 2007). But apart from politically or ideologically/religiously motivated actors, there are opportunistic ones as well, such as organized crime groups, or, in our context, pirates, exploiting openings created by weak or failed states. Even a cursory look at the Mexican syndicates' actions tells us that Hammes is right to argue that although "these groups have been around for centuries [...] commercially available weapons [now] allow them to overmatch all but the most well-armed police—they are even a match for the armed forces of some nations" (Hammes, 2007). This reminds us of Arquilla and Ronfeldt's (2001) concept of Netwar, but it also resonates well with the Royal Navy's assumption quote above that today's irregular threats are more complicated, less geographically constrained, and less containable than before. In this 4GW setting, the state has lost its monopoly on waging war that it has enjoyed since the Treaty of Westphalia, with "a return to the world of cultures, not merely states, in conflict" as the logical consequence in the view of some scholars. This assertion probably is the most striking one, making the 4GW concept quite distinct from MIA and CIC.

As the case of Somalia alone demonstrates, a 4GW indeed seems to be upon us already—just like many military thinkers claim, and as at least one al-Qaeda operative, the above mentioned Ubed al-Qurashi, conceded. Thus, in a sense, the concept of 4GW seems to work very well in the present maritime/naval context. However, it does so not so much because of the definitional value it adds to older concepts like AW but because of its ambiguity and breadth, and by lumping together all sorts of GAP, be it maritime crime or land-based violence or politically motivated or not. As such, at its previous stage the concept has much in common with the equally nebulous and hard to verify (if that is possible in human science at all) "nexus" idea: that the collusion of interests between terrorists and organized crime groups to conveniently lump them together for the purpose of fighting one of our modern "wars" against them such as the war on terrorism, the war on drugs, or maybe even the war on poverty. Without any proper definition of the concept, it seems also a bit premature to speculate on a Fifth Generation Warfare (5GW), as Hammes does. Quite ironically, there is much more evidence that 3GW, defined as conventional war between nation states, might make its comeback around the mid-century at the latest—if 3GW was ever gone.

3GW RELOADED: A CAVEAT

In our opinion, the notion that conventional war and conventional naval war is a thing of the past seems to be a very Western and post-modernist one. However, and unfortunately for our future, this view is not shared by two of the most formidable challengers to the West, especially when it comes to guaranteeing the free flow of old-fashioned, carbon-based energy, oil, and gas. These two (potential) challengers are the PRC and the Indian Union. Both states deem themselves to be modern, not post-modern—the latter moniker being quite an arcane and incomprehensible one for them in any case. This is why both states still deem conventional wars to be an extension of politics by other means and not as the obviously politically incorrect course of action as seen by the West.

With regard to the maritime/naval sphere, this means that both nations seem to be embroiled in an arms race, especially (but not only) a naval one. For both China and India, this is not only about national prestige: both nations are nuclear powers now, which equals a blue-water-navy status at the turn of the nineteenth century. Rather, both nations pursue some realpolitik objectives that among others forces them to build up high-sea fleets: ensuring and safeguarding the free and uninterrupted flow of energy is a *sine qua non* for their uninterrupted economic growth, which translates into the imperative of protecting their sea lines of communication (SLOCs). The Chinese Malacca Dilemma is well known, and so is their "String of Pearls" strategy to protect their extended SLOCs. The Indian Mahan-inspired naval posture is less well known outside of a rather small group of specialists, despite their new naval/maritime doctrine published only a couple of years ago. Obviously, in Anglo-American foreign offices, India still is seen as "one of us" (UK, Australia), or neglected in a more or less benign way (United States). Unfortunately, this could not be further from the truth: while China as the usual bogeyman could well be part of the solution, India as the "largest democracy," and, by extension, supposed to be our natural ally, could equally well be part of the problem. As Churchill famously quipped, "nations don't have friends, they have interests..."

During the last couple of years, both China and India invested billions of dollars in Iranian oil and gas fields. This under-reported fact illustrates the rising importance of carbon-based energy at a time when Western nations are full of talk about alternative energy sources, which may or may not be available in the next two decades. If they are not, which is to be expected, Western energy firms and Western states will have to deal with powerful newcomers in the Persian Gulf. Western states will also have to deal with the fallout of a potential conventional cold or hot war between China and India, either about hegemony in the Indian Ocean/Persian Gulf, or/and about SLOCs.

In the author's view, staying out of the fray simply will not be an option, and weapons systems acquired to face a 4GW threat may not necessarily be suitable for a 3GW environment. For example, with regard to the U.S. Navy's Littoral Combat Ship, developed with a 4GW setting in mind, some observers argue that it "has been so drastically reduced in capability that it cannot adequately handle low- to medium-threat areas," and a 3GW environment, however, would be medium to high threat area. With respect to the Royal Navy, the planned Type 26 Global Combat Ship (T26 GCS) also seems to be optimized for an MIA environment as defined above, and not for conventional naval war: according to the Ministry of Defense's own description, the vessel, "which is due to come into service after 2020, will be used by the Royal Navy in combat and counter piracy operations and to support humanitarian and disaster relief work around the world" (Ministry of Defence UK, 2012).

CONCLUSION: THE PERILS OF "SWIMMING IN THE INSTANTANEOUSNESS OF POSTMODERNISM"

In a Spiegel interview on the future of Europe, philosopher André Glucksmann (2012) made an interesting point:

> *Globalization brings global chaos, and a global police force—which the United States played for a long time—no longer exists. The players may not be keen on war, but they don't exactly mean well by one another. Everyone is playing his own game. In this anarchic confusion, Europe has to assert itself and face up to threats offensively. Putin's Russia, which wants to regain parts of what it lost, is a threat. China, a bureaucratic slave state, is a threat. Militant Islamism is a threat. Europe has to learn to think in terms of hostility once again. (German philosopher) Jürgen Habermas, for example, doesn't see this when he says that well-intentioned cosmopolitanism can unite everyone in global citizenship.*

He is quite right. Europe indeed has to relearn to think in terms of hostility, as unpopular as this may be at the moment. According to Indian and

Chinese estimates, the build-up of high-sea fleets on par with or better than Western forces will be finished around 2025–2030. Since this still is more than a decade in the future, there is a lot of time to prevent a worst-case scenario from coming true. According to the adage "*si vis pacem, para bellum*": if you want peace, prepare for war. However, this is exactly what Western navies are not doing. Instead, on the basis of shrinking budgets and mushrooming costs for (naval) weapons systems, Western navies are tempted or forced to take the easy way out, trying to rationalize their very raison d'être on the basis of today's GWoT (or whatever it is called at the moment). Again, in the opinion of this author, this will have a similar effect as the Maginot Line thinking before the WW II: ignoring the real threat, while preparing for the wrong one.

CHAPTER

5

Port and Border Security: The First and Last Line of National Security Defense

Andrew Staniforth

A NEW ERA

On the morning of Tuesday, September 11, 2001, millions of Americans were preparing for their commute to work. In the Eastern United States people awoke to nearly cloudless skies and for those heading to an airport, weather conditions were perfect for a safe and pleasant journey. Among those traveling that morning were Mohammed Atta and Abdul Aziz al Omari. They both arrived at the Portland International Jetport and approached the First Class counter of American Airways. Atta and Omari produced first class tickets which, Atta had bought online two weeks earlier with a Visa card. Allowing for the well-documented effects of events on the recollection or reconstruction of testimony, the memories of the many witnesses to that fateful day are instructive. At the check-in desk the ticket agent Michael Tuohey became suspicious recalling, "The first thing I knew was that these were $2,500 tickets, you don't see those too often really. Even at the first class check-in counter you don't see them that often."

Tuohey went on to reveal that he asked Atta and Omari two standard security screening questions which included, "Has anyone known to you asked you to carry anything on board the aircraft today? And have any of the items that you're traveling with been out of your control since the time you packed them?" As Atta and Omari responded to the questions the ticket agent was immediately troubled by the expression on Attah's face describing that; "He had the most hateful look on his face. He had the most angry look on his face and I had never got a feeling like this. I looked at him and I'm thinking, my God I sense anger. I said to myself if this does not look like an Arab terrorist then nobody does." Despite the ticket agent's reported suspicions he thought better of it and let Atta and Omari through the check-in process stating that; "I actually gave myself a mental slap saying they've got ties and jackets on and look like proper businessmen." Moments later Atta and Omari passed slowly through the airport's security arrangements that had become so benignly familiar to frequent flyers over the past 20 years.

Portland International Jetport closed circuit security television captured both men at the central security screening zone collecting their personal valuables having passed through the electronic detectors. The images revealed that both Atta and Omari by this time had removed their ties and smart sports jackets. Flight 175 to Boston departed on time at 6 A.M. for the 45 minute journey to Boston's Logan International Airport. Atta and Omari boarded their flight with hundreds of unsuspecting fellow passengers. As the aircraft rose into the Portland sky the final stages of al-Qaeda's "Planes Operation" was underway.

Atta and Omari arrived at Boston, at 6:45 A.M. They made their way to a connecting flight, American Airlines Flight 11 bound for Los Angeles. At the same time Satam al Suqami and brothers, Wail al Shehri and Waleed al Shehri, checked in and boarded this same flight scheduled to depart at 7:45 A.M. All five men passed through the mechanical and methodical layers of airport security. Atta, the leader of the al-Qaeda cell and the only terrorist trained to pilot the plane, took his seat in business class together with Omari and Suqami. The al Shehri brothers had reserved adjacent seats in first class. By the time Flight 11 pushed back from the gate at 7:40 A.M. these five men had quietly defeated all the security measures that America's civil aviation security system then had in place to prevent hijacking and terrorist attacks since the 1970's.

American Airlines Flight 11 provided nonstop service from Boston to Los Angeles. Nine flight crew and 81 passengers were on board, nothing unusual or suspicious had been reported to the crew; this was a routine flight, business as usual. Nobody knew that five terrorists were among the passengers who intended killing themselves and everybody onboard. Just 15 minutes after takeoff, at approximately 8:14 A.M., the terrorists had begun their attack by stabbing passengers and cabin crew to gain access to the cockpit. In a violent struggle the fate of the passengers and crew now lay entirely in their hands. The terrorists sprayed a chemical irritant in the first class cabin to force passengers and flight attendants to the rear of the plane. As Atta took control of the aircraft he announced to the passengers and crew that; "We have some planes. Just stay quiet and you'll be okay. We are returning to the airport" moments later followed by another announcement in which he said; "Nobody move. Everything will be okay. If you try to make any moves you'll endanger yourself and the airplane. Just stay quiet."

At the time of Atta's announcements, American Airlines cabin crew Betty Ong and Madeline Sweeney were able to contact their reservations office in North Carolina via an air phone. This was the first of several occasions during 9/11 that flight attendants and passengers reported events on board that provided United States investigators with detailed information on how the hijacks were conducted.

Approximately 20 minutes into the flight Ong reported that; "The cockpit is not answering, somebody's stabbed in business class and I think there's mace—that we can't breathe—I don't know, I think we're getting hijacked." For the next 25 minutes Ong and Sweeney provided detailed descriptions of the suspected hijackers including their seat numbers. Ong reported that the aircraft was "flying erratically." Sweeney also said that; "something is wrong. We are in rapid descent; we are all over the place." Sweeney was directed to look out of the window potentially to identify where they were. Sweeney stated that; "We are flying very, very low. We are flying way too low." Seconds later she said "Oh my God we are way too low."

At 8:46 A.M., American Airlines Flight 11 crashed into the North Tower of the World Trade Center in New York City. All on board, along with an unknown number of people in the tower, were killed instantly. Al-Qaeda had begun its first wave of attacks on America, in America, marking the dawn of a new era in ports and border protection.

SECOND WAVE

In the immediate aftermath of the 9/11 attacks governments across the world were quick to review and enhance their existing aviation security regimes, but their efforts did not dissuade al-Qaeda operatives from progressing further deadly and determined attacks. On December 21, 2001, British citizen Richard Reid attempted to board an American Airlines flight from Paris to Miami. Reid was stopped and examined at Charles de Gaulle airport by security authorities; his disheveled appearance raised sufficient concern for officials to question the purpose of his travel. Further suspicions were raised when authorities identified that he had unusually paid for his ticket in cash and had not checked in any baggage for the 10 hour flight. Port authorities questioned him for so long that he missed the flight. Upon his release from examination, Reid was undeterred and again attempted to board American Airlines Flight 63 from Paris to Miami the following day. On this occasion his passage through security was successful, although at the boarding gate his long hair, beard, and scruffy appearance drew attention. Fellow passenger, Annie Joly, recalled that; "I was immediately struck by how bizarre he looked." Despite unsettling travelers and having been extensively questioned the previous day, Reid made his way to seat number 29J and the aircraft departed on schedule. During the flight, American Airlines attendant Hermis Moutardier was clearing away meal trays when a passenger tugged at her sleeve and said he could smell smoke. Just 13 weeks after the 9/11 attacks the sense of anxiety among the 197 passengers and crew on board Flight 63 was obvious. Moutardier spotted the hulking seated figure of Reid, a 6'4" man weighing well over 200 pounds, leaning against the cabin window trying to strike a match. She thought he was trying to light a cigarette and when she challenged him, Reid mumbled an apology. Moments later Moutardier glanced over to seat 29J and again saw Reid bent double, trying to strike another match. Moutardier grabbed at his shoulder and was about to remonstrate with him when she noticed that he had one of his black baseball boots gripped between his knees and was trying to ignite a fuse protruding from its tongue. Moutardier shouted for assistance and fellow passenger, American college student Monique Danison, described the incident saying the flight attendant screamed "Oh, my god! Somebody help me! When someone screams the way she did, you know something bad is happening." While Reid was restrained by passengers and crew (one of them holding his ponytail) Moutardier poured a bottle of water over him to douse the smoke. Reid was further restrained by a series of passenger's seat belts and was sedated with drugs from the onboard medical kit. For the rest of the tense flight the captain warned that Reid might have accomplices onboard. As passengers and crew guarded their prisoner, Flight 63 was diverted to Boston, escorted by U.S. Air Force fighters.

On arrival at Boston's Logan airport, Reid was arrested and taken into custody. He was denied bail on December 28, 2001. On January 16, 2002, a U.S. federal grand jury indicted Reid on nine counts, including the use of a weapon of mass destruction and attempted murder. Reid pleaded not guilty to all charges. The indictment alleged that Reid had received terrorist training at camps in Afghanistan run by al-Qaeda. At the time of his arrest it was perhaps all too easy to dismiss Reid as a bungling amateur, but as the investigation into the foiled shoe bombing continued information came to light that would focus on Reid's terrorist recruitment and radicalization in London. UK security forces would come to learn that his story was about more than just one failed and inept terrorist plot targeting security vulnerabilities at European airports.

Richard Reid was not born into an Islamic family. His grandfather was a Jamaican immigrant to Britain and his father, Colvin Robin Reid, met and married Lesley Hughes, a white woman who was the daughter of an accountant and magistrate. Reid was born in London in 1973, by which time his father was in jail for car theft. Reid's father spent a total of 20 years behind bars stating that "I've seen the inside of most of London's prisons. I was no great example to my son." Reid's parents divorced when he was 11. He left school at 16 and by then he had drifted into the south London world of street crime and car theft. At 17, after mugging an elderly person, he was jailed for the first time.

In the next few years Reid was in and out of prison, following in his father's footsteps as a petty local criminal. When Reid next met him he found that his father had converted to Islam while in prison in the 1980s, and his father suggested something that seemed to make sense to Reid. His father told him that "Muslims treat you like a human being, plus, they get better food in prison." The next time Reid was convicted and given a custodial sentence he followed his father's advice and converted to Islam. At first Reid did not align himself with extremist groups but, on leaving prison in 1994, he gravitated to mosques in Brixton, London, and eventually the Finsbury Park Mosque where he came under the influence of the radical cleric Abu Hamza. Hamza had the knack of making misfits like Reid feel good about themselves. By 1998, violent jihad had become Reid's chosen path. He took the name Abdel Rahim and told his parents he was going overseas where he was later taught how to make bombs by al-Qaeda's Abu Khabab al-Masri. The red-bearded al-Masri, who became al-Qaeda's number one poisons expert and bomb maker, operated from the Darunta training complex near Jalalabad in eastern Afghanistan. Al-Masri's activities for al-Qaeda were regarded as such a threat that he attracted a \$5million bounty placed on him by the United States. In the late 1990s al-Masri ran a specialist course in explosives and toxins inside the Darunta camp to which only an elite few, selected for high profile missions, were admitted. Reid, who had never previously appeared to excel at anything, was able to stomach the tough regime of the al-Qaeda camps in Afghanistan; indeed he appears to have thrived. He was precisely what the terrorist handlers were looking for—zealous, stoical, uncomplaining, and easily led. In a setting where brighter, stronger, and fitter candidates from Finsbury Park were reportedly begging to come home after a fortnight, Reid's mentors congratulated him on being a better Muslim than such shallow men. Continue to do as he was told, they said, and he would become a legend, and achieve all the glories that go with martyrdom.

While being taught terrorist tradecraft in Afghanistan, Reid was also able to engage with other like-minded individuals. These included the Frenchmen, Zacarias Moussaoui, who was later arrested by the Federal Bureau of Investigation (FBI) in Minnesota during August 2001, after instructors at a flight training school he was attending reported his suspicious behavior. Moussaoui would later become the only person charged and convicted with complicity in the 9/11 attacks, pleading guilty to terrorism conspiracy. Moussaoui is currently serving a life sentence in a Colorado prison but, during his sentencing trial in 2006, he revealed that he and Reid were supposed to have been part of the al-Qaeda attacks on U.S. targets. To add credence to these claims the FBI 9/11 investigation revealed that Moussaoui had worshipped in Brixton while studying in London. Investigators also secured evidence in the case against Reid, which confirmed that the day before he boarded Flight 96 in Paris he had written a letter to his mother bequeathing his belongings to Moussaoui. For those in authority who initially considered Reid to be a feckless petty criminal were shaken by the possibility that violent jihadists from different parts of the world had been working together against the established order, and that there was increasing evidence of some kind of orchestrated effort beginning to weave all the terrorist cells into an international destructive tapestry, and, more worrying, that their intended targets were commercial flights packed with innocent civilians.

INDEPENDENT REVIEW

In direct response to the increased terrorist threat from al-Qaeda and their affiliates, the British government appointed Lord Carlile of Berriew QC CBE as Independent Reviewer of Terrorism Legislation (IRTL). In more than nine years as IRTL he spent a considerable proportion of his time on ports and border security. This is perhaps a mundane part of the Reviewer's routine, but its importance struck him very quickly. As he stood behind Special Branch officers at UK airports he realized how many extremely speedy judgments fall upon them, with a complex body of intelligence and law sitting on their shoulders. He also observed the questioning of passengers and stated that "I came to understand the intrusion faced by almost always innocent passengers, but the critical importance of the national security framework in which we all travel. In seaports and on ferries, I became conscious of the porous and fragile nature of our border controls. At huge freight terminals, I saw the opportunities for terrorist and other seriously criminal acts with the potential for irreparable public damage, and the subtlety required to meet those challenges in a proportionate as well as strictly lawful fashion." At and around the Channel Tunnel, the vital link between the UK, France, and all European Member States, Lord Carlile observed:

> *The importance and fair potential of automated number plate recognition yielding actionable intelligence at large executive jet hubs, and on small landing strips randomly distributed around the United Kingdom, I shared the challenge of policing the economically important executive and private aviation market, memorably described to me by a police officer in late 2001 as "the soft underbelly of ports policing.*

The private aviation sector was rightly identified as being in need of additional security measures, especially those centers offering flight instruction. A key component in the success of the 9/11 al-Qaeda Planes Operation lay in terrorists' ability to overpower and take control of commercial aircraft. This provided an opportunity for the suicide pilots to fly toward identified targets. Some of the flying skills had been acquired in flight training establishments within the U.S. general aviation sector. The United States has the largest flight training industry in the world. Students are attracted by the country's climate and its location, but fundamentally they are attracted to the competitive rates charged by their flight instructors. As a result large numbers of prospective flying students arrive in the United States from all corners of the world. Seeking flight training was therefore not unusual and the 9/11 suicide pilots blended into the large diverse and transient flight-training population. They were well placed to conduct their activities beneath the radar of U.S. intelligence and law enforcement agencies.

U.S. authorities had already debated suicide hijacking of aircraft prior to 9/11, but these were dismissed on the grounds that there would have been no opportunity for dialog with the hijackers. This dismissal was based on the notion that any suicide hijackers would wish to communicate to authorities to make demands, or outline their intentions or motivations. U.S. authorities also discussed the possibility of a hijacked airliner coming from overseas and crashing into the Pentagon. It was, however, decided that there would be sufficient time to identify the target and scramble interceptors to shoot it down before it reached the U.S. coastline. No one imagined that four aircraft could be forcibly commandeered in flight within U.S. air space and used to such devastating effect.

The 9/11 terrorist cell thoroughly researched its use of aircraft and the security regimes in place at ports and borders. The Commission highlighted that a large extent of the damage at the World Trade Center was initially caused by considerable amounts of burning aviation fuel. It is believed that all the hijackers targeted internal flights that were to cover long distances in the United States thereby maximizing the amount of fuel on board at the time of impact. In addition, several of the hijackers conducted hostile reconnaissance on internal U.S. flights where considerable information was gained to support their attack planning. On these reconnaissance flights the hijackers carried rudimentary weapons such as box cutters to identify what they could easily get past airport security. They also selected specific aircraft and assessed the exact time at which it would be best to take control of the flight. The 9/11 plot was carefully planned and executed. Thousands of dollars in cash was paid to training providers of flight simulators. Two terrorist cell members sought flying lessons but were not interested in commencing with small aircraft; instead they wished to learn how to fly

Boeing commercial jets. A number of different flight instructors stated that the suicide pilots wished to focus upon controlling the aircraft in flight but took no interest in take-off or landing procedures.

As a result of 9/11 and the use of commercial aircraft as weapons, the aviation industry has undergone an unprecedented period of reviewing security measures. It is clear that al-Qaeda continues to seek large-scale spectacular attacks using aircraft. The threat level within the global aviation sector remains high serving to underline the need for increased vigilance from authorities not just within the commercial scheduled and charter sectors but also within the general aviation business including flying schools and associated training. While the catastrophic events of 9/11 served to rapidly develop security protocols at ports across the globe, al-Qaeda remained undeterred from pursuing its vaulting ambitions to use commercial aircraft as a vehicle to deliver its deadly attacks. Five years on from 9/11, another al-Qaeda terrorist cell across the Atlantic developed new ways in which to breach border security measures introduced to protect innocent travelers.

TRANS-ATLANTIC TERROR

Overnight on August 9, 2006, large numbers of police officers were deployed across High Wycombe, London, and Birmingham as part of an operation to disrupt a major terrorist plot. A total of 24 terrorist suspects were arrested and questioned in relation to the commission, instigation, and preparation of an alleged trans-Atlantic terrorist attack of 9/11 proportions. The police and Security Service (MI5) had been monitoring the activities of British-based al-Qaeda-inspired terrorists for several years and the executive action taken overnight was the result of the largest operation ever conducted by the Metropolitan Police Service and MI5. A potentially catastrophic terrorist attack had been disrupted and intelligence and evidence had been painstakingly gathered under Operation OVERT. Despite the number of arrests that morning the security forces could not be confident that they had managed to capture all of the terrorist conspirators in their raids. In order to warn the public of, and protect people from, an imminent attack, the threat level was raised to "Critical," following a decision made at an emergency meeting of the Cabinet Office Briefing Room chaired by then Home Secretary, Dr. Jon Reid. Speaking outside New Scotland Yard to a gathered press, then Deputy Commissioner of the Metropolitan Police Service, Paul Stephenson, stated that "We cannot stress too highly the severity that this plot represented. Put simply this was intended to be mass murder on an unimaginable scale." The Deputy Commissioner went on to reveal more about the terrorist plot stating that

> *We are confident that we have disrupted a plan by terrorists to cause untold death and destruction and to commit, quite frankly, mass murder. We believe that the terrorists' aim was to smuggle explosives on to aeroplanes in hand luggage and to detonate these in flight. We also believe that the intended targets were flights from the United Kingdom to the United States of America.*

Other developments that following morning on August 10 included the arrival of U.S. air marshals, introduced to protect civil aviation following the events of 9/11, which had been sent to the UK to provide extra security to U.S.-bound trans-Atlantic flights. The U.S. Department of Homeland Security had increased the threat level applied to incoming commercial flights originating from the UK to "Red"; the first time it had been raised for external flights.

Airports suffered a day of confusion and thousands of holiday makers faced unprecedented security passenger measures. Heathrow Airport was thrown into chaos. Heathrow Airport authorities made an announcement to the press in an attempt to inform and pre-warn passengers traveling later that day to "Arrive with no hand luggage, bring only those items into the cabin in a clear plastic bag and still be prepared for delays. You will not be permitted to take any liquids, toiletries or cosmetics on board the flight."

Many passengers appeared prepared for long delays, canceled flights, and restricted luggage. On a day that perfectly illustrated the competing considerations facing counterterrorism authorities, a passenger queuing at Heathrow stated that "Honestly I think it is one of the safer days to fly because of the extra security in place," while another accepted the security measures saying, "If we have to wait we have to wait, we know they are doing the right thing." At the same time, some airline executives believed the government measures that morning were misplaced—more public relations than a practical response to the threat. The Chief Executive of Ryan Air, Michael O'Leary, said that "The London airports are at the point of break down, the airlines are cancelling 20–30% of their schedules, we're stranding thousands of British passengers all over Europe all for security measures introduced by government that are ineffective in terms of increasing security." In the weeks and months that followed, the British public would come to learn more about the realities of the terrorist plot as the police carried out extensive searches of buildings and vehicles, focusing on a 354 acre wood near High Wycombe. The huge scale of the searches and resources required were a significant logistical challenge, with more than 200 police officers on site every day. A variety of evidence was gathered including potential bomb-making equipment and chemicals, computers, telephones, and portable storage media such as memory sticks, CDs, and DVDs. Operation OVERT was gathering pace and investigators and prosecutors building their case against the conspirators were soon to identify the drive and motivation behind the terrorist plot.

The covert surveillance activity under Operation OVERT served to identify three primary terrorist cell members: Abdulla Ahmed Ali, aged 27, the leader of the terrorist cell; Tanveer Hussain, aged 28; and Saad Sarwar, aged 29 from High Wycombe. All three men were well acquainted. Ali was an engineering graduate who chose not to go into industry after leaving City University in London in 2002, but instead decided to pursue business opportunities in Pakistan. In contrast, Sarwar turned down a place at university in Chichester, Sussex, because he became homesick, and his second attempt to progress undergraduate study at Brunel University, Bristol, first with a sports science course, then earth sciences, also failed because he found the work too challenging. Ali and Sarwar were, however, to meet in 2003 when they both went to deliver aid to refugee camps on the Pakistan and Afghanistan border. In the aftermath of 9/11, the U.S.-led military response in the Bush administration's "War on Terror" resulted in ground troops from the coalition forces pouring into Afghanistan. As a result many Afghans decided to flee from areas of conflict, compounding a decades-old refugee crisis. In support of the worsening situation, the Islamic Medical Association, a charity shop located in Clapton, East London, raised money and collected equipment to send to the refugee camps. Ali and Sarwar were volunteers, distributing such aid in what seems to have been a sincere humanitarian endeavor, but their shared experiences at the refugee camps radically altered their worldview. Ali, married with a son, was shocked by the appalling conditions in the camps, where he witnessed many people dying. Sarwar would also be dismayed at what he had seen at the camps, later believing that the aid work in which he was engaged was an ineffective way of helping. Both Ali and Sarwar were angered about the situation and decided to tackle what they believed was the root cause of the death and misery they witnessed at the camps—Western foreign policy. The anger felt by Ali and Sarwar turned them against the UK and the United States and both began to relate to the anti-Western rhetoric of radical Islamism. Ali and Sarwar started to move in Islamist circles, which were increasingly calling for attacks on Britain. In particular, by March 2005, Ali was suspected of communicating with Rashid Rauf, a primary influential al-Qaeda figure who had fled to Pakistan from Birmingham in 2002 following the issue of a warrant for his arrest in connection with the stabbing of his uncle. Security forces suspected that Rauf had come to be engaged in numerous al-Qaeda plots including the London

7/7 suicide bombings and the failed 21/7 bombings in London during 2005. As the Security Service continued monitoring those individuals preaching violent extremism and calling for violent jihad against the UK, Ali became of significant interest.

When Ali returned from Pakistan in June 2006, the Security Service and antiterror police were sufficiently interested in him to conduct a covert baggage search of his luggage before it passed through to the arrivals hall. Inside his luggage they found an unusual powdered soft drink called Tang, and a large number of batteries. The discovery was extremely suspicious and served to attract more attention from MI5. The police and the Security Service needed to identify the purpose of the batteries and the drink bottle for they suspected, while under the influence and instruction of Rashid Rauf, that Ali had made the dangerous transition from al-Qaeda sympathizer and supporter to al-Qaeda attack planner. In the coming weeks, the Metropolitan Police Service and MI5 mounted Operation OVERT, which later captured Ali sending e-mails to Pakistan. These e-mails were coded messages, thinly veiled communications to his jihadist contact, believed to be Rauf. The e-mails talked about acquiring Calvin Klein aftershave for a business opportunity, but in reality the plotters were purchasing hydrogen peroxide, a legitimate but potentially lethal component of a home-made explosive. Investigators discovered that the quantities of aftershave referred to in the e-mails bore a striking similarity to the quantities of chemicals being amassed. Mobile surveillance of Sarwar captured him buying household items that could be used to make bombs and to increase the scope of the surveillance operation, MI5 covertly entered the east London flat used by Ali to install audio and visual recording devices. Ali's family had recently purchased the £138,000 two bedroom flat at 386a Forest Road for cash. MI5 were alarmed by what they saw when they entered the flat as it appeared to be a bomb factory, but one unlike any they had seen before. It did not have large containers filled with bubbling chemicals like the 7/7 bomb factory discovered at Alexandra Grove in Leeds, although it did contain signs that the residents were constructing devices. On August 3, 2006, MI5's tiny camera and microphone inside Ali's flat recorded Ali and Hussain constructing home-made improvised explosive devices from drink bottles. The covert cameras also captured Hussain drilling a hole in the bottom of a soft-drink bottle in order to replace its contents with concentrated hydrogen peroxide without breaking the seal on the cap. Combined with AA batteries filled with hexamethylene triperoxide diamine and concealed in a disposable camera in hand luggage, the component parts of the bomb were to be assembled mid-flight and detonated over U.S. towns and cities. MI5 captured further clues to the trans-Atlantic plot from their intrusive surveillance, recording the conspirators reviewing numbers, and talking of "18 or 19." It was not known whether this meant 19 devices, 19 targets, or even 19 co-conspirators but then, mobile surveillance teams observed Ali for over two hours in an Internet cafe, researching flight schedules. A jigsaw of extraordinary evidence was finally coming together. Ali's e-mail traffic continued and then the orders came back to push ahead with the project or "presentation," as Ali described it. The attack plan had entered a new phase as the plotters now had to find some willing volunteers to join their conspiracy and commit acts of martyrdom on board the trans-Atlantic flights. The tipping point at which executive action would necessitate the arrest of the conspirators was closing in but—as so often happens in even the best laid plans—events beyond the control of MI5, the police, or government, would dramatically change the carefully planned course of Operation OVERT.

The major covert terrorist investigation mounted against the trans-Atlantic plotters had captured considerable evidence indicating the scale of murder and destruction that they intended to bring about if not stopped. Postponing the arrest of all the suspects called for political courage by Home Secretary Dr. Jon Reid and operational leadership by MI5 and the Metropolitan Police. At one critical moment, the Home Secretary told Jonathan Evans, Director General of the

Security Service, that "If this goes wrong I'm out of a job, you're out of a job, and the government will fall." Across the Atlantic U.S. intelligence officials were rightly anxious about the possibility of another al-Qaeda attack and were concerned that any delay in arresting the plotters may allow time for them to complete their plans. Although unable to determine the timing of arrests in Britain, the Americans brought pressure to bear on the Pakistani authorities to arrest Rashid Rauf. It was extremely important that the arrest of all conspirators, in Pakistan and in the UK, was carefully coordinated so as not to cause a chain of events that would lead to the disclosure that the arrest had been made and leading to the potential escape of the terrorists. Unfortunately, Pakistani authorities decided to arrest Rauf on August 9, 2006, when he was stopped by local police in a taxi in Rawalpindi. The premature arrest brought unwelcomed pressure upon the senior managers leading Operation OVERT. Deputy Assistant Commissioner Peter Clarke stated that "We were at a critical point in building our case. If word got out that Rauf had been arrested, evidence might be destroyed or scattered to the four winds. More worrying still was the prospect of a desperate attack." The arrest phase of Operation OVERT was swiftly brought forward and put into action that same day. When Ali was arrested, found in his possession was a computer memory stick containing details of seven trans-Atlantic flights due to take off from Heathrow Terminal 3 within 2 1/2 hours of each other. For a large part of their combined flight times to cities in the United States and Canada, all the aircraft would have been airborne simultaneously. There would have been little chance of surviving the attack, which would have brought mass death and destruction to passengers, crew, and anyone that happened to be beneath them as they fell from the skies.

As the Operation OVERT post-arrest investigation continued, police recovered a series of videotapes containing "martyrdom" videos, which Sarwar had hidden in his garage. Contained within the videos was the martyrdom recording made by Ali. Wearing a headscarf and a black robe, while jabbing his finger repeatedly at the camera, Ali declared angrily:

> *Sheikh Usama warned you many times to leave our land or you will be destroyed, and now the time has come for you to be destroyed and you have nothing but to expect than floods of martyr operations, volcanoes, and anger and revenge and raping among your capital and yet, taste that what you have made us taste for a long time and now you have (to) bear the fruits you have sown.*

Another recovered prerecorded martyrdom message showed Hussain in which he warned of further attacks, declaring his regret that he could only be a suicide bomber once. He stated:

> *For many years, you know, I dreamt of doing this, you know, but I didn't have no chance of doing this. I didn't have any means to do this. You know I only wish I could do this again, you know come back and do this again and just do it again and again until people come to their senses and realize, realize you know, don't mess with the Muslims.*

With such evidence in their possession, the prosecution's case was building, but there were still many challenges to overcome.

The Crown Prosecution Service, Counter Terrorism Division (CTD), working alongside counterterrorism police investigators and operatives from the Security Service, prepared the prosecution case for Operation OVERT, described by then head of the CTD, Sue Hemming, as the "largest case of its type." The head of the CTD liaised with her U.S. counterparts both from the perspective of seeking assistance and because the intended targets of the attacks were in the United States. This brought a new challenging dimension to UK counterterrorism. Here was an active terrorist cell conducting their attack planning in the UK to deliver mass murder to our allies in the United States. The UK could not become a staging post or training ground for home-grown al-Qaeda-inspired terrorists. The duty and responsibility of the UK government had been extended, not simply to protect its own citizens from the excesses of Islamist violence but to ensure that British terrorists were not exporting death and destruction overseas. Of the 24 suspects who were arrested on August 9, 2006, 17 were formally charged. Eight men, including Ahmed Abdullah Ali, Saad Sarwar, Tanveer Hussain, Ibrahim Savant, Arafat Khan, Waheed Zaman, Umar Islam, and Mohammed Gulzar, were charged in connection with the trans-Atlantic plot. The trial began in April 2008 and ran for six months. After deliberating for over 50 hours on charges of conspiracy to kill by blowing up aircraft the jury failed to reach a verdict, but they did find Ali, Hussain, and Sarwar guilty of conspiracy to murder. Later, in September 2009, a second trial of the eight men, excluding Gulzar but with the addition of Donald Stewart-Whyte, found Ali, Sarwar, and Hussain guilty of the plot. In July 2010, Ibrahim Savant, Arafat Khan, and Waheed Zaman were also found guilty at Woolwich Crown Court, London, and sentenced to life in prison for conspiracy to murder. They must serve a minimum of 20 years in prison before being eligible for release. The Home Secretary, Theresa May, welcomed the guilty verdicts and commended the efforts of police, Security Services, and the Crown Prosecution Service in bringing the men to justice. The Home Secretary said, "I welcome the verdicts in one of the most significant terrorist plots the UK has ever seen. National security is our top priority and this result reflects the continued and serious threat we face."

SECURING THE BORDER

Terrorist plots that successfully managed to defeat all port and border security measures put in place after 9/11 caused great anxiety and concern among national security practitioners. The suicidal component of terrorist tactics used to devastating effect to target commercial passenger jets was, in truth, only part of the diverse range of terrorists threats to port and border security controls. During 2006 the International Institute for Strategic Studies in the UK published a report titled The Military Balance. It estimated that "one in ten of the 20,000 insurgents fighting the conflict in Iraq were foreign-born." It was also revealed that up to 150 radicals from Britain had travelled to Iraq to join up with a "British Brigade" that had been established by al-Qaeda leaders to fight coalition forces. It appeared that the flow of young men from Western Europe to Iraq was increasing. The supply chain of volunteers to join the ranks of the insurgency was not just restricted to the UK. In France, Pierre de Bousquet de Florian, the head of the French domestic security service, revealed that 15 young French men remained in and around Iraq that they were aware of and of these at least 9 had been killed. This raised two key issues for the UK government: (1) how were these men traveling to Iraq and receiving their training and (2) while it was understood that many of the volunteers may die in the theater of conflict in Iraq, some may well survive their experiences and return to the UK with military training and hardened combat experience. How would they use their training and combat experience when returning to their communities? It was considered by authorities that such skills and experiences presented a threat to the cohesion of local communities and created additional security risks further fueling the radicalization and recruitment of young, vulnerable, and impressionable young men to violent extremist causes.

The passage of British citizens through UK ports with violent extremist perspectives to swell the ranks of the insurgents in Iraq was a worrying trend. Border authorities had to become alive to this threat and prevent those intent on fighting coalition armed forces from departing. During March 2007 the British government set out a new philosophy of border control stating that

> *Border control can no longer just be a fixed line on a map. Using new technology, particularly biometrics, and new approaches to managing risk and intelligence, we must create a new offshore line of defence, checking individuals as far from the UK as*

possible and through each stage of their journey. Our aim is to make legitimate travel easier, yet prevent those who might cause us harm from travelling here.

While the British government grappled with measures to counter the travel of would-be insurgents, airports came under direct physical attack.

At approximately 3:11 P.M. on Saturday, June 30, 2007, a dark green Jeep Cherokee loaded with propane cylinders was driven into the glass doors of the Glasgow International Airport terminal. Initial reports indicated that protective security steel bollards had prevented the vehicle from entering the terminal. The vehicle contained two men, Kafeel Ahmed and Bilal Talal Samad Abdulla, who were both arrested at the scene. Ahmed suffered 90% burns to his body and died at Glasgow Royal Infirmary as a result of his injuries several days later. Although people assisting police officers at the scene of the attack suffered injuries, miraculously, no other casualties arose from this attempted suicide attack.

The attack came a day after the Metropolitan Police were also engaged in a counterterrorism investigation following a controlled explosion carried out on a car also packed with gas cylinders in the Haymarket area of London on June 29, 2007. An ambulance crew had reportedly seen smoke coming from a green Mercedes at 1:30 A.M. that morning, which had been parked near to a nightclub. Two major incidents within the space of 48 hours raised concerns that the UK was the target of a potential series of coordinated terrorist attacks. The attack was the first to specifically target Scotland providing evidence that no community was immune from the global reach of international terrorism.

Images of the unfolding events at Glasgow Airport were beamed around the world by 24 hour news channels within minutes of the attack and airports were again plunged into chaos and the attack triggered an immediate response from HM Government. As part of a series of measures Prime Minister Gordon Brown asked Lord West to conduct a review security. During 2007, as a direct result of the failed terrorist attacks in London and Glasgow in June that same year, Lord West, Minister for Homeland Security and Counter-Terrorism, conducted a review of UK security within the UK. The review specifically focused upon the protection of our strategic infrastructure, stations, ports and airports, and other crowded places. There were three key findings from the review which included;

1. A need for a new "risk-based" strategic framework to reduce vulnerability of crowded places
2. Focused effort on reducing the vulnerability of the highest risk crowded places by working with private and public sector partners at a local level
3. New efforts to "design in" counterterrorism security measures are needed, building on good practice from crime prevention

During a keynote speech at the Government Conference on Homeland and Border Security during July 2008, Lord West stated that

The protection and security of the UK is always on my mind—I have been hugely impressed by the hard work and dedication of the men and women in our police and security services and of the many others—in government, private and voluntary sectors—who contribute to the safety of the UK and the wider international community. Countering terrorism is not just about the work of the police, security service and intelligence agencies—it relies upon everyone to play their part. We need to build more resilient enterprises and a resilient society where our values and freedoms can survive and thrive in the face of this complex and serious terrorist threat. To do this, and to make sure that our response is proportionate and necessary, is an enormous challenge, but I am confident that we shall continue to succeed. By working together we shall ensure that violent extremists and terrorists find neither support nor sanctuary anywhere in the UK.

Unbeknownst to UK security forces at the time Lord West was making his speech, a new enemy, al-Qaeda in the Arabian Peninsula, had recruited and radicalized a young Nigerian student studying in London who was prepared to martyr himself for the extreme Islamist ideology espoused by al-Qaeda. On December 25, 2009, 23-year-old Nigerian-born Umar Farouk Abdulmutallab boarded Northwest Airlines Flight 253 at Amsterdam's Schipol airport. In possession of only hand luggage for a trans-Atlantic flight and with no coat, Abdulmutallab evaded all the sophisticated security screening in place at one of Europe's premier international aviation hubs as he successfully concealed an improvised explosive device (IED) in his clothing. Abdulmutallab took his reserved seat, 19A, which was directly located over the wings and fuel tanks of the aircraft as it pushed back from the gate and headed toward the U.S. city of Detroit, Michigan. As the aircraft made its descent, Abdulmutallab attempted to detonate his IED consisting of a primary and secondary explosive sealed in a plastic package and in the crotch of his underwear. As the device ignited it failed to detonate but alerted passengers, and crew extinguished flames coming from Abdulmutallab who was hidden beneath a travel blanket after describing to fellow passengers he was suffering from a stomach upset in an attempt to conceal his actions. Abdulmutallab was restrained on board the flight and was later arrested by U.S. authorities upon arrival in Detroit. On January 2, U.S. President Barack Obama addressed the world's media stating that "Abdulmutallab joined an affiliate of Al-Qaeda, and that this group, Al-Qaeda in the Arabian Peninsula, trained him, equipped him with those explosives and directed him to attack the plane headed for America."

Yet again this failed attack demonstrated the continued intention of al-Qaeda to use commercial passenger jets to deliver death and destruction on an unimaginable scale. It provided further evidence of their continued desire to mount spectacular terrorist attacks. Ports and borders across the globe were under relentless attack from al-Qaeda-inspired operatives, which presented a major vulnerability in the struggle against international terrorism. State action was required to plug the gaps at porous border controls. The British government, accepting the recommendations from Lord West's report together with an internal Cabinet Office review, established the Border Management Program. This is a cross-government initiative aimed at developing and implementing closer and more effective joint working in order to strengthen border security while minimizing the impact of legitimate traffic. The strategic objectives of the program were to do the following:

- Improve intelligence sharing in support of border agencies
- Jointly identify and manage risks
- Provide a more effective border control
- Minimize the impact on legitimate traffic and business partners

In addition, during November 2007, Prime Minister Gordon Brown announced a wide range of measures to counter terrorism, which included a new approach to strengthen UK borders. These measures included the creation of the UK Border Agency (UKBA) in April 2008, which brought together the work of three key agencies: the Border and Immigration Agency, UK Visas, and HM Revenue and Customs. The new UKBA had 25,000 staff present in 135 countries, and its strategic aim was to protect the UK with a simple system of triple checks which included the following:

1. New offshore controls including pre-arrival screening
2. A stronger UK border
3. Tougher checks at UK orders

The then Minister of State for Borders and Immigration, Liam Byrne MP, set out the plans of British government in the 2008 UKBA report, A Strong New Force At The Border stating that

> *The purpose of the UK Border Agency is clear—to secure our border and control migration for the benefit of our country. That means we will protect our borders and our*

national interests. The public will notice border controls that are stronger. They will see our customs and immigration staff in a single uniform. They will see a primary checkpoint at passport control where they will be asked about the travel documents they have and the goods they are carrying. They will see our new force with the purpose, the power and the punch to protect our border in the 21st Century.

During 2010, UKBA published a five-year strategy for enforcing UK immigration rules and addressing immigration and cross-border crime. The strategy—protecting our border, protecting the public—set out how UKBA will continue its efforts to cut crime and make the UK a hostile environment for those that seek to break UK laws or abuse UK hospitality. The strategy is divided into four mutually reinforcing objectives:

1. **Deter:** Strengthening UK resilience against immigration and cross-border crime
2. **Disrupt:** Breaking up criminal activities and organized criminal groups.
3. **Detect:** Identifying and locating those responsible for criminal activity and the smuggling of illicit goods
4. **Deal:** Taking action against those engaged in all levels of criminal or noncompliant activity.

When combined, the four pillars form the basis of what was considered to be a renewed approach to tackling criminal activity, including terrorism in all of its forms. The five-year strategy was primarily designed to develop UKBA's law-enforcement capability as well as increasing its collaborative approach with partners to tackle immigration and cross-border crime. The UKBA faced a series of key challenges, none more so than the simple volume of traveling passengers. Aviation is a major UK industry, carrying over 235 million passengers a year and over 2.3 million tonnes of freight. To manage the sheer volume of passenger and freight movements, UKBA had to develop new tools, and their operational expertise was essential to the development of the Freight Targeting System, which now provides real-time risk assessment on freight arriving in the UK, allowing officers to target high-risk goods. The introduction of the e-Borders program, which enabled the effective screening of individuals and goods that cross the UK border, had more impact. It alerted UKBA to those passengers of interest prior to their arrival in, or departure from, the UK. For the very first time e-Borders allowed UKBA to check every person traveling into or out of the UK against a range of watch lists, thereby identifying criminals and immigration offenders. e-Borders has collected details on 141 million passenger and crew movements since 2005 and thus far has resulted in the 4,800 arrests: including 104 arrests for rape/sexual assault and 33 arrests for murder alongside the seizure of over 150 fraudulently used British passports and the identification of 950 people with an adverse immigration history being refused entry to the UK.

UKBA now deploys a series of tools to increase its capacity and capability to detect people and property of interest who pose a threat and risk to the UK. One of the most valuable tools in its armory is the Schengen Information System (SIS), a pan-European database designed to enhance judicial and police cooperation among European Member States. SIS provides UKBA with real-time data on over 27,000 people subject to a European Arrest Warrant (nearly 50%of which relate to violent/sex crimes, drugs, or terrorism offences), missing persons, persons assisting the judicial authorities, and over 42 million alerts on lost or stolen property. This system has greatly enhanced UKBA's ability to detect people and property of interest to UKBA and overseas partners.

Despite making a significant contribution to the protection of UK national security, UKBA has, over recent years, been subject to a series of criticisms across a number of its operating responsibilities. During 2011 several reports published by John Vine, Independent Chief Inspector of UKBA, criticized its efforts in relation to convict deportations, revealing that more than 5,000 foreign criminals awaiting deportation remained in the UK. The report suggested that the number who were not deported or could not be deported at the end of their sentence was rising, adding to the financial burden upon the public purse. UKBA came under increasing pressure during 2012 when it was revealed that large queues at border controls were being experienced by passengers at key hub international airports. A further inspection by the Chief Inspector of UKBA revealed that the Border Force—part of UKBA's frontline operations with overall responsibility for securing the border, which delivers customs and immigration functions at airports, seaports, and the Channel Tunnel—had suspended important security checks without permission. In addition the Chief Inspector's report also identified that UKBA had spent millions on new technologies but chose not to use them; that it was led by managers who did not communicate with their staff; and had that it had sent reports to ministers that were inaccurate, unbalanced, and excluded key information. These reports made a series of recommendations about how to improve the operation at the border, all of which were accepted by the Home Secretary, Theresa May. The Home Secretary did not, however, believe that the answer to the very significant problems exposed in the Chief Inspector's reports at UKBA was just a series of management changes and so significant reform to the Border Force was approved.

On February 20, 2012, Home Secretary Theresa May announced to the Houses of Parliament that the UKBA Border Force needed a whole new management culture, revealing that she had appointed Brian Moore, Chief Constable of Wiltshire Police, as the interim Head of the Border Force. The Home Secretary went on to reveal that

> *There are many hard-working and dedicated members of staff in Border Force. They want to get on with their work securing our border, and I want to make clear that this Report is in no way a criticism of them. But, as the Home Affairs Select Committee and its Chairman have argued consistently, there is no getting away from the fact that UKBA, of which the Border Force is part, has been a troubled organization since it was founded in 2008. From foreign national prisoners to asylum seeker backlog to the removal of illegal immigrants, it has reacted to a series of problems instead of positively managing its responsibilities. With a new chief executive and a plan for comprehensive change, I believe that UKBA is in better hands for the future. But I also believe that the extent of the transformational change required—in the Agency's case working functions and in the Border Force—is too great for one organization. I can therefore tell the House that from 1 March 2012, the UK Border Force will be split from UKBA and will become a separate operational command, with its own ethos of law enforcement, led by its own Director General, and accountable directly to ministers.*

The UK Border Force continues to provide the necessary operational capacity and capability to protect the UK border from all manner of threats. The public rely on them to stop dangerous people from entering the UK, to prevent illicit, dangerous goods such as drugs and weapons from reaching their communities, and to take effective action against those that break immigration laws. The role and function of UKBA and the Border Force at UK ports and borders continues to be an essential part of the UK's security apparatus, but the swift change to border security architecture reveals the complexities and challenges that governments encounter when attempting to deliver safety and security at borders in response to a diverse range of contemporary threats.

ALL HAZARDS APPROACH

The protection of ports has, over recent years, quite rightly focused upon the determined threat from contemporary international terrorists wishing to expose and exploit vulnerabilities in border security. With That said, there are numerous hazards for port practitioners to counter that directly impact national security, the most pressing are the threats from serious and organized crime. Globally, the United Nations (UN) estimates that the most powerful international organized crime syndicates each accumulate in the region of $1.5 billion a year. The international drug market alone is estimated to be worth £200 billion. The UK's National Security Strategy also notes that cyber crime has been estimated to cost up to $1 trillion per year globally. Europol's Organized Crime Threat Assessment continues to highlight the regional threat posed by organized crime. The biannual report, which assesses current and expected trends in organized crime affecting the European Union, explores how a new criminal landscape is emerging, marked increasingly by highly mobile and flexible groups operating across boundaries and criminal sectors. It shows that organized crime is changing and becoming increasingly diverse in its methods, group structures, and impact on society.

It therefore remains essential for all port security practitioners to understand that organized crime groups attempt to control the market for their commodity or service and respond to changes in demand, price, or risk. They use unscrupulous means to satisfy market demand, which provides a way of understanding their motivation. Organized criminals also respond to law enforcement action and react to legislative or enforcement changes. For example, when specific commodities were protected from UK value-added tax fraud, criminals quickly found new commodities to attack the tax system. The capabilities of organized criminals expand as they gain experience. Many cultivate criminal contacts, which enable them to be more flexible, for instance, in their use of routes and methods, which helps increase profits. They are early adopters of new technology. Technology has made more markets and victims accessible more quickly. Organized criminals also seek to exploit one-off opportunities such as major events. Half of all organized crime groups are involved in more than one type of crime. Most organized crime groups are involved with money laundering to make their money appear legitimate. To launder money, organized criminals often turn to cash-rich businesses like bars, fast-food outlets or car washes, competing unfairly against those legitimately running such businesses.

While the laundering of criminal cash and the importation of drugs are key challenges for security forces engaged in the protection of ports and borders, the trafficking of human beings remains a primary concern to governments and is a serious crime that demeans the value of human life. Human trafficking is the acquisition of people through the use of force, coercion, deception, through debt bondage or other means with the aim of exploiting them. Men, women, and children can fall into the hands of traffickers either in their own countries or abroad. Trafficking occurs both across borders and within a country; it is not always visible—exploitative situations are frequently covert and not easily detectable. During the sitting of the UN in Palermo, Italy, in 2000, the Protocol to Prevent, Suppress, and Punish Trafficking in Persons, Especially Women and Children, supplementing the UN Convention Against Transnational Organized Crime was adopted. This protocol, more commonly known as the Palermo Protocol, provided the very first internationally recognized definition of human trafficking as the following:

> *Trafficking in persons shall mean the recruitment, transportation, transfer, harbouring or receipt of persons, by means of the threat or use of force or other forms of coercion, of abduction, of fraud, of deception, of abuse of power or of a position of vulnerability or of the giving or receiving of payments or benefits to achieve the consent of a person having control over another person, for the purposes of exploitation. Exploitation shall include, at a minimum, the exploitation of the prostitution of other forms of sexual exploitation, forced labour or services, slavery or practices similar to slavery, servitude or removal of organs.*

Sadly, children are among the most vulnerable victims of this increasingly organized crime. Sometimes they are sold by family members or families are in debt bondage to traffickers and their children are put into forced labor or domestic work where they are vulnerable to sexual or physical abuse. Children may be abducted or handed over by their parents in the belief that they may have a better life and access to education. Children are also vulnerable to being used in criminal enterprises, working in cannabis farms or pick-pocketing gangs. Some are unaccompanied asylum-seeking children who can be preyed upon by those who exploit them to enable others to acquire state benefits. Security forces at ports continue to tackle human trafficking through prevention and disruption mechanisms achieved by dismantling criminal networks, constructing robust prosecution cases, and confiscating assets that are the proceeds of crime.

The challenges of countering human trafficking provide evidence that ports and borders are under relentless attack from all types of hazards. In this age of economic uncertainty nations across the world must act quickly and effectively to be able to address new and evolving threats to their security. In an increasingly interconnected, interdependent, and globalized world, individual countries of the Western world remain a vital link in the global economic network, which presents great opportunities and potential vulnerabilities. It therefore remains the first duty of government to ensure that its nation is secure by not only protecting all of its citizens, but also safeguarding its economy, infrastructure, and wider territory from all major risks that can, if left unchecked, affect them directly. A vital component in protecting national security and economic prosperity continues to be the successful delivery of a strong and resilient border.

The diverse range of security challenges encountered at borders also provides significant opportunities to lawfully gather intelligence and evidence on those individuals, groups, and states who seek to exploit, disrupt, or destroy the free and democratic way of life in the West. Yet the increasingly commercializd and economically driven focus of our ports of entry provides a challenging environment in which law enforcement and intelligence agency practitioners must operate. All in authority must never forget that the safety of the traveling public and the wider security of its citizens must continue to take precedence at all times and that the development of a strong, united, and resilient border, with agencies and governments working together, will ensure increased security for all. New threats will no doubt emerge in the future and those intent upon defeating security measures at borders will continue to create new and innovative solutions to carry out their unlawful activities. Therefore, in the post 9/11 era, ports and borders across the world remain the first and last line of national security defense.

SECTION 2

The Public, Communication, Risk, and National Security

CHAPTER

6

Risk Communication, Risk Perception and Behavior as Foundations of Effective National Security Practices

M. Brooke Rogers, Julia M. Pearce

INTRODUCTION

Any event that impacts upon the ability of individuals, communities, organizations and nations to carry on as normal constitutes a threat to national security. As a result, the management of risk is a central role of governments around the world. The UK's Blackett Review of High Impact Low Probability Risks states, "One of the key drivers in a major crisis is to avoid collateral impacts and restore 'normality' as quickly as possible" (Government Office for Science, 2011, p. 24). Whether you believe in the possibility of returning to "normal" or embrace the concept of "the new normal" (McArdle et al., 2012, p. 754), the success of government interventions before, during, and after a crisis relies on the cooperation of the public. Risk communication is a primary tool for achieving understanding and cooperation from society. Therefore, risk communication is an essential part of risk management and forms an important cornerstone of the foundation of effective national security practices.

This chapter will argue that risk communication, risk perception, and behavior must be recognized as foundations of effective national security, and that policies and practices that do not incorporate the concepts underpinning effective risk communication are bound to fail. In doing so, this chapter will provide a detailed explanation of the primary drivers and, thus, the primary targets, of all risk communication: risk perception and behavior. It will explain how risk perceptions influence the psychological and behavioral responses that play a key role in determining health, social, and economic impacts of extreme events. It will then unpack the relationship between risk communication, risk perception, and behavior. This will entail an exploration of the underlying issues informing expert and public variations in perceived risk, the identification of likely behavioral reactions to extreme events, and an illustration of the ways in which effective risk communication can mitigate negative impacts by reducing unnecessary care-seeking and increasing compliance with official guidance. Finally, this chapter will suggest steps that can be taken to better enable practitioners and policy makers to develop risk communication messages and emergency response programs capable of informing public perceptions of risk and positive health behaviors.

RISK COMMUNICATION: A PILLAR OF NATIONAL SECURITY

Risk communication is increasingly recognized as a primary tool within an effective security arsenal. Depending on the nature and evolution of the risk being addressed, risk communication can be used to reduce anxiety during times of crisis, to manage awareness in order to maintain engagement, or to raise awareness of issues when engagement is low (Government Office for Science, 2009, p. 2). As a result, risk communication can play a variety of roles in preparing for, responding to, and recovering from a broad range of natural and technological, accidental, and malicious disasters. Overall, risk communication is believed to improve resilience across organizations and communities.

The increased focus on the use of communication to improve our ability to prepare, respond, and recover from extreme events can be seen in a variety of government documents (see Government Office for Science, 2009, 2011; Cabinet Office 2011a), national and international research programs (e.g., PIRATE, CIE Toolkit, PRACTICE, Resilient Futures), and independent government reviews. These reviews reflect upon the success or failure of governments to respond to crises ranging from infectious diseases such as pandemic flu (Hine, 2010), to extreme weather events such as flooding and hurricanes (U.S. House of Representatives, 2006; Cabinet Office 2008a), to social unrest (HMIC, 2011), and, finally, to terrorist attacks (The 9-11 Commission, 2004; Coroner's Office, 2011; Home Office, 2012). The independent findings and lessons learned repeatedly stress the need for improvements in the communication of risk. Why is the effective communication of risk important enough to lead governments from around the world to emphasize the need to communicate and engage with members of the public?

THE IMPORTANCE OF EFFECTIVE RISK COMMUNICATION

Evidence gathered from real-world emergencies and disasters, government training exercises, and academic studies prove repeatedly that getting the messages right can mean the difference between a successful response and a failed response. In short, the effective communication of risk can be a matter of life or death during a crisis situation. This is because risk communication messages inform the public psychological and behavioral responses to risks, and as a result, help determine the subsequent likelihood of disease or injury (morbidity) and survival (mortality) rates in response to a crisis (Rogers et al., 2007).

The relationship between risk communication, morbidity, and mortality rates was evident in the reported evacuation experiences of the World Trade Center (WTC) during the terrorist attacks in New York City on September 11, 2001. While the evacuation of the WTC is largely viewed as a success, resulting in a survival rate of 99% for those below the crash site, there were instances where communication had the potential to put lives at risk (Glass and Schoch-Spana, 2002; Proulx and Fahy, 2003; Sheppard et al., 2006). For example, survivors report that, shortly after the first airplane struck Tower 1, a message was issued over the public announcement system informing individuals that, "...Tower 2 was secure and that it was safe for occupants to return or stay in their offices" (Proulx and Fahy, 2003, p. 30). This information was correct at the time (approximately 9:00 A.M.), although Tower 2 was struck by an aircraft three minutes later. Behavioral responses to the message varied: some occupants in Tower 2 started their evacuation prior to the announcement and continued to evacuate after hearing the announcement, others disregarded the instruction to return or stay in their office and began to evacuate, and others turned back in order to return to their offices upon hearing the announcement (Proulx and Fahy, 2003). The behavioral response to the risk messaging resulted in different health outcomes for different individuals caught up in the same event.

Acknowledging Variations in Behavioral Responses to Risk Communication

The importance of understanding variations in behavioral response to risk communication cannot be underestimated. Individuals can instigate a number of major and minor behavioral changes in order to avoid real or perceived risks. Some of these changes are "spontaneous precautionary behaviours" (Rubin et al., 2010), undertaken without (and sometimes in the absence of) official advice. Examples of spontaneous precautionary behaviors include the self-evacuation of approximately 40% of the population within 15 miles of the Three Mile Island nuclear plant during a partial meltdown in 1979 (Glass and Schoch-Spana, 2002); the widespread purchase of face masks during the 2005 avian flu and 2009 swine flu pandemics; and the avoidance of public transportation after the 9/11 attacks in the United States, the 2004 Madrid bombings, and the 2005 London bombings (Gigerenzer, 2004; López-Rousseau, 2005; Fasolo et al., 2008). Some behaviors, such as the adoption of hand and respiratory hygiene advice during an infectious disease outbreak (Hine, 2010) will only have a minor impact on day-to-day life, while others such as social distancing, unwarranted self-presentation at health facilities, the unnecessary uptake of antibiotics, and the avoidance of travel, have the potential to impact the economy, as well as the ability of health services and society to respond to current and future threats.

Implications for Physical Health

While a number of spontaneous precautionary behaviors enable members of the public to avoid or decrease the likelihood of exposure to risk, some behaviors lead to exposure to additional risks. As a result, the way in which individuals perceive risks and the way in which they behave as a result of those perceptions can have serious implications for physical health.

Consider the relationship between the perceived risk of travel on public transportation after terrorist attacks and the resultant public behavior. Evidence for this relationship began to accumulate in December 2001, when David G. Myers noted that cancellations had crippled the aviation and holiday industry. He assumed that Americans were driving instead of flying and based on a mile-for-mile comparison in the United States, which demonstrated that Americans are 37 times more likely to die in a car crash than on a commercial flight expressed concern that "...the terrorists may still be killing us, in ways unnoticed" (Myers, 2001). Gigerenzer (2004) confirmed that, when compared with the 2000 statistics, the 2001 national revenue passenger miles showed a significant decrease from October to December 2001. Additionally, road traffic data indicated that Americans were driving further, as the monthly miles driven were significantly higher than in previous years. Gigerenzer went on to note significantly higher levels of road traffic fatalities during the months of October, November, and December 2001, which he calculated as resulting in an additional 353 road traffic fatalities. When put in context, the number of additional road traffic fatalities that came about as a result of this spontaneous avoidance behavior exceeds the number of lives lost (266) on the hijacked airplanes.

A similar trend has been observed in Europe, where "...dying as a passenger on a train is 20 times less likely than dying as a passenger in a car" (López-Rousseau, 2005, p. 426). An analysis of the Madrid train bombings in 2004 and the London bombings in 2005, however, demonstrates that avoidance behaviors resulted in an entirely different set of physical health outcomes for the Madrid and London public. For example, in spite of a significant decrease in the number of train travelers after the Madrid bombing, Spaniards did not replace train travel with car journeys. As a result, members of the Spanish public were not exposed to the additional risk associated with car journeys (López-Rousseau, 2005). Londoners avoided the underground and buses after the London bombings in 2005 to a greater extent than the Spaniards, but to a lesser extent when compared to the Americans. Instead of increases in the use of cars, however, Londoners took to the roads on bicycles and two-wheeled vehicles. Fasolo et al. (2008) did not find evidence of an increase in fatalities as a result of the Londoners' substitutions.

López-Rousseau (2005) and Fasolo et al. (2008) suggest that the role of cultural factors must be taken into consideration when discussing variations in public behavioral responses to extreme events. For instance, variations in the extent of behavioral change between the three countries could have been influenced by the differences in scale and number of fatalities as a result of the attacks. Other issues, such as a desire to avoid London's congestion charge for all private vehicles entering London and the lack of a car culture in Spain could have influenced the level of travel substitutions. Finally, both the Spanish and UK publics have had previous experience with terrorist attacks, which also has the potential to mediate the public reaction (López-Rousseau 2005, Fasolo et al., 2008).

Implications for the Ability of Systems to Respond

It is evident that public perceptions of risk and their resulting behavioral responses can create additional risks for members of the public. The interactions between risk perceptions and behavior can also impact the ability of governments and government agencies to respond to disasters, thus creating additional risks for national security. While many emergency response plans focus on the potential of public behavior to overwhelm their resources and ability to respond, threats to emergency response systems can arise from over-response, under-response, and the more likely, unpredictable, mixed responses to official advice, or a lack, thereof.

The potential of members of the public to overwhelm a systems ability to respond through over-response was illustrated by the 1987 radioactive incident in Goiania, Brazil. In this case, government failure to communicate about an accidental release of radioactive material resulted in the healthcare system becoming overwhelmed when more than 112,000 individuals sought medical examination for an incident that resulted in 4 deaths and 260 contaminations (IAEA, 1998; Acton et al., 2007; Rogers et al., 2007). Similarly, the Sarin attacks in the Tokyo subway (1995) tested the ability of systems to cope with unpredictable public behavior. In this, case, "5,510 of those who flooded hospital emergency departments were "psychological casualties"—that is, they experienced physical symptoms without direct exposure to nerve agent" (Lemyre et al., 2005 p. 2). More recently, the World Health Organization had to offer guidance on how to avoid radiation in response to the "panic" buying that took place during the Fukushima nuclear accident in Japan in March 2011. In this instance, members of the public were buying potassium iodide tablets, which prevent the absorption of radioactive iodide-131 into the thyroid gland (McCurry, 2011).

Conversely, in the case of the 2009 swine flu pandemic, in spite of the fact that the UK government's communication strategy successfully built up high levels of public awareness and understanding of pandemic flu (Hine, 2010), reported levels of engaging in protective behaviors and likely acceptance rates for vaccines were low (Rubin et al., 2010). These low levels were attributed to a number of factors (e.g., levels of worry about the possibility of one's child catching swine flu or personally catching swine flu, perceiving the government to be well-prepared for swine flu, etc.), including the perception that too much fuss was being made about the risk of swine flu (Rubin et al., 2010). The 2009 swine flu pandemic was not as severe as it could have been, but it issued a warning to governments that communication about future outbreaks of infectious disease must address the low uptake of recommended behaviors observed during this outbreak.

Finally, it is important to recognize that, when discussing the ability of systems to respond, we must not plan for a simple over-demand or under-demand for health or emergency services. This is because the public behavioral response to extreme events is often mixed, as demonstrated by the public response to the Anthrax attacks in the United States (2001). Public behavioral responses to the Anthrax attacks ranged from the "panic" purchasing of gas masks and ciprofloxacin, to multiple reports of calm and orderly behavior as individuals waited in line for their turn to be tested or given prophylaxis, to reports of individuals in affected states reporting little or no concern about contracting anthrax (Glass and Schoch-Spana, 2002). These variations in public behavior in response to a shared event (anthrax) did not generate enough of an impact to cripple the system's ability to respond. However, they did provide evidence that it does not take a large percentage of the population to change their behavior(s) in order to increase the demands on the system in quite a large way. This highlights the need for the development of resilient, adaptable response plans and procedures built upon a strong understanding of the relationship between risk communication, risk perceptions, and behavior.

Inspiring Trust, Influencing Response

One constant trend in the midst of all of these complex behavioral reactions to extreme events is that the likelihood of the success or failure of risk communication is strongly and consistently mediated by the levels of public trust (Glass and Schoch-Spana, 2002; Earle, 2004; Kasperson and Palmlund, 2005; Rogers et al., 2007; Pearce et al., 2012a). This is due to the fact that the ability of governments and government agencies to respond to disasters is limited by the extent of public trust in the efficacy of the response and in those tasked with communicating the response. Building and maintaining trust is vital, as trust is considered the primary route to cooperation and once lost it is very hard to regain (Wynne, 1992; Rogers et al., 2007; Pearce et al., 2012a,b).

Fear that the public will panic has repeatedly led authorities to withhold information following extreme events. Rather than reassuring the public, this has resulted in an increase in unnecessary care seeking of the sort seen following the 1987 Goiania radiation accident. Trust was clearly an issue during the anthrax incident in the United States in which behavior ranged from over-reactions to under-reactions. The variation noted in public behavior during and after the anthrax attacks must be "…seen in the context of conflicting reports from experts about the nature of the threat, as well as vague and nonspecific government alerts about additional possible attacks, the level of public concern appears measured and reasonable" (Glass and Schoch-Spana, 2002, p. 222). Additionally, public confidence in the Japanese government rapidly declined in the wake of the 2011 Fukushima nuclear accident when the hydrogen explosion made the severity of the crisis clear despite government attempts to downplay the seriousness of the situation (Funabashi and Kitazawa, 2012). Organizations tasked with risk communication are also learning that the use of modern technologies and 24/7 news broadcasts to document and respond to crises are hindering attempts to downplay the seriousness of events in an information-rich media environment.

Lack of trust may not only lead to an increase in unnecessary care seeking, but may also lead to refusal to adopt recommended protective health behaviors. For example, following the 2011 anthrax attacks in the United States, failure to effectively communicate reasons for lack of consistency in testing and treatment regimes for groups exposed at different times, or to address ongoing concerns about history of racial discrimination in vaccination programs led to particularly low uptake of prophylactic vaccination among at-risk African American postal workers (Blanchard et al., 2005). While the consequences on that particular occasion were not fatal, failing to persuade the public to comply with targeted vaccination programs following a catastrophic outbreak of an infectious disease could determine the effectiveness of these interventions and subsequent mortality and morbidity rates.

RISK PERCEPTION: A FOUNDATION FOR UNDERSTANDING PUBLIC RESPONSES TO EXTREME EVENTS

In order to establish the likelihood that the public will adopt protective health behaviors, it is important not only to understand the extent of trust in those tasked with providing advice and treatment, but also to understand public perceptions about the risks associated with compliance or noncompliance with recommended behaviors. For example, the sharp decrease in uptake of the measles, mumps, and rubella (MMR) vaccine after a 1998 *Lancet* report linking the MMR vaccine with autism suggests that the risks associated with the vaccine were perceived by the UK public to be higher than the risk of nonvaccination, despite the fact that increased incidence of measles and mumps could result in severe and permanent injuries, as well as deaths (Bellaby, 2003). Additionally, the muted public response to the 2006 radioactive poisoning of Alexander Litvenenko in London resulted in a low public level of uptake of testing for individual levels of contamination in spite of the fact that the incident involved radioactive contamination in public areas, an issue believed to have the potential to cause extreme changes in public behavior (Acton et al., 2007). In this instance, the perception that the attack was a targeted assassination led to low perceptions of risk (Rubin et al., 2007).

To truly understand the relationship between risk communication, risk perception, and behavior it is important to realize that risk means different things to different individuals. It is also imperative for organizations who hope to engage in effective risk communication to recognize that a gap exists between public and professional perceptions of risk (Rogers et al., 2007).

Expert Perceptions of Risk

When experts discuss risk, they are attempting to quantify the amount of harm or the number of deaths that can arise as a result of exposure to a risk. Experts often discuss risks in terms of cause and effect relationships, and they use complex statistical methods in order to analyze large amounts of data in an attempt to identify the risks that are the most likely to impact public health (Slovic et al., 1981, 1986; 1998; Beck, 1999; Weidemann et al., 2003; Rogers et al., 2007). They are then tasked with communicating these risks to members of the public.

Public Perceptions of Risk

The low uptake of suggested precautionary behaviors following the 1998 MMR controversy (Bellaby, 2003) and the 2009 swine flu pandemic (Rubin et al., 2010) demonstrate that public perceptions of risk do not necessarily reflect the level of risk that experts are trying to communicate. Members of the public are interested in the science driving expert perceptions of risk, but expert communication of risk to members of the public often fails to take the social context of risks into account (Beck, 1999; Rogers et al., 2007). Table 6.1 illustrates the different factors that experts and members of the public focus on when making decisions about risks. Risks qualify as "dread risks" if they are believed to have catastrophic potential, fatal consequences, seen as uncontrollable, seen as inequitable, and have a high risk to future generations (Slovic et al., 1981).

Risk perceptions are made in dynamic, ever-changing environments. Information that influences one factor will have knock-on effects across the other factors and, thus, have the

TABLE 6.1 Variations in Public and Expert Perceptions of Risk

Expert Risk Perception Factors	Public Risk Perception Factors
Can I identify a clear cause and effect relationship?	Is the risk voluntary or involuntary?
Can I quantify the amount of harm?	Is the risk familiar or unfamiliar?
Do I suspect a hazard based on past experience?	Do I have control over the risk or do I trust the organizations who are supposed to be responsible for the risk?
Is there a possibility of an accident?	Is the risk fair or unfair?
Is there a possibility of exposure to the risk?	Is the risk natural or technological?
Is there evidence of damage?	Does the risk cause dread?

potential to change the overall perception of the risk. Risk perceptions will, in turn, inform the behavioral response to the risk.

BEHAVIOR: UNDERSTANDING LIKELY PUBLIC RESPONSES TO EXTREME EVENTS

Professional Planning Assumptions and Public Behavior

Variations in the perception of risk and levels of trust are not the only stumbling blocks hindering the effective communication of risk between experts and members of the public. In spite of the growing recognition of the importance of communicating with members of the public, many emergency planning assumptions and risk communication campaigns are built upon an imperfect understanding of the relationship between risk communication, risk perception, and public behaviors. As a result, emergency planning assumptions often fail to incorporate human behavior and are based on contradictory expectations.

The contradictions that can exist within emergency planning assumptions can be seen in focus groups, which examined UK and Polish health care responders' views of the public response to a hypothetical chemical event (Pearce and Rogers, 2011). Responders in both countries assumed public panic. For example, one respondent said, "Camera crews, it wouldn't take them long to get there, so it'll be on television within minutes and people around them will be starting to panic." However, as the event progressed, the same respondent suggested that, upon hearing official advice to shelter, members of the public would comply. He stated, "I think quite generally in any incidents like that they see a uniform and they sort of feel a bit better, 'I feel safe,' and you've just got to say something to most of them and they'll just do as you tell them." Clearly, it would be difficult for individuals who were in the middle of a panic to calmly comply with official advice. Which view is accurate? Will the public panic or will the public comply with official advice during an emergency? It is important to answer this question in order to ensure that emergency planning assumptions incorporate accurate data about the ability of the public to respond to disasters in their own right.

Assumptions of public panic in light of an emergency event are the norm, rather than the exception. Discussions about the ways in which members of the public might respond to an extreme incident are based on "...the assumption that the general public tends to be irrational, uncoordinated, and uncooperative in emergencies—not to mention prone to panic" (Glass and Schoch-Spana, 2012, p. 217). Assumptions of panic lead emergency planning processes to exclude members of the public. As a result, emergency planning assumptions fail to understand and plan for the variety of behaviors that can take place during an extreme event.

Is the assumption of public panic valid? There is some evidence for panic, particularly in situations where there have been fires in enclosed spaces where there is no visible means of escape (Glass and Schoch-Spana, 2002). However, overwhelming evidence suggests that when faced with disasters and emergencies, people become cooperative and panic is rare. For example, there were multiple reports of helping behavior and an atmosphere of calm during the evacuation of the WTC during the 9/11 attacks (Proulx and Fahy, 2003; Sheppard et al., 2006); first-hand accounts of pro-social behavior and calm during the 7/7 transport bombings in London during 2005 (Sheppard et al., 2006; Drury et al., 2009); and resourcefulness, civility, and mutual aid were reported during the response to the 1918 Spanish influenza pandemic (Glass and Schoch-Spana, 2002). What appear to be panic behaviors are therefore more likely to reflect a lack of choice, rather than poor decision making or a rational choice under the circumstances and in light of the available information. The continued focus on panic has the potential to obscure the broad range of possible behaviors identified in previous examples.

Public Behavioral Response Assumptions during an Extreme Event

Members of the public have also been known to make inaccurate assumptions about their responses to extreme events. In spite of the fact that compliance with official guidance is often poor (Lemyre et al., 2010; Rubin et al., 2010; Pearce et al., 2012a), members of the public tend to overestimate their levels of compliance with official advice during an emergency event. For example, Pearce et al. (2012a) assessed public intention to comply with official advice to shelter in place following a hypothetical chemical spill in the UK and Poland. Their findings demonstrated that members of the public reported high levels of intended compliance with official advice to shelter in place when at home (95% UK, 93% Poland), although this figure dropped when members of the public were asked to shelter in a public place such as the post office (30% UK, 70% Poland). However, once intended behaviors that were incompatible with compliance (e.g., collecting children from school, evacuation, checking on friends and loved ones) were taken into account, intended compliance was significantly reduced. In fact, this more accurate measure of intended compliance suggests that only 43% of UK respondents and 7% of Polish respondents would comply with advice to shelter at home. In the post office, only 10% of UK respondents and 5% of Polish respondents intended to comply with the advice to shelter. The key factors that influenced the public intention to comply in this study included the perceived ease of compliance, the intention to collect children from school, the individual's perception of their ability to respond, the perceived ability of the professional organizations to respond, the perceived cost of responding, and the level of trust that they had in authorities (Pearce et al., 2012a).

Inaccurate and uninformed assumptions about public behavioral responses to extreme events must be addressed on the policy-making level, the professional emergency response level, and on the public level. Assumptions of panic can have far-reaching consequences as they can lead to a less effective emergency response by obscuring the ways in which public behaviors, concerns, and anxieties can be modified by effective risk and crisis communication. Overestimations of compliance suggest that members of the public do not truly understand the emergency response processes and procedures that will take place during an extreme event. Risk communication messages and programs must be designed in order to address these gaps. What steps can be taken to better enable practitioners and policy makers to develop risk communication messages and emergency response programs capable of informing public perceptions of risk and positive health behaviors?

RISK COMMUNICATION IN PRACTICE

This chapter has demonstrated the importance of recognizing risk perception, risk communication, and behavior as foundations of effective national security and recognizing that policies and practices that do not incorporate the concepts underpinning effective risk communication are bound to fail. In short, all approaches to public risk communication, irrespective of the reason for communication, must be built upon an understanding of risk perceptions and likely behaviors. This is especially important in light of the fact that, "For many high impact risks we do not understand what the public actually expects in a situation, or how tolerant they may be of 'abnormal' risks during a crisis" (Government Office for Science, 2011, p. 24).

First and foremost, it is important to recognize that getting messages right can mean the difference between a successful or failed response. Risk messages influence risk perceptions and behavior and, in turn, influence the likelihood of exposure to disease, injury, survival, and death. Acknowledging that public behavior in response to extreme events often reflects the quality and availability of information is a crucial step toward reworking one of the strongest tools available to governments: communication.

The interactions between risk perceptions and behavior can impact the ability of governments and government agencies to respond to disasters. Members of the public can over-react, under-react, or demonstrate a mixed response to government attempts to communicate about and respond to an extreme event. All forms of response can lead to a threat to national security. In order to address these issues, organizations must concentrate on building, improving, and maintaining levels of trust before, during, and after an incident occurs.

Additionally, in order to enhance the efficacy of risk communication, government organizations should "...work more closely with risk communication experts and behavioral scientists to develop both internal and external communication" (Government of Science, 2011, p. 26) to produce clear messages that are based on an understanding of the differences between expert and lay perceptions of risk. Trust can be built by the provision of factually correct, consistent, regularly updated information (Rogers et al., 2007). Unpalatable information should not be withheld due to fear of provoking panic. The resulting loss of credibility and absence of clear direction are both associated with spontaneous protective behaviors that are likely to lead to behaviors that may overburden medical services and hinder emergency responses (Glass and Schoch-Spana, 2002).

Overall, risk communicators should focus on improving knowledge and understanding as well as managing public expectations. Well-informed risk communication can help practitioners achieve these goals by enhancing the likelihood that at-risk populations will take appropriate protective actions and decrease the likelihood that low-risk populations will overburden emergency response systems. Effective risk communication can also reassure those who are not at risk by reducing rumors and fears and maintaining public trust and confidence in agencies responsible for ensuring public welfare (Becker, 2004; Rogers et al., 2007).

CHAPTER 7

Promoting Public Resilience against Chemical, Biological, Radiological and Nuclear Terrorism

Kristian Krieger, M. Brooke Rogers

INTRODUCTION

Responding to the threats from terrorism has traditionally revolved around the notions of prevention and security. Responses therefore often target potential perpetrators, make extensive use of technology (e.g., surveillance), and are dominated by law enforcement and military activities. The prevention of numerous terror attacks vindicates the repeated focus on prevention and security. However, successful attacks, such as those with Sarin on Tokyo's subway (1995), the bombing of the World Trade Center in New York (2011), and the train bombs in Madrid (2004) and on London's transport system (2005), suggest that counterterrorism policies need to include measures that improve the resilience of societies and communities in order to enable them to respond and adapt in the instances where preventative measures fail to stop an attack. Resilience refers to the ability of individuals and communities to cope with and adapt to major perturbations, as well as to "bounce back"(Adger, 2000; Masten and Osofsky, 2010; Poortinga, 2012) after experiencing an extreme event. The broadening of counterterrorism activities to include policies and activities targeted at promoting a resilient society against terrorist attacks has substantial research and policy implications. Most fundamentally, it recalibrates the attention of authorities from the perpetrators, law enforcement, and the development of technology onto behavior of members of the public and the activities of emergency response authorities.

Notable advances in programs targeted at the development of community resilience have been made by the UK Cabinet Office with the launch of the first publicly available National Risk Register in 2008 (Cabinet Office, 2008a,b, 2012), the Community Resilience Web site (http://www.cabinetoffice.gov.uk/content/community-resilience), the Strategic National Framework on Community Resilience (Cabinet Office, 2011a,b), and the Center for Disease

Control's First Hours program (United States) (http://www.bt.cdc.gov/firsthours/). All of these programs make information publicly available for a variety of risks ranging from natural hazards to lower likelihood–high-impact risks such as chemical, biological, radiological and nuclear (CBRN) terrorism. In fact, the inclusion of the threat of CBRN terrorism in public discussions is a new risk communication trend driven by the knowledge that "terrorist networks have made no secret of their desire to acquire and use chemical, biological, radiological and nuclear (CBRN) weapons" requiring "a comprehensive strategy to try to stop them succeeding" (UKCO, 2008, p. 28).

In this chapter, we examine how public resilience against terrorist attacks using CBRN agents can be promoted. The case of CBRN is particularly interesting for the study of public resilience because the uncertain and dreaded nature of the hazard and its consequences suggests the potential for a particularly strong impact on the mental and physical health of individuals, as well as on their behavior. We argue that in order to mitigate the adverse impacts on health and ensure behavior that facilitates response and recovery to emergencies, policy makers, law enforcement, and emergency management authorities need to devise effective crisis and risk communication strategies. The communication strategies need to take into account a set of four factors that require the integration of insights from social psychology and other social sciences, as well as a thorough engagement with members of the public. These factors include the following:

1. Public perception of risk
2. Public perception of authorities
3. Public perception of response and self-efficacy
4. Quality and timing of risk communication

This chapter briefly discusses how counterterrorism policies and debate have traditionally been dominated by prevention and security logics and demonstrates that public resilience has recently attracted increased attention by policy makers and academics. It then explores the role of the public and the authorities' engagement with the public in promoting resilience against CBRN terrorism. This chapter also uses the case of CBRN terrorism to identify four factors that affect the public resilience against terrorist attacks, and concludes by reflecting on the institutional/policy implications of the four factors, as well as advocating for a comparative research agenda.

BEYOND PREVENTION AND SECURITY IN COUNTERTERRORISM: PROMOTING PUBLIC RESILIENCE AND MANAGING RISK

Counterterrorism is conventionally attributed to the policy field of security. Historically, security policies were primarily populated by the military and law enforcement agencies and targeted at the prevention of violent attacks, whether they originate from other countries or from within national borders. Counterterrorism policies were no exception. One case in point is the 1985 U.S. counterterrorism strategy in which the U.S. government promises to "resist the use of terrorism by all legal means" and "act in a strong manner against terrorists without surrendering basic freedoms or endangering democratic principles" (Bush, 1986, p. 7). Modern approaches to security policies indicate that prevention continues to form a central plank of counterterrorism strategies. For example, both the UK's CONTEST counterterrorism strategy and the European Union (EU) Counterterrorism Policy (EC, 2010) include the streams of "pursue" (to stop terrorist attacks) and "prevent" (to stop people from becoming or supporting terrorists).

However, counterterrorism policies have undergone a notable evolution over the past two decades. The aforementioned UK CONTEST and EU counterterrorism policies, for instance, also include streams named "prepare" (where an attack cannot be stopped, to mitigate its impact) and "response" (mobilizing the EU civil protection mechanisms), respectively. Similarly, governments have revised emergency response policies against terrorism and other catastrophic events: One example is the 2004 Civil Contingency Act (UK) that seeks to improve coordination among the diverse network of response organizations and planning for a wide range of events and emergency situations. Another example can be found in the United States: The mid-1990s saw the instigation of the Federal weapons of mass destruction (WMD) preparedness program designed to improve the nationwide capacity to manage the consequences of a chemical or biological weapons attack through federal assistance to state-level disaster management agencies. The WMD program had rapidly become one of the fastest growing federal programs of the late 1990s (Falkenrath, 2001). In Germany, the federal government established the Federal Office of Civil Protection and Disaster Assistance in 2004 in order to enhance preparedness and coordination for emergencies and crises. At the European level, the European Civil Protection Mechanism was established and revised in 2001 and 2007, respectively, through which individual member states can—when national resources are deemed inadequate—request assistance from the disaster management agencies of other EU Member States.

The incorporation of discussions and plans for preparedness and response in the case of extreme events such as terrorist attacks have also been reflected in the recent debates in the academic literature on security. Rasmussen (2001, p. 308), for example, argued that the new post-9/11 "strategic reality is better understood by the concept of risk society than by traditional notions of terrorism." The concept of risk society, as developed by Ulrich Beck (1992, 1999), points to the emergence of a society in which uncontrollable and unpredictable dangers are present. Beck even argued that these dangers cannot be governed due to the fact that these risks are beyond rational calculations. The particular challenge of governing terrorism, along with other low probability–high impact risks (Blackett-Review, 2011), is also acknowledged in recent policy initiatives. For instance, the EU counterterrorism policy recognizes that the "risk of terrorist attacks cannot be reduced to zero" (EC, 2010). However, as critics of Beck note, that does not mean that risks such as terrorism are ungovernable (Aradau and Munster, 2007). Rather, they are in fact already being governed through a wide range of mechanisms. Some of them resonate with the traditional idea of providing security and prevention, such as more extensive profiling and surveillance (Munster, 2004); others, however, accept the uncertainty of the state's ability to provide security and protection and aim to promote public resilience, such as insurance (Ericson and Doyle, 2004) and emergency preparedness and response (Anderson, 2010; Anderson and Adey, 2011).

In the context of this evolution in counterterrorism policy, this chapter focuses on how the state can improve public resilience via emergency preparedness and response. Policy initiatives such as the aforementioned UK Civil Contingency Act or the U.S. WMD preparedness program often focus on coordinating the diverse networks of actors involved in the response to complex emergencies (Anderson and Adey, 2011). However, less attention in policy practice and literature has been paid to the engagement of the authorities with members of the public and the role of the public itself in emergency management.

RESILIENCE AND THE ROLE OF THE PUBLIC

Although interest has increased in recent years, the historically limited interest of practitioners and scholars of security and emergency management in the role of the public to ensure that society "bounces back" and copes with a terrorist attack is surprising for two reasons.

First, members of the public are a primary target of terrorist perpetrators. Scholars have been arguing that terrorists are strongly interested in the psychological effects of their activities. This school of thought identifies terrorism as theater, as embodied by Brian Jenkins' (1975, p. 15) classical statement: "Terrorists want a lot of people watching, not a lot of people dead." While the latter part of the statement may no longer be true, Ursano et al. (2004) elaborated on Jenkins' statement and argued that terrorists pursue three aims: to foster pervasive fear and insecurity through specific acts of violence, to promote behavioral changes that disrupt public infrastructure, and to create a loss of confidence in public institutions. To prevent terrorists from achieving key objectives and ensuring that the public copes effectively with terrorist attacks, it is important to improve the public response in cognitive, emotional, and behavioral terms.

Second, members of the public are also key actors in emergency management. In contrast to the often assumed irrational and counterproductive panic among members of the public (Glass and Schoch-Spana, 2002; Proulx and Fahy, 2003; Sheppard et al., 2006), the actions of individuals can mitigate the adverse impact of CBRN incidents, benefit the authorities' response and recovery efforts, and should be viewed as a rational response to the available information about the threat and the evolution of the event. However, information and perceptions may also lead to behaviors that interfere with the efforts of the authorities (IAEA, 1998). From the perspective of the emergency responders, it is therefore of key importance to understand what drives human behavior in emergency situations and how members of the public can become partners in, rather than barriers to, effectively responding to terrorist threats and emergency situations.

Both the role of the public as a terrorist target and as an actor in emergency management are particularly relevant in the context of CBRN terrorism. Scholarly literature and case studies suggest that members of the public are particularly strongly affected by CBRN threats in terms of mental well-being and behavioral responses, making the deployment of CBRN agents attractive to potential perpetrators. Scholars point in particular to the emotional response of anxiety and fear as a consequence of the uncertainty about the nature and effect of the CBRN incident (Holloway et al., 1997; Hyams et al., 2002; Acton et al., 2007). As a consequence, Page et al. (2006) found that, in respect to chemical incidents, the idea of the hazardous agent can, at times, be more harmful that the agent, itself. They argue that even "the impact of episodes in which there is no actual environmental hazard at all but only the perception of such a threat can be as damaging as those in which there is at least some chemical exposure" (Page et al., 2006, p. 413). Another observation that underpins the unique impact of the CBRN threat on mental health is the phenomenon of a "mass sociogenic illness" (DiGiovanni, 1999), i.e., (massive) episodes of physical symptoms of acute injury triggered by unusual odors and rumors of contamination.

Beyond the effects on mental and physical health of members of the public, CBRN terrorism has behavioral and political implications. The Assessment of the Vulnerabilities of Modern Societies to Terrorist Acts Employing Radiological, Biological or Chemical Agents with the View to Assist in Developing Preventative and Suppressive Crisis Management Strategies (ASSRBCVUL) project provides a useful overview of wider implications of CBRN threats and incidents (Bastings et al., 2007; Figure 7.1).

As seen in Figure 7.1, while anxiety and mental health issues clearly belong to the first tier "health" of the model, CBRN incidents have important wider repercussions. For example, images of first responders in personal protective equipment displayed in the media after the 2001 anthrax attacks spread fear among the population, especially as the identity of the perpetrators and the nature of the agent continued to be unknown (Dobbs, 2001). This spread of anxieties about exposure can in turn lead to overburdened healthcare facilities because the early symptoms associated with some CBRN events are often difficult to distinguish from common flu-like symptoms. Scholars have called this phenomenon the "worried well" (Stone, 2007), and overburdened health facilities point to (in ASSRBCVUL's terminology) second-tier implications, affecting "societal services."

Furthermore, given the uncertainty surrounding deliberate, complex-agent CBRN incidents, the medical professions may not be able to identify the agent immediately or determine the long-term health consequences with confidence. This absence of assurances by medical professionals may, along with the possible *ad hoc* use of vaccines and antidotes that may not even have been approved by medical regulators, increase public distrust vis-á-vis the medical

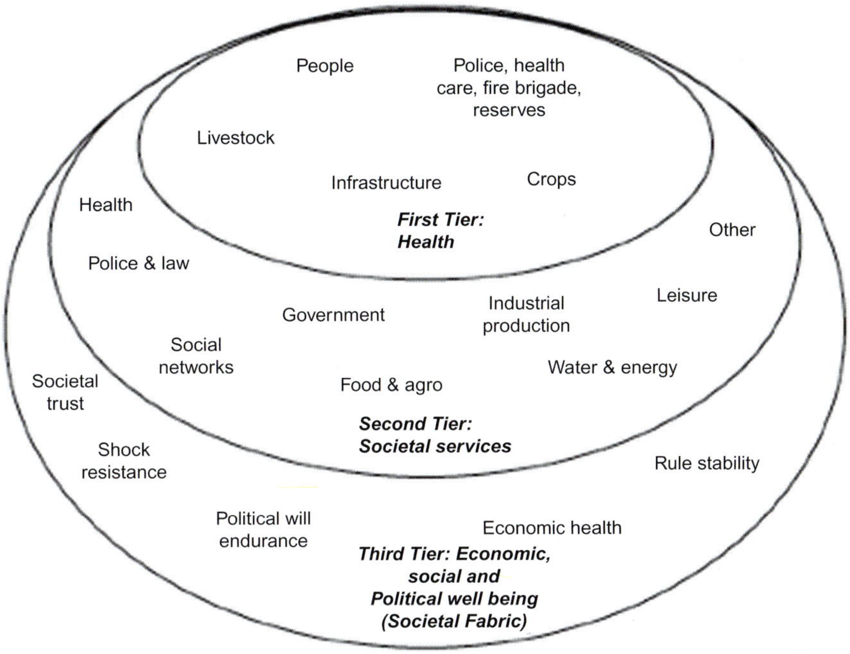

FIGURE 7.1 ASSRBCVUL's model of societal vulnerability *(Bastings et al., 2007).*

profession and government officials (Berezuk and McCarthy, 1992; Birchard, 1998). The loss of trust in emergency professionals and officials can in turn undermine political stability and endurance, thus rupturing the "societal fabric" of ASSRBCVUL's third tier. Moreover, CBRN incidents can impact the economic well-being of a society. Long-term psychological consequences, such as post-traumatic stress disorder, have the potential to affect the everyday life of victims (Whoriskey and Jenkins, 2001) and can therefore be expected to carry wider consequences for the economic activities of societies. Economic life will also be impacted as affected communities and their products and properties become stigmatized due to concerns or perceptions about contamination (Petterson, 1988). In short, CBRN threats and incidents can have significant adverse consequences on public health, as well as, more generally, on political and economic processes of a country. While some of this is related to direct physical harm and material damage, the emotional and cognitive, as well as behavioral responses by members of the public, can contribute significantly to the harm done by CBRN attacks. The following section focuses on the psychosocial and institutional factors that explain emotional, cognitive, and behavioral responses to CBRN threats and attacks. Understanding the drivers of these responses is essential to enhancing the resilience of the public against CBRN attacks.

PUBLIC PERCEPTIONS, RISK COMMUNICATION, AND THE PROMOTION OF RESILIENCE

Which drivers should policy makers and emergency planners focus on when designing interventions capable of shaping and informing the emotional, cognitive, and behavioral response of publics to CBRN threats? Four sets of factors have been identified on the basis of existing academic literature and case studies targeted at managing CBRN incidents, natural disasters, health behavior, and risk (Rogers et al., 2007; Krieger and Rogers, 2012; Pearce et al., 2012a,b). The factors and their effects are

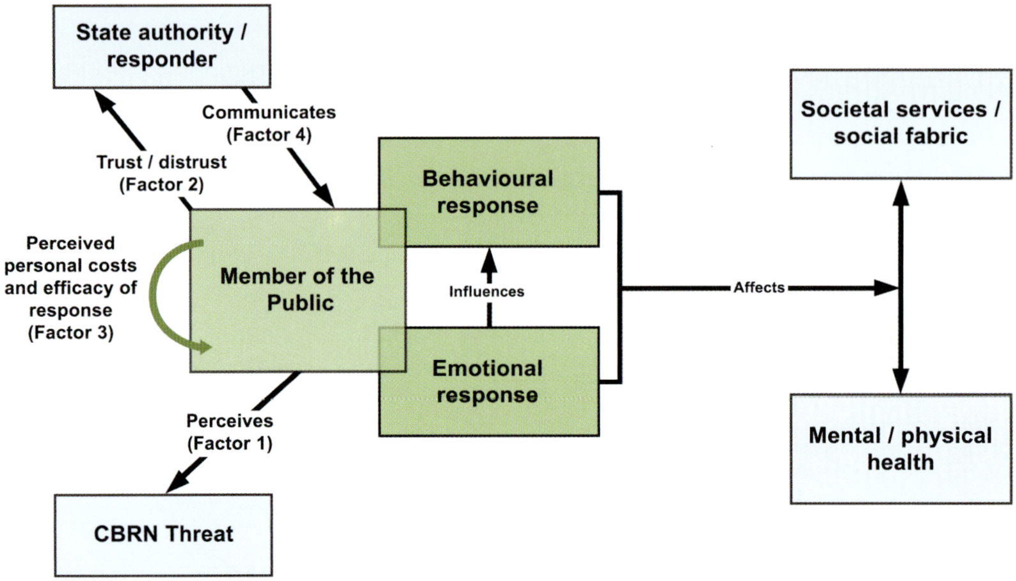

FIGURE 7.2 Four factors shaping emotional, cognitive, and behavioral responses to CBRN threats *(Krieger and Rogers, 2012).*

presented in Figure 7.2. The factor sets include the following:

1. Public risk perceptions of the CBRN threat
2 Public perceptions of the trustworthiness of authorities and policy makers in charge of emergency management and counterterrorism
3. Public perceptions of the efficacy and costs of responses, as well as individual's perceived self-efficacy
4. The quality of risk communication

The perceptions of and communication with the public affect emotional and behavioral responses to CBRN threats, which in turn influence the degree of individual and societal vulnerability/resilience to such threats. These four factors will be discussed in greater detail in the following sections.

Public Risk Perceptions of the CBRN Threat

The first set of factors concerns the public perception of CBRN threats and draws on arguments from social psychological research. Slovic et al. (2004) used the concept of "affect" to link risk appraisals by members of the public to negative emotions such as fear and anxiety and, thus, to subsequent effects on mental well-being and behavioral choices.

Often referred to as "risk as feelings," affect implies that a member of the public appraises risk, sometimes consciously, sometimes unconsciously, in a binary way, either as good or bad. This allows rapid assessment in a complex environment with multiple risks. This basic differentiation is often linked to past events that had an emotional significance, such as the images/experience of bloodied individuals wandering out of a train wreck. These emotional associations in turn shape the perceived benefits and costs in risk evaluation, a phenomenon called "affect heuristic" (Slovic et al., 1981; Slovic, 1987,). Some scholars have identified these past experiences and associations as "mental models" that can be uncovered through in-depth interviews (Morgan et al., 2001).

Social psychological research has, in the past decades, identified characteristics of threats that trigger emotional responses (Weisaeth and Tonnessen, 2003). CBRN threats resonate with many characteristics that trigger negative associations and lead to strong adverse impacts on their mental and physical health outcomes and behavioral responses. For instance, Hyams et al. (2002) noted that exposure to CBRN is involuntary; that CBRN risk is man-made and implies unfamiliar threats with unknown health effects; and the effects of CBRN events may also affect children and future generations, all of which increase public fear and anxiety. Franz and et al. (1997) argued that CBRN is perceived as particularly threatening because the events can harm large numbers of ordinary citizens in places generally considered safe, i.e., workplace or residential neighborhoods. These types of health effects (while often unknown) are also dreaded because CBRN agents can cause death and injury in strange and prolonged ways. In general, as Fullerton et al. (2003, p. 5) argued, terrorism can be distinguished from other risks by the "characteristic extensive fear, loss of confidence in institutions, unpredictability and pervasive experience of a loss of safety."

However, there are two important caveats to focus exclusively on risk perception as a factor to explain emotional, cognitive, and behavioral responses to CBRN threats. First, risk perception factors have been found to lack explanatory power in quantitative empirical studies (Sjoberg, 1996). Second, risk perceptions are shaped by the particular empirical context. In other words, risk perceptions may vary, depending, for instance, on the levels of public trust in institutions, factors influencing associated mental models (e.g., safety record of industry, history of incidents in a specific country/region), and ethnic and social classifications (Siegrist et al., 2005; Pearce et al., 2012b).

Public Perceptions of the Trustworthiness of Emergency Managers and Policy Makers

The second set of factors concerns public trust in institutions. Social science literature on trust normally interprets it as a key foundation to any social interaction. More specifically, it is described as a lubricant of social interactions (Tyler and Degoey, 1996) and an important instrument to reduce social uncertainty and complexity (Luhmann, 1979). Moreover, as a key prerequisite for social capital, trust is also seen as essential for a healthy economy and democracy (Putnam, 1993; Fukuyama, 1995). These general descriptions of the role of trust in society already point to the concept's relevance for risk research in general and the field of CBRN risk in particular. Given the uncertainty and the wider political and economic effects associated with many risk issues, it comes as no surprise that risk scholars have taken an increasing interest in the issue of trust.

Rosseau et al. (1998) described trust as a psychological state that entails the acceptance of vulnerability on the basis of positive expectations of the intentions and behavior of the other. Trust in risk-managing institutions has been associated with many factors, such as competence, integrity, and accountability and other values such as honesty, fairness, and caring (Poortinga and Pidgeon, 2003).

Risk researchers suggest that trust is a crucial factor that shapes public risk perception and the effectiveness of risk communication (Meredith et al., 2007; Rogers et al., 2007). One way of demonstrating how different levels of trust influence the impact and response to CBRN events is that different trust levels can be associated with different levels of concern as a result of lack of familiarity with a threat and the resultant information needs of the public. A public perception factor such as lack of familiarity carries less weight if members of the public trust the organizations in charge of managing a risk (Kasperson et al., 1992; Pidgeon et al., 1992). On the contrary, distrust has been linked to stigmatization of risky technologies (Flynn et al., 2001), as well as the social amplification of risk after risk-management failures (Freudenburg, 2003). In addition to shaping risk perceptions, different levels of trust can also be expected to lead to different levels of compliance with the authorities. Meredith et al. (2007), for instance, discovered in the event of a bioterrorist incident, decisions about vaccination, fleeing the scene, and adopting a wait/watch and see attitude could be linked to levels of trust of the public in the responding authorities and their communications.

While (promoting) public trust is therefore important for the effective management of CBRN events, it is important to note that trust can aggravate as well as enhance CBRN impact and management performance. For instance, "blind" trust (in the sense of uncritically accepting what authorities announce) may worsen the performance of CBRN management because critical feedback by the public can encourage authorities to improve their communication and public engagement (Poortinga and Pidgeon, 2003).

Finally, degrees of skepticism and trust visá-vis institutions can vary significantly between groups and countries, resulting from factors such as different types of individual interactions with the responsible institutions, historical risk management records of an agency, and the wider political cultures in which members of the public and responsible agency are embedded (Poortinga and Pidgeon, 2003; Krieger, 2012; Pearce et al., 2012a). Sjoberg (1996), for instance, noted that in contrast to U.S. studies on risk perception, Swedish research found that trust is not such an important factor in determining risk perception. Pearce et al. (2012a) identified variations in levels of trust in emergency responders and official messages about a chemical incident in Poland and the UK, and Pearce et al. (2012b) identified variations in levels of trust between the UK and German populations in response to a radiological emission device. These varying levels of trust have implications about the ways in which members of the public respond to extreme events.

Public Perceptions of Response Costs and Efficacy

Public perceptions of costs and efficacy of responses, as well as of the individual's own efficacy of undertaking protective behavior, are the third set of factors that affect the impact and effective management of CBRN incidents. This factor is based on the principles of Protection Motivation Theory (PMT) from the field of health promotion behavior (Rogers, 1975; Maddux and Rogers, 1983; Pearce et al., 2012b).

Maladaptive individual behavior has also been recognized as a problem in the health behavior and promotion literature (Floyd et al., 2000; Milne et al., 2000). This literature addresses the puzzle of why individuals do not follow the recommendations for health-promoting behavior provided by medical experts, but engage in potentially harmful or negligent behaviors. An example for often disregarded, health-promoting behaviors are routine physical examinations, monitoring blood pressure routinely, exercising aerobically, eating a well-balanced diet, and abstaining from smoking to prevent cardiovascular diseases. Others include preventive behaviors against HIV (e.g., use of condoms) and cancer (e.g., abstention from smoking).

The majority of health-related literature assumes that protective, health-promoting behavior is motivated by the perceived threat and a desire to avoid the adverse outcome. PMT, however, identifies two mental processes: the threat appraisal and the coping appraisal. Coping appraisal is concerned with the perceived response efficacy (the individual's belief about whether the recommended response will be effective in reducing the threat to the individual), self-efficacy (beliefs about whether the individual is able to perform the recommended protective behavior), and the response costs (how costly performing the response will be to the individual) (Floyd et al., 2000; Milne et al., 2000). Meta-analytical reviews of PMT (Floyd et al., 2000; Milne et al., 2000) establish that coping appraisal has stronger explanatory power (higher significance ratios) for health-related behaviors than threat appraisals, confirming some of the doubts expressed about the empirical validity of the psychometric risk perception theory (Sjoberg, 1996).

The focus on perceived efficacy of the response and oneself is a recent addition to the CBRN response literature. For example, survey results for the UK and Polish public show correlations between perceptions of self-efficacy and response efficacy and intended compliance with recommended behavior following a hypothetical chemical spill (Pearce et al., 2012a).

Quality of Risk Communication

The fourth and final set of factors concerns the effectiveness of crisis and risk communications (Wessely et al., 2003; Glik et al., 2004; Glik, 2007). Risk communication by public authorities before, during, and after a CBRN incident has been argued to significantly influence the impact of such incidents: Well-designed and well-timed messages can reduce the uncertainties surrounding the nature, exposure, and consequences of particular hazardous agents, as well as provide effective behavioral guidance to the public. Glik (2007) suggested that high-quality risk communication before, during, and after CBRN incidents has to take into account a wide range of factors, including characteristics of the audience (e.g., "emotional" state and socioeconomic and demographic group), context (e.g., local culture; history), source (e.g., trusted and familiar authority), channels (e.g., multiple channels and mass media), content (e.g., simple, clear, and comprehensive), and timing (e.g., rapidly and evolving with the emergency).

This is particularly important because crisis and risk communication can go wrong in many ways. For instance, inconsistent, ill-timed, and incomplete messages can increase uncertainty and decrease trust in institutions (Lofstedt et al., 2011). Even well-designed and well-timed message have pitfalls; they may reassure the public too much, undermining incentives to engage in protective behavior (Rubin et al., 2009).

Communicating with the public is particularly important to organizations and community groups involved in the development of greater public preparedness and resilience in the event of lower likelihood–high-impact extreme events such as CBRN incidents. CBRN events normally occur with little warning, the threat is invisible and unknown, the perpetrators and their plans are often unknown, and they have the potential to cause "dread." As a result, response organizations should expect members of the public to show higher frequencies of information-seeking behavior in comparison to the information-seeking behaviors experienced during natural disasters such as flooding where the hazard and its consequences are well known, unintentional, and often come with some warning.

What are the implications of these four factor sets for emergency managers and counterterrorism practitioners in terms of promoting public resilience? How can academic research further advance the public resilience against terrorist acts?

CONCLUSIONS: PUBLIC RESILIENCE, CBRN, AND HUMAN FACTORS

Promoting public resilience has become an integral part of counterterrorism strategies in many countries. This broadening in counterterrorism strategies acknowledges the limitations of approaches that focus too narrowly on prevention and security, and reflects the importance of thinking about the potential of extreme events to adversely affect the mental well-being of members of the public, as well as every day, economic and political life in a society. This chapter focused on public resilience to unconventional CBRN terrorism. Terrorist attacks such as the use of Sarin in Tokyo and anthrax in Washington, D.C. illustrate the terrorists' willingness to use CBRN terrorism, as well as the adverse psychological and behavioral consequences associated with such attacks. CBRN attacks clearly constitute a unique challenge to public resilience as the unknown, unfamiliar CBRN agents can cause uncertainty and dread, thus making preparedness, response, and recovery objectives difficult to achieve.

We argued that there is a set of four factors that can be used to explain the impact on and response behavior of members of the public, such as public perceptions of risk, public perceptions of the responders and authorities, perceptions of response and self-efficacy, and quality of existing risk communication. The fourth factor, risk communication, is the key to understanding how emergency responders and counterterrorism policy makers can promote public resilience because risk communicators can—if well-designed and implemented—reduce uncertainty and dread, as well as strengthen perceptions of efficacy and trust.

When communicating risk, it is essential for emergency responders and counterterrorism policy makers to take into account all four factors when promoting public resilience. Disregarding any of these factors carries the risk of failing to capture the full extent of public vulnerability to CBRN events and bears the previously mentioned pitfalls of risk communication. For instance, focusing solely on public risk perception of radiological threats may lead to the conclusion that people will be extremely worried and anxious because of the association of radiation with nuclear risk and the stigmatization of nuclear technology. However, the degree of public anxiety is likely to be significantly altered by the degree of trust in authorities responsible for nuclear and radiological risk management (Bord and Connor, 1990; Pijawka and Mushkatel, 1991–1992). Similarly, failure to take existing risk communication messages or perceived self-efficacy into account may result in an incomplete prediction of the levels of anxiety and worry felt by individuals and their behavioral choices.

Moreover, risk communicators do not only need to make sure that their messages are short, simple, comprehensive, and consistent, they must also ensure that the messages address the diversity and diverse needs of the public(s); the diversity of the cultural and institutional contexts in which members of the public are embedded; and the history, memories, and associations triggered by particular incidents of different risk communication target groups. This requires substantial engagement with the public ahead of a crisis and in-depth research at national and local levels to learn about the diversity of existing mental models and existing levels of trust in particular institutions, as well as response efficacy and cost assessments.

CHAPTER 8

From Local to Global: Community-based Policing and National Security

Fahimeh Tabatabaei, Seyed Mohammad Reza Nasserzadeh, Simeon Yates, Babak Akhgar, Eleanor Lockley, David Fortune

INTRODUCTION

In 1829 Sir Richard Mayne wrote:

> *The primary object of an efficient police is the prevention of crime: the next that of detection and punishment of offenders if crime is committed. To these ends all the efforts of police must be directed. The protection of life and property, the preservation of public tranquility, and the absence of crime, will alone prove whether those efforts have been successful and whether the objects for which the police were appointed have been attained.*

In attaining these ideals, there is a heavy dependency on the approval and cooperation of the public, which has been determined by the degree of esteem and respect in which the police are held. One of the key principles of modern policing in Britain is that the police seek to work with the community and as a part of the community.

The origin of the British police lies in older historical customs for securing order through appointing representatives. In Saxon times "the people" were "the police." This system involved dividing people into groups of 10, called "tythings," each with a tything-man as a representative and into larger groups, made up of 10 tythings, under a "hundred-man" who was responsible to the "Shire-reeve", or Sheriff, of the County.

The tything-man system survived in part because of the arrival of the Normans and feudalism. The tything-man became the parish constable responsible to the Sheriff, who was the Justice of the Peace. This system expanded over British history often comprising one unarmed able-bodied citizen in each parish, who was appointed or elected annually to serve as parish constable. In towns, other specified groups of citizens known as "The Watch," and often paid for by guilds, guarded and patrolled public spaces. In the eighteenth century the greater movement of the population to large urban spaces saw the gradual break down of these systems. It was this need to re-create a public group to undertake the "policing" of public space that provided the basis for the formation of what has become the UK police force. The original concept of policing by the consent of the community remains at the core of UK policing policy and rationale; however, the specifics of actual police action may play out or be perceived.

Policing through the ages has therefore continually adapted and evolved. Importantly, in the

UK context policing has nearly always tended to have a focus on community and citizens as the foundation. Such a community and citizen focus is not only "fit for purpose" when dealing with minor community issues, but it is essential for addressing current issues in terms of terrorism and organized crime (things that are commonly used in the same sentence as the term "national security"). "Community and Neighborhood" and "Local Area" policing are the "default" positions for all police forces in the UK. Although specialist and reactive police units (such as serious crime, counterterrorism, and public order units) can consume significant resources at key times or periods of specific concern, community-based policing remains key.

Partnerships and professional working relationships with many local and national government, health, and social service agencies have developed such that "policing communities" is not the sole responsibility of the police force. Many different agencies and vitally the community itself must be involved to ensure safety and security of citizens and communities alike. A continued challenge in this multi-agency approach is the lack of interoperability between agencies' data storage systems and the poor flow of information between agencies. This issue is equally replicated between those agencies working at the serious and organized crime/terrorism end of the policing spectrum.

POLICING BY CONSENT, COMMUNITY, AND PREVENT

All policing is intelligence driven. While some of that intelligence is gathered covertly by various government agencies and law-enforcement operations, the majority comes from the community. Trust, understanding, and communication are required to get community information, to turn it into community intelligence, and to use it to make informed policing decisions. Without these three key elements those that choose to undermine society and provide misinformation, mistrust, and misunderstanding of the actions of policing. Within the UK the link between community-based policing and counterterrorism strategy is the PREVENT agenda.

Looking specifically at terrorism and the prevention of violent radicalization it is vital that the community understand, support, and can act as advocates, particularly in areas of low trust and community confidence. The PREVENT strategy is not about convincing people that terrorism is wrong; it is about gaining community support and trust. Radicalization happens when terrorist ideologies are not challenged. Within communities young people are most at risk of radicalization and possibly the most difficult to connect with by police and law-enforcement bodies, as they often have great mistrust and misunderstanding of policing and law-enforcement bodies. However, the actual numbers of young people who are at risk of radicalization are very small. Often this small target group may be citizens of the country, they may feel they do not belong, and have no stake in the country in which they are living. While the current focus of this preventative work is on al-Qaeda, it must focus on all forms of radical extremism including extreme right wing and nationalist groups.

So the strategy and actions are focused on a very small group of people. The community has the best chance of identifying those key individuals and articulate a counter narrative that balances extremist and violent views. One good example is the case of Taimour Abdulwahab Al-Abdaly, who, in December 2010, killed himself in a bomb attack in Stockholm, Sweden. His behavior and beliefs had raised concerns at a mosque he attended in the UK, and those beliefs had been challenged by religious leaders resulting in expulsion from his mosque. However, those religious leaders did not think it appropriate to refer this information to either the police or other authorities, perhaps due to a lack of trust and not having good clear and open lines of communication with the police.

Policing and community policing in particular is about being a catalyst for change. It is about supporting the community to stand up for their rights and beliefs. When this is done, the community can then prevent the extremist views and narratives of a very small minority cannot gain traction with community discourse. The community can also look to challenge accompanying criminal activity.

Policing today also faces the challenge of increasing demand and public expectation at a time of decreasing resources caused by the economic crisis. These "economic hard times" can create an environment where terrorism and serious and organized crime can flourish, meaning police and law enforcement must meet these new and evolving challenges head on. These issues also pose a threat to national security and economic stability. Poly-criminal organized crime gangs are entrepreneurial in nature, exploit the same weakness as terrorist networks, and indeed may be linked to them. National security policies, police, law enforcement, and intelligence organizations cannot win the fight against them alone. Without the support and contribution of the community they will face an uphill struggle.

The Challenge of New Forms of Networked Communities

In the contemporary networked society "community" has a range of definitions beyond that of local place and space defined social groups. The challenge is to work with new and emerging definitions of community, such as virtual, industry sector, geographic, or common interest focused. They will often hold key pieces of information and intelligence that can make a difference and help prevent and detect crime. Without the trust, confidence, and ability to communicate that information to the police and law-enforcement bodies the police will be attempting to piece the intelligence "jigsaw together" often with some of the vital pieces missing.

The challenge for the police service in this globalized context is that of evolving to manage an increasingly e-connected citizen, and "police" the e-community in new ways that deal with the new threats and criminal opportunities that exist via the Internet and e-enabled world. Communities are often physically and socially disconnected, and yet are electronically connected. Increasingly people have concern for e-friends across the globe, but do not know their neighbors who live on the same street. We have a new definition of community, but this new e-community still requires policing.

Policing today and tomorrow is different and there is a growing need to deal with the online behavior of citizens and criminals/terrorists. Citizens need to be protected and policed in a way that protects them in the same manner as the police in the "real world." The challenge is to deliver traditional policing in a way that is fit for today's and tomorrow's world.

DEVILRY THROUGH THE DARK WEB: PREVENTING ONLINE RADICALIZATION

The question therefore is how to police a networked society, holding onto the ideals of policing by consent, and bringing in the learning from PREVENT strategies in location-based communities. The obvious context here is the challenge of online radicalization of individuals or groups. In fact, the use of cyberspace to disseminate radical material and messages has become the predominant means that extremists use to recruit individuals to their cause. This is increasingly so as more global efforts have been made to monitor and disrupt more conventional means and venues of recruitment based in local communities strategies such as PREVENT, or through denying access to specialist events and locations (e.g., foreign training camps).

Individuals, particularly young, computer-literate males, are becoming self-radicalized through access to increasing amounts of sophisticated online material promoting and justifying extreme views and actions. The detection of these self-radicalized individuals is difficult, and detecting the threat of particular individuals becoming self-radicalized even more so. Therefore, this chapter aims to explore how to identify factors that might highlight safe online behavior—to have or to read about extreme views is not often criminal in most Western democracies—from behavior that may indicate radicalization and intentions to act criminally as a result of such views.

Mandel (2010) defined a "radicalized" person as someone who holds an opinion about a certain topic (e.g., animal rights, Nazism, or religion) that gradually starts to deviate from normative opinions and for whom accepting opposite opinions becomes difficult. Very often they then seek to act against social norms as directed by their opinions or justified by them. In many cases this is violent action. By that time, their actions may have caused serious harm and the chances of deradicalization will be reduced, which is evidenced by the fact that they feel strongly enough to cause harm.

In the context of online radicalization, some authors (Chen et al., 2008; Lenselink, 2011; Geeraerts, 2012) differentiated between cyber terrorists and cyber extremists. They believed that the most important aim of terrorist use of the Internet is spreading ideology and "most terrorists can also be considered extremists, but not vice versa." For example, Neo-Nazis or religious fundamentalists may be viewed as extremists by the majority of society but not necessarily as terrorists. In addition, the concept of radicalization refers to the opinions, views, and ideas by which individuals have led to act according to an overt message but not necessarily in a terrorist manner (Chen et al., 2008; Royal Canadian Mounted Police, 2011; Geeraerts, 2012).

Three aspects of the Internet used to radicalize people have been found to be problematic:

1. Extremists may use the Internet to express and impose their ideological messages, and potential recruits may get access to powerful pictures and films to support their political views.
2. Joining formal organizations is facilitated by the Internet for potential recruits, and they can find individuals with similar ideas without any danger.
3. Virtual locations, online community spaces and places, afforded by the Internet create new social environments in which ideas and opinions that might be considered socially unacceptable are "normalized" for that group or context. In other words ideas, speech, and opinions that might be socially sanctioned in public interaction; for example, racist speech, go unsanctioned and may be applauded by a specific online community. Such support and encouragement of extreme ideas and behaviors is not specific to Internet contexts. It could and does just as easily take place in such contexts as private meetings, camps, or events. The new affordances of the Internet include the global reach of the medium and its disconnection from surveillance in location by the police or the community (Guangrong, 1998; Kshetri, 2005; Stevens and Neumann, 2009; Institute for Strategic Dialogue, 2011; Geeraerts, 2012).

There is also a threat that the Internet will be the location for acts of extremist terrorism through acts of cyber crime or cyber terrorism, which may be spawned by the Internet. Such acts may need little resource and impact many areas of social life from the political through to the military, and may have economic, social, and material consequences (Guangrong, 1998; Kshetri, 2005; Chen et al., 2008; Mainz, 2009; Yunos and Hafidz, 2011, p. 20).

Weimann (2004) studied the role of the Internet in the radicalization process as enhancing communication (sending and receiving messages, searching for information and like-minded people, and maintaining anonymity); diffusing propaganda (posting resources into multiple languages and linking to related forums, videos, and content of group Web sites); and recruitment and training (massive and dynamic online libraries to train, interactive services, that is, games, multimedia content, and online registration). Similarly Chen et al. (2008) conducted several experiments on cyber terrorism activities in major Web sites and blogs such as YouTube and Second Life and popular hosting service providers such as blogspot.com and wordpress.com. Their findings showed that extremist groups abuse the cyber space to achieve their goals (Chen et al., 2008;Yunos and Hafidz, 2011, p.20).

The power of mass communication for social and political change via social media has been argued at length (Thompson, 2011). Today, social media particularly social networks that provide knowledge sources such as instant messaging (e.g., IRC, AIM, MSN, Jabber, Skype), sharing sites (e.g., Flickr, Picassa, YouTube, Plaxo), blogs (e.g., Blogger, WordPress, LiveJournal), wikis (e.g.,Wikipedia, PBWiki), microblogs (e.g., Twitter, Jaiku), Web- based social networks (e.g., MySpace, Facebook, Ning), and collaboration networks (e.g., DBLP) (Malik and Malik, 2011) are all tools available to social political actors of any persuasion from the most liberal to the most extreme. The key issue is not simply that such media are used by extremist groups—so was the printing press—but how to identify, block, challenge, and counter such use so as to minimize the opportunities for individuals and groups to be radicalized (Correa, 2011).

Self-radicalization Process

According to Sageman (2008), a radicalization process consists of four interchangeable steps that reinforce each other. The process starts with "moral outrage." This is then "interpreted in a specific way" by the person or others (e.g., through interactions on online forums). The third step requires that it "resonate with own personal experience" in which like-minded people interact with others. They therefore give each other further self-confidence by justifying their ideology, actions, and beliefs to each other. Finally they may be "mobilized through a network."

The Internet provides many routes through these four stages or through just general use, but more often through engagement with specific materials and specific online contexts. These are often ones where countering voices and alternate perspectives are absent or denied (Zhou et al., 2005; Institute for Strategic Dialogue, 2011; Lenselink, 2011). An integral part of the radicalization process are the real-world relationships, as social network theorists, such as Sageman (2008, p. 121) claim:

> *...most online participants also have friends who share their views and desires but do not spend so much time on the internet. Terror networks consist of a mixture of online and offline elements, and their respective in-person and virtual discussions mutually influence each other.*

Online self-radicalization and self-recruitment will rarely occur if there is no relation to the outside world. This highlights the continued importance of the real-world social and community relationships. Direct human contact and ties of friendship and kinship can never be completely replaced by the Internet; they can only be supported and facilitated by the Internet. This is supported by the findings from research on the recruitment of the Islamist militants in Europe, where few of them had been radicalized or recruited merely on the Internet (Stevens and Neumann, 2009; Institute for Strategic Dialogue, 2011).

IMPLICATIONS OF ONLINE BEHAVIOR FOR NATIONAL SECURITY

Online radicalization is therefore not wholly disconnected from other forms of community-based radicalization. It should be considered as part of the real-world events and developments and their integration into cyberspace and vice versa. As a result, a more comprehensive approach that shows the real-world nature of the problem has to be accounted for and integrated into any policy aimed at tackling online radicalization. Thus, solving the radicalization problem by any simple policing or technology-based process is not possible (Stevens and Neumann, 2009).

So, In recent years, governments have utilized the three methods of "removing," "filtering," and "hiding" to overcome the abovementioned problems. Removing means abstracting the unwanted content from the Web, filtering means restricting users' access and controlling the exchange of information, and hiding means manipulating search engine results to make undesirable content more difficult to find. These issues are of course pertinent to many other aspects of Internet behavior that individuals, groups, or governments are concerned about, from the availability of pornography, information on drugs and drug use, through to the infringement of copyrighted materials.

- **Removing:** How is removing possible? All Web sites depend on a "hosting" company, which provides online storage, and a "domain name provider," which supplies the Web site addresses through which the site can be accessed.

A government can ask the hosting company to "take down" content or ask the domain name authority to deregister the relevant domain, and the undesirable content will be removed from the internet (Stevens and Neumann, 2009).

- **Filtering:** Since most Internet users are connected to the Internet through a small number of Internet Service Providers, filtering, which tries to control the flow of information between computers, is made possible.

Of course many of the filtering technologies that are currently in use are either too crude or too expensive to operate (Stevens and Neumann, 2009). Internet filtering is a widespread phenomenon triggered by governments. Almost every country and even those with a deep commitment to "democratic principles and civil liberties filters or censors access to Internet content in some way ...However, the location, quantity, and manner of the filtering vary greatly" (Initiative OpenNet, 2004).

It is noteworthy that countries usually rationalize the laws that allow filtering by invoking one of two broad themes: upholding "community standards" and ensuring "national security." Filtering, however, makes enforcement easier for governors.

> *It seems state's control over Internet infrastructure; service providers, content providers, or users may be adequate to enforce its determinations through filtering.*
>
> ***Initiative OpenNet, 2004***

However, when radical thoughts result in violence and other criminal behavior, society can be put at stake by the actions of individuals and groups. Before this happens, authorities need to take lawful steps to target the illegal behavior (Royal Canadian Mounted Police, 2011).

- **Hiding:** In this method, governments interfere with the process whereby search engines produce their results in order to make extremism-related Web sites less visible.

If a search engine is manipulated to drop certain Web pages or rank them significantly lower than others, this is in fact, hiding them from public view. Certain Web pages or requests for banned keywords are dropped in one method, and in another method, search engine results are altered, so that appropriate sites appear higher than inappropriate ones (Stevens and Neumann, 2009).

In addition to law enforcement agencies, various stakeholders such as the Project for Research of Islamist Movements and the Search for International Terrorist Entities (SITE Institute) explore the Internet to find and analyze the radicalized content. Nevertheless, approaches that are based on keywords may result in many false positives or topic content. Moreover, the unstructured and informal nature of content (abbreviations, colloquialism, and transliterations) adds fuel to the fire of the problem complexity. Therefore, manual search for detecting, collecting, and analyzing such content is not only time-consuming but also infeasible (Correa, 2011).

From a security and intelligence viewpoint, self-radicalized individuals are harder to detect and monitor compared to those who belong to a recognized radical group. Although such individuals may have informal social connections and acquaintances with similar individuals and to individuals who may have radical influence, these loose associations and networks are more difficult to detect and analyze than formal organized groups.

Reviewing studies that seek a method to identify Web sites and radical contents resulted in Table 8.1. It indicates that combating the radicalization needs more extensive research in the near future.

TABLE 8.1 A Review of Studies about Dark Web Detection Methods

Researcher	Context	Method
Stevens and Neumann (ICSR; 2009)	Strategies to combat online radicalization	An in-depth review of action strategies to counter extremist groups
Thompson (2011)	Use of social media to radicalize individuals	A deep description to the role of social media
Qin et al. (2007)	Terror campaigns analysis on the Internet (point of views: technical sophistication, content richness, and Web interactivity)	Dark Web Attribute System (DWAS)
Institute for Strategic Dialogue (a working paper; 2011)	Online self-radicalization to violence	An in-depth review of technical ways for governments
Chen et al. (2008)	Dark Web forums analysis	Measurement online radicalization using automated approach to sentiment analysis
Correa (2011)	Solutions to detect and analyze online radicalization	Deep review on Web mining techniques to detect dark content (i.e., Web text mining/Web structure mining)
Lenselink (Thesis; 2011)	Online radicalization (patterns of social interactions)	Semi-automated method (using Web crawler technologies)
Bermingham et al. (2008, 2009)	Online radicalization	Combining automated lexical analysis and sentiment analysis
Sureka et al.	Extremist videos, users, and hidden communities on YouTube	Web mining, a semi-automated system using social network mining
Glass and Colbaugh (2011)	Security informatics applications (focusing on social networks)	Text classification using tow algorithms: a semi-supervised sentiment classifier and a sentiment estimation algorithm for social media content
Memon and Larsen (2006)	Terrorist networks analysis	Investigating Data Mining (IDM)
Tsfati and Weimann (2002)	Terror on the Internet	Deep review of content types and target audiences of terrorist/extremist Middle Eastern Web sites

CONCLUSION

As data become a critical resource across the world, efficient access to data, data sharing, information extraction from the data, and information utilization are also urgently needed. There have been many efforts to not only integrate the various data sources scattered across several sites, but also to extract information from them to the form of patterns and trends (Del Carmen and Alonso, 2003; Thuraisingham, 2005). Nevertheless, a large amount of work is still needed to integrate the semantics of various systems and applications (Thuraisingham, 2005). Today all prominent active extremist groups have established at least one form of presence on the Internet. Their activities are very dynamic, as their Web sites appear and vanish in a few days (Lenselink, 2011). The rise of social media has had a further significant impact on the challenge of discovering such content (Correa, 2011).

Just as with the development of the PREVENT agenda in the context of community policing there is a need to develop a range of approaches to identifying and countering online extremist views with credible voices. There is therefore a need to identify a common profile of self-radicalized individuals, and also the process of self-radicalization. This will help to detect and monitor vulnerable individuals and their network of associations, to prevent radicalization, disrupt the self-radicalization process, and to provide an opportunity to deradicalize individuals before they cause harm.

The Internet has become a significant instrument for extremists and terrorists. They increasingly utilize online environments to inculcate their content and ideology and propaganda and to recruit and widen their communication area. However, knowledge of such individuals and group activities is not mature. In particular, presenting short-term solutions that are manual or semi-automatic have not been effective. A mixed approach linking technical solutions with community-based policing and countering of online radical content is needed. In Section 3 we have noted some key technical and knowledge management techniques and methods that might allow the police, governments, and communities to address the security challenges faced by networked societies. We have also noted in this part of the book (Section 2) that addressing threats to communities and to national security must also be tackled by understanding the social, political, and cultural behavior of those involved whether using new digital media or not.

CHAPTER

9

The Role of Social Media in Crisis: A European Holistic Approach to the Adoption of Online and Mobile Communications in Crisis Response and Search and Rescue Efforts*

Marco Manso, Bárbara Manso

INTRODUCTION

Worldwide crises have drawn significant attention to the role information communication technologies (ICT) play in search and rescue response efforts within the national security framework, particularly the implementation of civil defense and emergency preparedness measures and the use of analytical intelligence methods and tools to detect and act upon external and internal security threats to a nation. With a century old research history, the sociological study of emergencies has shown that ICT expanded the reach of disaster sociology (Palen et al., 2007), which is now focused on the emerging trend of increasing citizen involvement through online (social) media and mobile communications (computer-mediated interaction) by providing, seeking, and brokering information and connecting those within and outside the geographical area of a crisis with broad implications to response efforts. Still, first responders (FRs) and public protection and disaster relief (PPDR) organizations, upholding their traditional command and control models, have not easily adapted to this new area of global and digital crises.

*This chapter is developed from the extended paper "The Role of Social Media in Crisis: A European Holistic Approach to the Adoption of Online and Mobile Communications in Crisis Response and Search and Rescue Efforts," by Marco Manso and Bárbara Manso, and presented at the 17th ICCRTS, Fairfax, VA, June 19—21, 2012.

The iSAR+approach described herein proposes the integration of ICT tools and functionality (Web portal, PPDR support tools, and mobile applications) to enable additional communication channels between PPDRs and citizens. In the process, PPDRs will be able to rapidly generate high awareness levels, during and after a large emergency or crisis, by simply exploiting the bidirectional communication potential provided by the mobile technology and social media that citizens have widely adopted and integrating it into existing PPDR crisis management and response systems.

This section is structured as follows. First, there is a presentation of lessons learned from past crises, then the role ICT tools and social media platforms play in crises is acknowledged, the multidimensional approach to the problem is introduced, and finally the iSAR+platform enabling the involvement of new media users in the effort to leverage citizens' security in crisis situations is proposed.

LESSONS FROM PAST CRISIS SITUATIONS

Large-scale natural disasters and human-induced emergencies, or simply crises, disturb routines and raise significant public scrutiny and undivided media attention; a reality that reinforces today's global emphasis on ICT. Orlikowski's theory on the coadaptation of technology and society, highlighting the creative interdependence of social evolution and technological innovation, has its ultimate token in society's addiction to the Internet and mobile technologies, displaying unprecedented information and communication capabilities (Sutton et al., 2008). Today, mobile and online social media are dramatically changing geopolitics, economic contexts, and the competitiveness of businesses that are transforming societies, a catharsis that is also present in the potential for an improved response to disasters and crisis response efforts.

Throughout history, overwhelming natural disasters and human-provoked actions and their associated response efforts have left their mark, providing valuable lessons for crisis management systems, plans, and organizations. We analyzed several of those lessons, trying to understand how current crisis response efforts are carried out by FRs and PPDRs and how citizens become involved in those efforts. We focused mostly on the use of mobile technology and online (social) media.

The analysis reveals that, in crises, amid the collapse of all critical infrastructure, communications withstand extreme damage or destruction. Landline phone networks are intermittently available or unavailable, with the surviving emergency service stations rapidly becoming overwhelmed by the incoming volume of calls, as occurred in the 9/11 terrorist attacks (McKinsey et al., 2002), the Madrid train bombings, and the London tube explosions. A different case was registered in Norway, where 112 operators dismissed all calls that were not related to the Oslo bombings, even though they were reporting the shootings at the island Utoya, Norway. This case proves that emergency systems may fail for reasons other than technology. Fiber-optic connectivity and mobile telephone networks are also affected but, in most crises, they exhibit a more resilient performance, quickly restoring the ability to exchange short message service (SMS) and text messaging communication (ICT4 Peace, 2011). Overall, the scarcity of communications access hampers information availability and, consequently, crises' situational awareness and response efforts. This was fairly evident both in the 2004 Indian Ocean tsunami and the 2010 Haiti earthquake.

ICTs have proven their worth in providing both PPDRs and the public the scarcest resource in crisis situations: information. In a society grown accustomed to immediacy and instantaneity, ICTs and their strong real-time emphasis are fundamental to gathering and delivering information, such as alert and emergency notifications and also enhanced situational awareness. And yet the majority of official PPDR channels do not include online social media, choosing instead to practice a unidirectional information dissemination model toward the public, which is often hampered by scarce information details and empathy. The traditional TV and radio broadcasts become official communication channels, a choice often criticized. In the California wildfires and the Norwegian massacre, those media were heavily attacked for sensationalism and irresponsiveness to the local communities' needs (Sutton et al., 2008). Communicating with the public is always a challenge, a difficulty highlighted in the 2005 London tube attacks, because the UK's Data Protection Act prohibits sharing personal data without the consent of those concerned, thus limiting what information officials could give agencies and families about victims' identities and status (Committee for Disaster Medicine Studies, 2007). Indeed, crises do question the balance between individual privacy and public security.

As a result of the repetitive lack of official information, and because high levels of preparedness are rare in societies (an exception is the Chilean people), citizens choose the rich contents provided by mobile and online social media technologies when requesting assistance, looking for information, and trying to locate missing relatives and friends. This is why, over the long span of the Chinese SARS outbreak crisis, 120 million text messages were exchanged (Palen and Liu, 2007) and, immediately after the 2010 great Japan earthquake, Twitter became the emergency service with 1,200 tweets per minute, and why Mission 4636 received 1,000 SMSs per day after the Haitian earthquake. Moreover, the growing global phenomenon of citizens' journalism has been instrumental in providing eyewitness accounts and first reports from the affected areas, thus contributing to the enhancement of the global situational awareness in crisis situations and suspicious activity report actions. This happened in natural catastrophes like the 2004 Indian Ocean tsunami, the 2005 Hurricane Katrina, and the 2010 Haiti earthquake, as well as in the United States, London, and Madrid terrorist attacks. Volunteering across the globe and the use of ICT tools have prompted citizens' initiatives like the South East Asia Earthquake and Tsunami Blog, the Digitales por Chile, and the Ushahidi platform to become useful sources of information during the evolution of these crises, the relief assistance that ensued, and the tracking of missing persons.

Notwithstanding the benefits of the amazing Facebook, Twitter, flickr, YouTube, instant messaging, SMS, multimedia messaging service, and e-mail technologies, it is also relevant to be mindful of their misuses and abuses. Not only are they extremely efficient instruments for defamation, slander, rumor, and misguidance, but they have also proven to serve as tools of change during the 2011 Middle East upheavals to organize protests and during the 2011 UK riots to coordinate looting activities. Today, there is an ongoing debate about whether these tools should be further monitored or controlled, a matter that the Chinese government already resolved during the 2003 SARS epidemic when it used Operation Golden Shield to prevent public unrest (Gordon, 2007).

The harsh reality of the crisis response efforts analyzed indicates that PPDRs seldom have the ability to fully implement the official national emergency plan that is carefully created to provide an effective and efficient response to crisis events. Albeit most European countries' emergency plans show concern for informing the general public and the media (to be able to follow authorities' instructions and adopt autoprotective measures), there is no reference to Internet-based tools (institutional Web site or e-mail) or to social media (Facebook or Twitter) as additional means for informing and contacting citizens, reinforcing the model's underlying unidirectionality.

In conclusion, reports and evaluations from past crises and European emergency response plans have discovered the following:

1. PPDR emergency services rapidly become overloaded or inoperational, including the emergency call service, disabling citizens to reach PPDRs.
2. Short-text messaging (via SMS cellular, Twitter, or Facebook) has proven to be a resilient and efficient way for citizens to

communicate (e.g., 5530 TPS during the Japan earthquake) and save lives in the process (e.g., Mission 4636).

3. PPDRs do not regard social media as a valid authoritative tool, and they do not exploit the potential to locate, reach, and communicate with citizens via mobile technology. In fact, PPDRs still follow the structures, models, and approaches defined well before the Information Age.

The straightforwardness of the classic response approach to the natural and human-provoked crises presented here tackles the diversified aspects that convey today's crisis response efforts and citizens' involvement, particularly through the use of mobile technology and social media communications. Together with the statistics that prove the growing global trend of online and mobile adoption, these examples are true accounts of the current state of the art in the use of new communication/social media in crises. Nevertheless, it is worth noting the few exceptions that demonstrate the potential of that use, even when the example for these new ICT tools for crisis management is provided by civil society.

THE ROLE OF ICT TOOLS AND SOCIAL MEDIA IN CRISIS

From Haiti's earthquake to Japan's tsunami, from Hurricane Katrina to the California wildfires, from the Virginia Tech shootings to the Norwegian island massacre, from the upheavals in the Middle East to the English riots, peer-to-peer communications through mobile phones and social media, text and instant messaging applications, blogs, wikis, and Web fora have become the designated ways for citizens to be involved and active in society, during and after crisis events.

Citizens' engagement, strongly enabled by social media and mobile technology, is supporting the dissemination of information, often critical and accurate, into the public sphere by providing eyewitness accounts, sending alert messages, exchanging evacuation and rescue experiences, searching and publishing event-related information, volunteering goods and services, and collecting donations. More than mere information distribution tools, these technologies connect people and provide information, establish collaboration mechanisms, create informal networks, and build no-boundary communities. With a strong emphasis on real time, new mobile and online technologies have significantly improved the capability of affected citizens and victims to help each other and themselves, with their messages enabling improved situational awareness among PPDRs and FRs, guaranteed by the gathering of a wide variety of data and information—an activity coined as crisis informatics.

Indeed, several initiatives, mostly private, leverage the real tributes of citizenship and volunteering toward crisis response efforts, whether on Web-based crisis management systems, mobile applications for emergencies, location awareness technology in crises, RSS feeds, social networking platforms, or Web citizenship on security. Web-based systems for crisis, such as Ushahidi and Sahana, contribute to the permanent monitoring of the evolution of crises, enabling crowd-mapping functions, report tracking on maps and calendars, alert services, and the interaction with multiple sources of information (text messages, e-mail, tweets, Webforms). Also, Google's Crisis Response is a free portal service applied in past crises (Chile, Haiti, and Japan) to enable donations, alerts, and infrastructure status reports, as well as satellite imagery and the Google Person Finder, whereas OpenStreetMap provided free worldwide geographical data created by a community of volunteers. These Web-based systems started off as project-oriented initiatives and have since evolved to accommodate the requirements of several other crises. Other projects remain confined to the specifics of determined events, namely the KatrinaHelp Wiki, dedicated to the Katrina hurricane, or DigiCel's Mission 4636, launched as a free phone number to meet the urgent needs of the Haitian people (medical care, food, water, security and shelter) through SMS messaging. Even the U.S. Department of Homeland Security launched the Haiti Social Media Disaster Monitoring Initiative to assist the Haiti response and recovery effort, creating a situational awareness vehicle to monitor the publicly available from online forums, blogs, and Web sites and message boards to collect critical information.

Web-based crisis systems are also present at official headquarters, such as in the United Nations (UN) and the European Union (EU). They aim solely to facilitate and support crisis response efforts for civil protection, which is a well-recognized national responsibility. Hence, the DG Humanitarian Aid and Civil Protection operates the European Community Mechanism for Civil Protection, comprising the Monitoring and Information Center tool, which provides all participating Member States permanent access to a platform of national civil protection resources, and the Common Emergency and Information System tool, which is a Web-based alert and notification application that facilitates emergency communication among participating states. They provide details of required assistance and offer help and make available the status information on emergencies through an online logbook. The Global Disaster Alert and Coordination System is a cooperation framework between the European Commission and the UN to strengthen the worldwide network of crisis information providers and users to provide reliable and accurate alerts and impact estimations upon major disasters and improve the cooperation of international responders in the immediate aftermath of natural, technological, and environmental crises.

Addressing ICT tools for crises, it is quite clear the relevance of mobile applications for affected citizens to contact family and authorities, to send and receive text messages, or to access platforms like Ushahidi or Google to know and provide critical up-to-date information. Today, mobile technology supports advanced functions for improved user experience, a benefit exploited by applications published by the Federal Emergency Management Agency (FEMA), the North Dakota State University, Ushahidi, and the Pacific Disaster Center, providing citizens useful crisis-related information and upholding built-in bidirectional communication. A notable initiative undertaken by the North Yorkshire Police Authority (NYPA) resulted in the first police mobile app to be launched in the UK, now shortlisted for the Good Communication Awards of 2011. The NYPA application offers numerous functions, including news, contact numbers, alerts, and online maps, and it is well integrated with Facebook, Twitter, and YouTube platforms, accessible by different smartphones (iPhone and Android) and tablets (iPad). Figure 9.1 portrays how the NYPA application is used to disseminate information and collect feedback from citizens, and it has been accessed more than 5,000 times (North Yorkshire Police Authority, 2011).

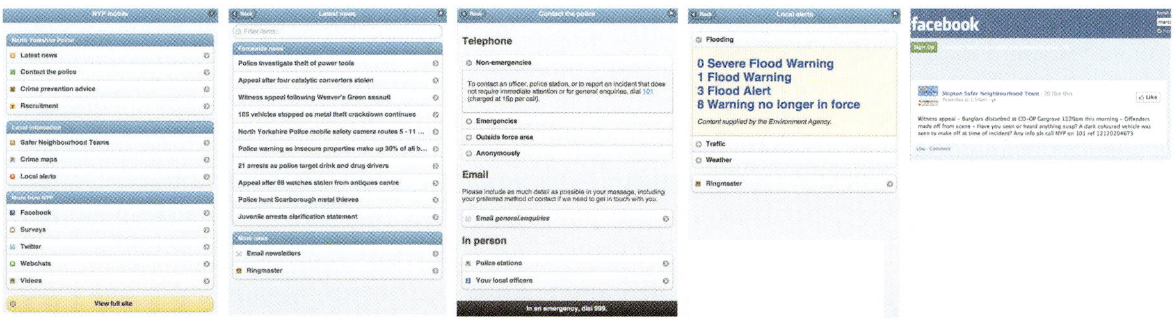

FIGURE 9.1 NYPA mobile application.

The mobile application Gaia GPS has a special version dedicated to the Haitian Disaster Relief, interfacing with OpenStreetMap to retrieve map and satellite information (and associated geolocalized data), to provide latitude–longitude coordinates location and guidance to waypoints and along tracks. Indeed, location awareness services are valuable for crisis response efforts, such as search and rescue actions, particularly in automated location-based services for directed messaging and alerts generation. In this context, the governmental approach has been careful and even conservative, considering not only the legal framework involved but also the ethical considerations. Nonetheless, mobile phones can be located using the equipment's own capabilities (GPS receiver) and/or existing communications infrastructure (cellular and Wi-Fi), a functionality widely accessible through the Google Latitude service, through which citizens share their location and visualize it over a map. Apple followed this example and is now offering the Find My Friends' service, allowing users to share their location, and the Find my iPad service, for users to locate their Apple devices (iPad, iPhone, and MacBook). Additional commercial solutions exist by using SMS or e-mail to locate mobile devices in cooperative and noncooperative modes and to provide automatic alerts when a phone enters or exits a zone, when a phone is nearby, and when a phone is turned on (Deveryware's TeamLoc, WhereAreYou, and PermiLoc applications).

The cooperative mode enables users to provide their location as explicit content and "opt out" at anytime, whereas the noncooperative mode uses the phone's beacons for real-time location. This method requires legal authorization to obtain data from telecommunication operators, thus it is primarily used for law enforcement and security purposes. The benefits of these features are relevant in crisis and search and rescue response efforts for their lifesaving potential, provided suitable legislation and ethical principles are safeguarded.

Next is a final word for social networking sites and their relevance in crises. Initially created to connect those with common relationships or interests, these sites have become a ubiquitous reference and, as a result, also present in times of crisis: there was a peak of 5530 TPS during the 2011 Japanese earthquake and YouTube has extensive video footage of the Haitian and Japanese crises. Notably, these social media platforms successfully attained mobile reach, enabling all users to easily post text, messages, or videos. Table 9.1 presents some of the most popular social networking sites worldwide and its use by citizens.

Moreover, citizens prove to be highly promiscuous in launching and contributing to innovative security-related initiatives, such as the portals and Web sites created by volunteers in the aftermath of major crises to coordinate assistance and help rebuilding efforts. Examples include the 2007 Virginia Tech shooting Wikipedia page composed by 1,500 individuals and the wide collection of tweets posted by citizens to assist firefighters and citizens during the 2007 California wildfires. However, the presented ICT tools, platforms, and services require communications to be accessible to those located

TABLE 9.1 The Main Use of Social Networking Platforms

Social Networking Mobile App	Main Use for Citizens
Facebook	The most used social media worldwide, with the ability to share messages, images, and video EU alternatives are Hyves.nl (in The Netherlands) and Nazka-kazka.pl (in Poland)
Twitter	Real-time microblog with short text geotagged messages connecting those that share interest in specific topics
YouTube	Online video publisher that supports videos upload (captured by smartphones)
flickr	Online image publisher that supports geotagging images Alternatives are Picasa and Panoramio (photos of the world)

within affected areas, and after large crises, and it is often the case that communications infrastructures are damaged or destroyed. In Haiti and New Orleans the option was to rapidly deploy communications networks via cellular and/or IP-based connectivity (typically Wi-Fi), assisted by satellite links when necessary. Table 9.2 summarizes the rich ICT tools ecosystem applicable to crisis response efforts, with the potential to promote the bidirectional communication between PPDRs and citizens.

Built on online social and mobile technologies, this new citizen-centric approach evolved from the basic citizens' journalism for raising awareness on the benefits of a bidirectional communication between citizens and authorities, with eyewitness accounts, first-hand observations, and photos and video directing FRs to where they are most needed and guiding citizens away from harm's way. Because of this better decisions are made and the existing situational awareness gap is closed.

TABLE 9.2 ICT Tools Ecosystem

Type of Communication	ICT Tools
Web-based emergency and crisis platform	Google Crisis Response; Sahana; Ushahidi
Mobile applications for crisis	Disaster Recover Log; FEMA Mobile Application; Gaia GPS; NYPA Mobile Application
Mobile location	Google Latitude/Apple "Find my iPad" Deveryware's TeamLoc, WhereAreYou and PermiLoc applications for location permission in cooperative and noncooperative modes
Voice calls	112 service calls supported by telecommunications operators VoIP (e.g., Skype and Freedom4Wireless service)
SMS/text messaging	Telecommunication operators' service; Mission 4636; Twitter; Facebook
Images	flickr; Panoramio
Video	YouTube
Communications infrastructure	Telecommunications perators and ISP TEKEVER and THALES solutions for MANET and hybrid networks

However, despite the growing global trend of using mobile technology and social media platforms, visible in the lessons learned from worldwide crises, and the empowerment enabled by the new ICT tools dedicated to crisis management and response, FRs and PPDRs continue to have an impaired ability to gather, process, and distribute information, which is reinforced by the likely inoperability, overload, or destruction of local communication infrastructures. During such circumstances, PPDR organizations struggle to develop situational awareness to establish priorities, develop a realistic response plan, and properly assign resources. Thus, decisions are often based on partial, limited, and incomplete information and, even, intuition. This state of low awareness and high uncertainty ought to be avoided and prevented, so expected performance, concerning the efficiency and effectiveness of crisis response efforts, is not compromised. It is therefore clear that a challenge emerges.

Main Challenge

To enable PPDRs and citizens to rapidly generate high levels of situational awareness upon the occurrence of a large emergency or crisis event.

Considering current European economic–financial constraints, it is unfeasible to advance solutions that require significant resource demand or overload, whether in personnel and/or equipment, to cope with crises, independently of the associated negative impact. The following are two key factors that, properly exploited, would bring strong benefits to crisis response efforts, without significant added cost.

1. A high level of adoption and use of mobile technology by citizens
2. Proactive online behavior of citizens to produce and consume information in crises

These factors should be understood as an unavoidable main opportunity.

Main Opportunity

To exploit the citizens' (1) high-level of adoption and use of mobile technology and (2) their proactive online behavior of information production and consumption.

In crises, citizens are the *in situ* first sensors. Empowered by the new communication media (mobile phones with cameras, text messaging, and Internet-based applications connecting to social media platforms), citizens are capable of receiving and transmitting large quantities of data, of capturing high-resolution imagery and video, of retrieving geolocation information (GPS and orientation), and of running sophisticated applications. Experiencing the urgent craving for information, citizens frequently exchange information with PPDRs and are often confronted with the inefficiency of the voice-based emergency service (112 for Europe, 911 for the United States) caused by infrastructural damage, lack of capacity to handle large demand, or the official unwillingness to provide continuous information. Consequently, citizens found online social media to be useful tools in crises, always enabling information sharing, assistance requests, and contacts to loved ones. Figure 9.2 illustrates current communication flows between citizens and PPDRs (the dashed gray line represents communication channels likely to be nonoperational in crises). For years PPDRs have relied on broadcast media and local signs to keep citizens informed; more recently, Internet-based services (RSS and e-mail in near real time) have been adopted, but most PPDRs still remain disconnected from online and social media platforms.

Hurdles to the Use of New Communication/Social Media in Crisis Situations

Indeed, FRs and PPDR organizations, and their traditional command and control models, do not easily adapt to the new trends of the global, digitally enabled social arena to worldwide crises.

Barrier (PPDRs)

PPDRs distrust online social media as a credible information source and a viable communication tool with citizens in crises.

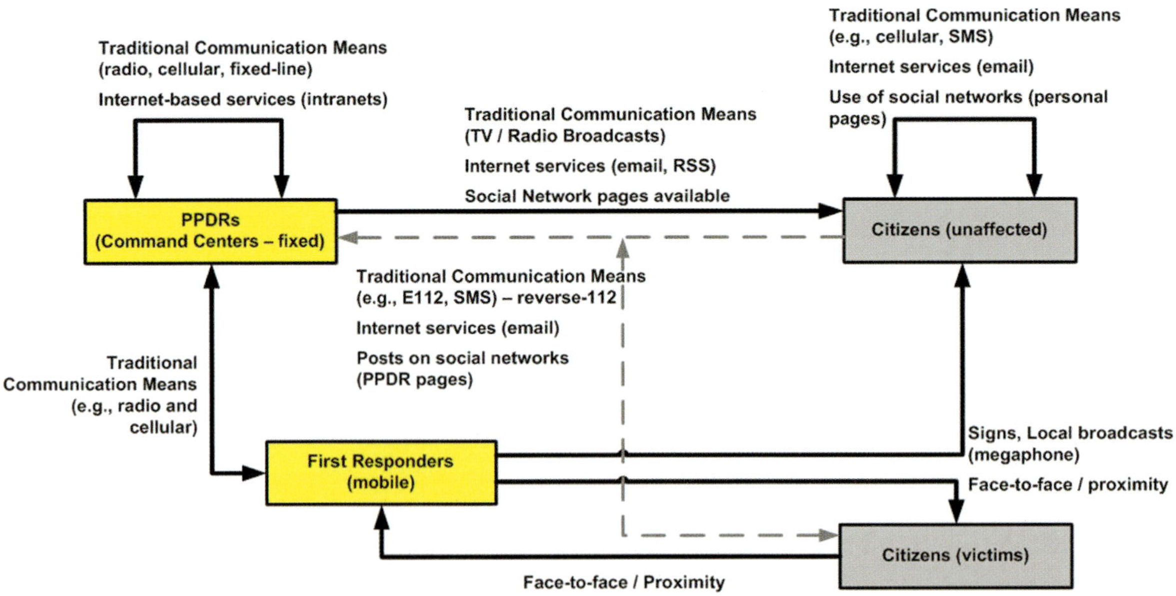

FIGURE 9.2 Current communication flows between PPDRs and citizens.

PPDRs have centralized governance, with clearly defined roles and well-established processes. Bach and Kaufman determined that the lack of trust between officials and the public is a significant obstacle to overcome when it is required to create bonds and establish the necessary confidence to act on the basis of another's words, actions, and decisions (Woodcock, 2009).

Barrier (PPDRs)

Popular online social media platforms cannot be used as a formal PPDR tool for they do not uphold EU ethical principles and legal framework.

Apart from Bach and Kaufman's mistrust, there are also control, ownership, privacy, and security issues and information overload, accuracy, verification, and (source and content) validation problems, as well as concerns with the responsibility or accountability of ICT providers and the possibility of citizens' overzealous behavior. On the other hand, the fear of commercial exploitation of private data, the use of misinformation as a weapon and, even, of political and ethnical pursuit, once people's location information becomes part of the public sphere is legitimate. Overall, it is comprehensible that the absence of a thorough ethical and legal framework to address the issues raised by the use of social media in crisis response efforts also contributes to PPDR organizations' general skepticism toward the integration of new social media as an additional public communication tool.

Barrier (PPDRs)

Introducing change to PPDRs entails new risks and it is a process to be carefully structured, managed, and tested before implementation.

Likewise, the long-time established PPDR organizational culture and tradition will probably act as an additional obstacle to change, especially considering that the integration of social media in PPDR processes will certainly provoke a significant impact on existing structures, roles, functions, procedures, information workflows, and technology, as well as add a new risk to a critical activity that serves, protects, and saves lives. The absence of superior endorsement, sometimes prompted by unfamiliarity, and the lack of expertise, resources, and budget are major hurdles to consider in the analysis, especially at a time when public safety budgets shrink and there are fewer resources to address greater needs. Although the impact of not adhering to the new mobile and social communication media is unlikely to be considered, the impact of the integration of social media in crisis response efforts is certainly reduced if the proposed change is a process carefully structured, managed, and tested before implementation.

Old Habits are Difficult to Change and they are not Exclusive to PPDR Organizations

Barrier (citizens)

Citizens use the online platforms to which they are accustomed. Forcing a change of habits and the adoption of an unfamiliar platform will likely result in failure.

Citizens also have recurrent loyalty behavior toward specific social media networking and, in the face of a new PPDR social media platform, there could be some difficulty in rapidly aggregating a significant number of followers. Undoubtedly, to have a PPDR social media platform segregated from the online circles widely adopted by the public, such as Facebook or Twitter, would be a high-risk enterprise.

To overcome current hurdles and respond to these interesting challenges, our approach delivers a platform that addresses the organizational, human, ethical and legal, and technological dimensions of constructing an integrated, unified pathway toward meeting the needs of PPDRs and citizens in crises. When time is of the essence, information overload is a problem and actionable intelligence is a requirement.

Our vision, the iSAR+Way, is presented in the next section.

THE iSAR+WAY: APPROACHING A MULTIDIMENSIONAL PROBLEM

To surpass PPDRs' state of low awareness and high uncertainty, which compromises efficiency and effectiveness when responding to crises, it is possible to explore the opportunity to involve citizens in crisis response efforts as information providers. Today, citizens are equipped with advanced mobile and online communication platforms (including social media) and are capable of providing critical crisis information that otherwise would be impossible to collect by FRs and PPDRs. This opportunity becomes clearer in the face of the crisis-related information that citizens have posted in public social media networks in recent crisis events, as opposed to the channels provided by PPDRs; in fact, in the digital era, there is still no 112 equivalent on the Internet. However, to enable the engagement of citizens and PPDRs in crisis response efforts, encouraging the use of the new communication media requires more than one other PPDR technological system; it requires a profound change in PPDR communication principles, as presented in Table 9.3.

With the integration of the new (mobile and online) communication media, the communication scenario between citizens and PPDRs could be as depicted in Figure 9.3. We consider that the challenge of incorporating new social media and mobile tools to significantly engage citizens and PPDRs and contribute to the exchanged information flow requires a multidimensional analysis, involving the organizational, the human, and the ethical and legal framework as well as the technological dimensions.

Therefore, achieving the iSAR+Way requires tackling the following dimensions:

- **Organizational**: This dimension focuses on the PPDR organizations and their culture, roles, processes, competences, training, and technologies to explore how to adapt these organizations to work with social media platforms, building their trust in online networking platforms, and thus facilitate the process of introducing new prototypes into the organizations' operational processes.
- **Human**: This dimension comprises the citizens' perspective on the acceptance and adequate employment of state-of-the-art mobile and social media communication technologies in crises, concerning human factor analyses, message delivery channels, and message content.
- **Ethical and legal**: This dimension deals with the PPDR platforms' requirement to abide by the ethical principles and legal framework, developing an ethics-by-design project approach, reflecting the ethical and legal framework concerns and waivers that emerge from the debate on the boundaries of privacy rights and public security, regarding the integration of online and mobile technology in response efforts.

TABLE 9.3 PPDR Communication Principles: Today and With the New Communication Media

PPDR Communication Principles	Today	With New Communication Media
Information exchange policies	Strictly defined; top-bottom approach Unidirectional (from PPDRs to citizens)	Not predefined; bottom-up approach Bidirectional (between PPDRs and citizens)
Control of information	Fully controlled by PPDRs	Control free and community driven
Information trust and quality	Office sources with adequate verification and validation	Anonymous or unofficial sources and self-correcting
Citizens' inclusion	Seen as auxiliary source and reactive	Seen as primary source and proactive
Communications target	General public	Tailored according to user profile
Approach	Need-to-know; command and control	Need-to-share; connect and collaborative

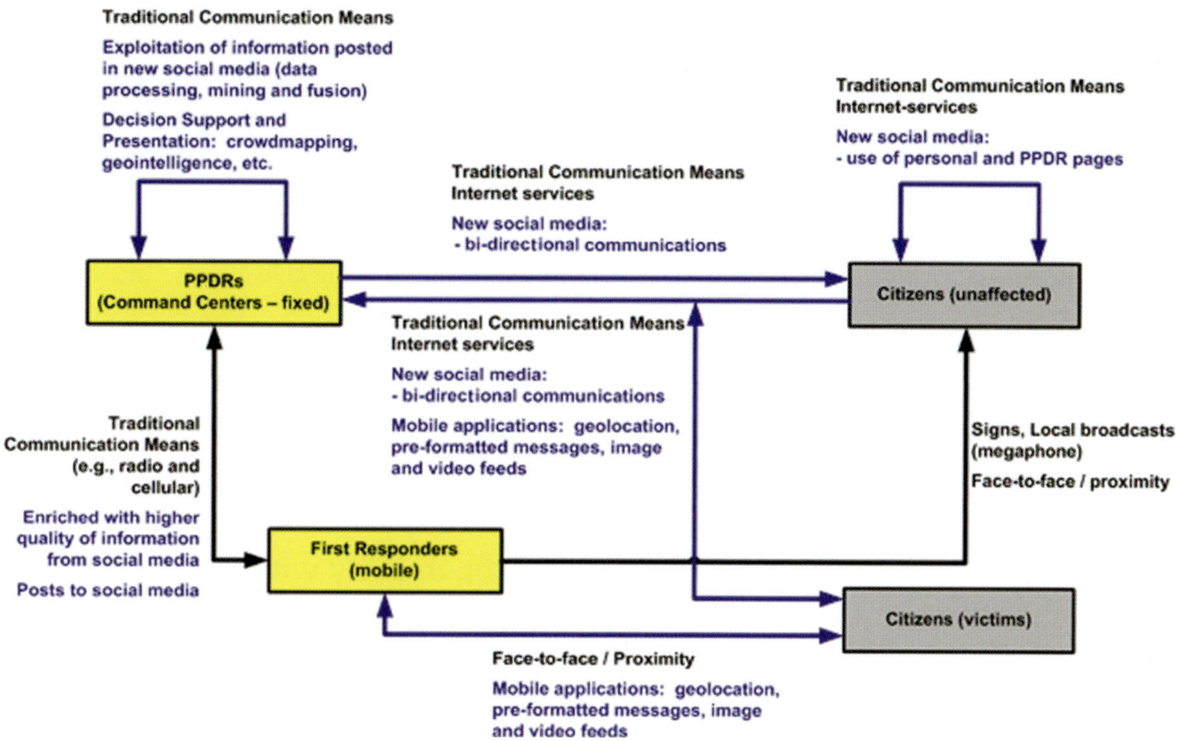

FIGURE 9.3 Enhanced communication flows with the new mobile and online social media.

- **Technological:** This dimension entails the integration of current ICT tools for crisis, equipment, communications, information processing technologies, and current standards into the PPDR platforms to explore the real potential of ICT and mobile technologies in what concerns crisis and search and rescue response efforts.

It is our conviction that this unified perspective, derived from a multidisciplinary alignment, is a guide to future policies, processes, organizations, individuals, and technologies to support existing and newly extended social dimensions of crisis and search and rescue response efforts. Upholding ethical and legal principles, the iSAR+Way considers the citizens and human factors approach and also the involved organizations' operational culture and processes, as it proposes the integration of existing ICT tools (including mobile applications and social media) already field tested in crises and the research and development of technical and technological improvements for advanced added-value services for citizens and PPDRs. The proposed platform and associated services for iSAR+are presented next.

The iSAR+Platform and Services

The proposed iSAR+platform is based on two different spaces. The PPDR space comprises the applications, services, and data storage specifically fit for PPDR needs, such as the "fusion center" and "risk maps." Because of ethical and legal concerns toward protecting citizens' privacy and upholding public security, part of the information within the PPDR space is not disclosed to the public. With respect to the involvement of the general public, the PPDR space includes traditional unidirectional broadcast media (television and radio) and new bidirectional interfaces connected to social media and the i112 portal. Moreover, it includes pluggable interfaces to external crisis systems and other portals alike. The citizens' space consists of the (open source) applications and services available for the general public, with information deemed of an open and public nature but still relevant for the platform's iSAR+purposes.

As perceived from Figure 9.4, both spaces interact in numerous and bidirectional forms. To deliver its technological vision we suggest a number of technological components that are introduced next.

The i112 Portal (for Citizens and PPDRs)

We propose an iSAR+portal that enables a bidirectional and rich communication channel between citizens and PPDRs. Citizens are able to upload crisis-related information, in multiple formats, optionally geotagged: short messages, e-mail, posts, images, and video. Additionally, the iSAR+platform provides useful services to citizens in crisis (people finder, useful contact numbers, and geolocalized information of shelters, safety houses, and food distribution centers) and operator-assisted feedback (reply-to) messages. Existing ICT tools for crises are embedded in the i112 portal, such as Ushahidi and Sahana, as well as social networking functionality and links to external social networks. With respect to the PPDR and FR component of the i112 portal, authorities benefit from an additional channel to provide useful crisis-related information (recommendations, guidelines, warnings, situation updates) to citizens, continuously interacting with them, in their time of need and despair. The i112 portal will also include areas with exclusive access for the PPDRs and FRs, where they find functionalities to ease crisis and search and rescue response efforts.

Interoperability with Existing Social Media and ICT Tools for Crisis

In this component, iSAR+integrates existing social networking platforms and ICT tools applied to crises as follows:

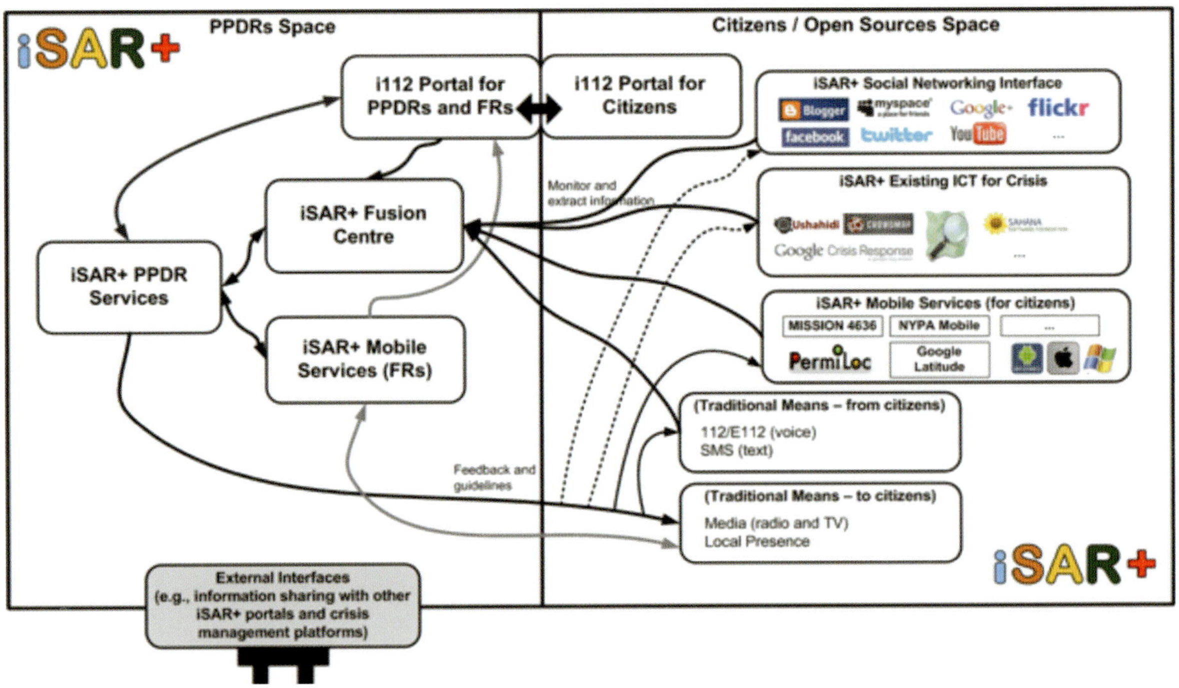

FIGURE 9.4 iSAR+platform overview.

- For the social media networking component, and because the applicable service providers host these platforms, an information extractor is placed in accordance to ethical and legal recommendations, enabling automatic or manual extraction of information into the iSAR+fusion center.
- For the existing ICT tools for the crisis component, two options are proposed. First, open-source platforms (as Ushahidi) are embedded in the i112 portal and an iSAR+service is provided by PPDRs and hosted within PPDR servers (see previous component). This option grants PPDRs full platform control and the possibility to create built-in events and external interfaces (a built-in information extractor). Second, an information extractor is placed to extract information from existing deployed platforms (Google Crisis Response, OpenStreetMap, Ushahidi) hosted by third parties, assuming bidirectional links exist (PPDRs may publish information in third-party platforms). Both options assume active platform use by PPDRs, with the purpose of developing enhanced situational awareness and improved communication with the public.

A mechanism to extract public information to feed the iSAR+fusion center is proposed (see next section) to prevent information overload. This technique is often referred to as open source intelligence, and the information extractor includes an anonymizer filter to ensure citizens' data privacy.

The iSAR+Fusion Center

The iSAR+fusion center is the platform's component for storing, processing, mining, and fusing large quantities of (open) information, so that actionable information is presented to FRs and PPDRs. It monitors information from all iSAR+components, including the interface with existing social media platforms (the open sources' space). Candidate technologies portrayed include: text analysis (automatic extraction of relevant information from data that is heterogeneous, malformed, and noisy); image and video analysis (detection, identification, and recognition of specific objects and/or situations in images and videos); and information mining and fusion (combines information from multiple sources and processed information assuming imperfect sources of information and generates alerts based on detected trends).

iSAR+PPDR Services

To deliver increased situational awareness for PPDRs, we identify several relevant services: crisis detection and monitoring (using social networks as sensors to detect, provide, and maintain information relative to ongoing crises); mapping/crowdsourcing (information over online maps to reveal location and citizen presence during crises, depending on individual express consent, and helpful resource coordinates such as hospitals, shelters, and assistance centers, an information further enriched with text, photos, and video); visualization and decision support tools (application of statistical techniques to spatial data for real-time pattern or signature detection, use of charts, sociograms, and smart search for information and mapping and crowdsourcing capabilities); missing people tracking (includes missing people's information, publicly accessible and constantly updated, as provided by Sahana; Prutsalis et al., 2010); risk maps (probability of the occurrence of a crisis with consequences or impact analysis, requiring the integration of geographical information systems or spatial maps with hazards, exposures, and vulnerability functions to obtain quantitative factors to be visualized on maps, often represented as threats; Yan, 2010).

iSAR+Mobile Services

The iSAR+services targeted to mobile users include the following:

- **SMS services gateway**: Enabling send and receive SMS (event-related information, warnings, alerts) from mobile phones located within and outside the crisis area. The service benefits affected citizens, deployed FRs, PPDRs, and the general public.

- **Mobile applications**: Running on the most popular mobile platforms (iPhone and Android) to provide access to the i112 portal and deliver critical services for citizens in crises: automatic localization of mobile phones, receiving crisis-related information specific to a geographical location, easy upload of crisis-related images and videos, and a preformatted SMS capability (to send SOS requests and receive relevant crisis-related information).

iSAR+Interfaces: The iSAR+Network

Crises cross borders and nationalities, thus the exchange of information between distributed iSAR+platforms is of the utmost importance. With each platform managed by its respective national authority, the exchanged information will include data on affected non-national citizens. This novel distributed approach is easy to implement and serves as an alternative to a single iSAR+centralized architecture and control authority. We propose this information exchange between iSAR+platforms through an iSAR+interfaces module.

CONCLUSION

The iSAR+approach encourages the use of mobile technology and social media to improve crisis communications between citizens and PPDRs, and contributes to the information gathering and sharing processes as well as enhancing situational awareness by exploring two main behavioral trends observed:

1. The ubiquitous presence of mobile devices in society
2. Broad adoption and use of social media by citizens to share information, even in crisis situations

These new communication channels—mobile technology and social media platforms—between PPDRs and citizens in crises have proven to be, in past events, more resilient and effective than traditional channels, namely the 112 emergency number, due to the inability to cope with the high number of incoming calls) and broadcast media, criticized for being too generic. Additionally, the proactive exploitation of new mobile technology capabilities, such as the automatic localization of smartphones, contributes to a dramatic increase in the situational awareness levels of PPDRs, thus enabling better planning and resource allocation and, consequently, increased effectiveness and efficiency in response efforts.

iSAR+broadens the reach and richness of PPDR-citizens communication, highlighting specific benefits as follows:

- Faster response reaction times
- Improved interaction among prevention, detection, reporting, and rescue
- Improved performance of FRs, medical personnel, police, and law enforcement agencies
- Improved national security framework assisted by innovative analytical intelligence methods and tools

ABOUT THE CONTRIBUTORS

This section is the collective result of a multidisciplinary group of expert entities that were gathered to present their view on the use of social media and mobile technology in crises for purposes of disaster response and citizen security. Our view, created as iSAR+, has been selected by the EC for implementation on a time frame of 30 months. As part of this effort, we plan to conduct a number of open workshops in Europe and one in the United States.

The following people contributed to this section: Bárbara Manso and Marco Manso (TEKEVER, Portugal); Hernan Joglar (retired Colonel, Army of Chile); Marco Artioli and Davide Piombini (Bridge129, Italy); Emilio Mordini and Andrew Rebera (Center for Science, Society and Citizenship, Italy); Stéphane Schmoll (Deveryware, France), Kari Junttila, Esa Kokki, and Hanna-Miina Sihvonen (Emergency Services College, Finland); Richard E. Hayes and Grace Scarborough (EBR, United States); Dieter Rhode and Silke Schmidt (Ernst-Moritz-Arndt-University

Greifswald, Institute of Psychology, Department Health and Prevention, Germany); Michal Choras and Krzysztof Samp (ITTI, Poland); Egil Bovim and Øyvind Hope (National Center of Emergency Communication in Health, Norway); Jukka Koponen (North Savo Rescue Department, Finland); Dave Fortune, Simon Jones, Simon Nott, and Tom Stirling (NYPA, UK); Emmanuelle Villot (Pôle Pilote de Sécurité Locale, France); Terhi Kankaanranta and Vesa Muttilainen (Police College of Finland, Finland); Thomas Delavallade, Catherine Gouttas, and Olivier Sagnes (Thales Communications and Security, France); Michael Cooke (University of Dublin, Trinity College, Aerospace Psychology Research Group, Ireland), Taina Kurki and Matti Nykänen (University of Eastern Finland, Finland); Elizabeth Avery Gomez (University of New Jersey, United States); and Alessandro Zanasi (Zanasi and Partners, Italy).

CHAPTER

10

Emerging Technologies and the Human Rights Challenge of Rapidly Expanding State Surveillance Capacities

Mohammad Dastbaz, Edward Halpin, Steve Wright

They that can give up essential liberty to obtain a little temporary safety deserve neither liberty nor safety...

Benjamin Franklin (1759)

INTRODUCTION

We live in the shadow of rapidly developing and converging technologies, from the humble beginnings when the Internet and the World Wide Web emerged, out of developing network technologies and hypertext, to today's connected world where from the comfort of our laptops we have access to terabytes of data that up until recently were reserved exclusively for large corporations. We are living in the era of "smart devices" that we trust with our most intimate personal and financial information and carry with us wherever we go. The integration and embedding of these technologies in our daily routines are such that life without our smart devices is unimaginable, and a world without them looks something like a postapocalyptic Hollywood blockbuster.

The use of these technologies informs every dimension of modern lives. From learning and training, to commerce and re-engineering our business processes, these tools facilitate communication between governments and their citizens and even the way we are made aware of social and political issues. Yet we are only exploring the tip of the iceberg in terms of the width and breadth of second- and third-order impacts from the application of these technologies.

Typically such rapid growth and advancements in mobile information communication technology (mICT) and its "uses and misuse" have received wide and varied treatment in the literature. Perhaps one of the current, controversial, and most talked about aspects of the use or misuse is the unprecedented application of these technologies in state surveillance versus citizens' rights in the emerging world.

Jacques Ellul, in *The Technological Society* (Ellul, 1964), considered the emergence of technology philosophically before concluding "….. What good is it to pose questions of motives? Of Why? All that must be the work of some miserable intellectual who balks at technological progress." While we do not balk at technological progress of the expanding state surveillance technologies in this chapter, we do try to pose questions of motive and of "why?".

There is a well-trodden history to the development of the surveillance capacities within the state; however, for the purposes of this chapter we will attempt to look at what might be described as the new era of intelligence gathering and security. It might be easy to pick the post-9/11 period as the significant point of change, although the origins of the response to 9/11 pre-existed the tragedy that occurred. A large range of public domain U.S. military thinking predates 9/11 and shows the need for, and use of, advanced information-gathering techniques, including Vision for 2020 (United States Space Command, 1997), and Information Operations (U.S. Air Force, 1998). In Information Operations the question of why is answered clearly and resolutely:

> *Today, information systems are part of larger information infrastructures. These infrastructures link individual information systems through numerous and redundant direct and indirect paths, including space-based systems. There is a growing information infrastructure that transcends industry, the media, and the military and includes both government and nongovernment entities. It is characterized by a merging of civilian and military information networks and technologies. Collecting, processing, and disseminating information by individuals and organizations comprise an important human dynamic, which is an integral part of the information infrastructure.*
>
> *U.S. Air Force, 1998*

In a special issue of *TIME* magazine (Thornburgh, 2005), Massimo Calabresi poses an interesting thought about the nature of the technology and how it's used these days. He stated:

> *If anybody wanted to develop a global system for tracking human beings and collecting information about them, it would look a lot like the digital mobile device networks. It knows where you are, and—the more you text, tweet, shop, take pictures and navigate your surroundings using a smart phone—it knows an awful lot about what you are doing…*

He goes on to highlight the growing issue of U.S. federal and local law enforcement agencies requesting mobile providers to hand over their data for various requests. According to Calabresi, such cases are now very common.

Ed Markey, major U.S. cell carrier and provider, states that they received 1.3 million requests for cell-phone tracking data from federal, state, and local law enforcement officials in 2011. This, compared to the 3,000 wiretap warrants issued in the United States in 2010, shows the growing problems and issues related to the private and personal data held by mobile carriers.

While there could be justifications made about the use of these technologies and data in tracking criminals and bringing them to justice, there is very little doubt that such data can also be easily used to violate individual's rights and freedoms. The images of societies where the state is constantly tracking its citizen's personal and quite often very private data can be used to change people's behavior is nothing short of the *1984* Orwellian nightmarish society where everybody is watched by "big brother." Yet our current surveillance capability so far exceeds Orwell's nightmare dystopia; in 1984 there were no computers, no "dataveillance," no geolocation, and no communication, command, and control based on algorithmic rather than human intervention.

A BRIEF SURVEY OF EMERGING SURVEILLANCE TECHNOLOGIES

While it is outside the scope of this chapter to review all emerging surveillance technologies, we felt it would be appropriate to highlight some of the emerging key technologies.

Naomi Wolf (2012), in an article in the *Guardian* (August 15, 2012) titled "The New Totalitarianism of Surveillance Technology," quoting a software engineer from her Facebook community, highlights an interesting case on surveillance technologies that affects our day-to-day activities. She stated that, while visiting Disneyland with his partner, the software engineer goes on a number of rides and later notices that the theme park is offering him photos of him and his partner, with his credit card information already linked to the offer. He is baffled as he had not entered his name or any other information into anything in the theme park or indicated that he was interested in photos or had informed anybody from the theme park who his partner was. He then comments that on closer inspection this scenario could have only happened if Disney was using advanced facial recognition technology. He then went on to claim that Disney had recently shared data from facial recognition technology with the U.S. Army. Wolf (2012) further stated that News21 supported by the Carnegie and Knight foundations recently reported that the Disney sites are indeed controlled by the same facial recognition software in which the U.S. military is interested.

Biometric Technologies

Biometric technologies generally refer to the use of technology to identify a person based on some aspect of their biology. Fingerprint recognition is one of the first and original biometric technologies that have been grouped loosely under digital forensics. With the ever-growing number of video surveillance cameras mushrooming in large cities, the use of the data captured by these cameras has been at the center of a number of privacy and human rights storms. Following the 9/11 terrorist attack, the use of facial recognition, especially in crowded places, as a means of detecting possible threats has been debated widely. The way the technology works is straightforward. CCTVs in streets, public places, and office buildings record images 24/7, sophisticated algorithms then carry out a matching exercise with an existing database of images of potential "villains" or "targets." A match will trigger enhanced surveillance and possible future and further action. For the system to be effective, the matching database should be as wide and comprehensive as possible. It is not surprising to note that to put such a database together security agencies never (at least we cannot identify any evidence) consult or seek permission to keep people's records in their data centers. Furthermore routine phishing activities through the Internet and social networks provide a fertile ground for not only a simple one-dimensional set of data (photos and other personal data) but potentially three-dimensional datasets of associated friends, links, habits, and quite often current location. In early August 2012, Michael Bloomberg, Mayor of New York, and Ray Kelly (NYPD Commissioner) unveiled a new police surveillance infrastructure developed by Microsoft called the Domain Awareness System, which links existing police databases with live video feeds from a variety of different sources.

Furthermore, according to a Homeland Security newswire in the United States, billions of dollars are being invested in the development of various biometric technologies capable of identifying anyone anywhere in the world. These include iris-scanning and foot-scanning technology, and voice pattern ID, as well as facial recognition technologies (Wolf, 2012).

Location-based and Tracking Technologies

In a recent visit to Prague, in the city's main square, I decided to find a restaurant called Bily Konichcek, about which I had read good reviews. After several attempts at finding the restaurant through asking people failed, I decided to use Google Map to find it. The search result was quite interesting. Not only did Google map find the restaurant for me, it also found a person called Bily Konichcek, which was within the search parameters I had given. Was Bily Konichcek aware that his location was being broadcast to a total stranger looking for a restaurant? I doubt it.

Wang and Loui (2009) defined the working of the GPS systems as using constellations of GPS satellites that orbit the earth. These satellites then broadcast signals on radio frequencies that consist of the time of the message and orbital information. A receiver measures the transit times of messages from four satellites to determine its distance from each satellite, and thereby calculate its location. They further noted that

> *In the United States, law enforcement officials use GPS technology to track criminal suspects and parolees without their awareness. For example, they may attach to the individual's car a device such as Trackstick, ™ which is a GPS data logger integrated with GoogleEarth. Law enforcement officials argue that GPS devices fall outside the scope of laws regulating wiretaps and similar forms of electronic surveillance because they do not record conversations.*
>
> *Wang and Loui (2009)*

As well as GPS systems, we also now have a Global System for Mobile Communications (GSM), which provides a wealth of data including locations to mobile operators and providers. For example, GSM signals transmitted from mobile devices can be used to monitor a traveling car and its passengers.

Through-the-wall Surveillance Technology

In the early years following the turn of the millennium, military and law enforcement agencies began developing technologies that were capable of detecting human movements and positioning behind enclosed spaces and solid walls. The technology, loosely termed through-the-wall surveillance, used radar technologies aimed at providing vital information to security forces dealing with difficult emergency situations. The more peaceful version of the same technology can be usefully deployed in natural disasters to detect victims of earthquakes buried beneath rubble. The main drive behind the development of this technology was for "safe" surveillance of potential criminals and threats to state security. In defining the possible application areas of such technologies, each year correctional and law enforcement officers are injured because they lack the ability to detect and track offenders through building walls. While the early versions of this technology were not able to map building or room interiors or could not tell how many walls are between the user and monitor, the later development of the same idea now provides quiet sophisticated three-dimensional mapping of buildings using building blueprints.

Mobile Surveillance and Wireless Sensor Systems

Tseng et al. (2005) explored the possibility of incorporating the environment sensing capabilities of wireless sensors with video-based surveillance systems. The result is an Integrated Mobile Surveillance and Wireless Sensor System (iMouse) capable of detecting and analyzing unusual events.

The proposers of this system believed that such a surveillance system could enhance human life in areas such as healthcare, building monitoring, and home security, but clearly one can see security agencies and the military could also be interested in the mobile capabilities of such sensors, which are versatile and battery operated.

Virtual Reality, Surveillance, and Security Systems

Another technology having a major impact in the development of surveillance and security applications is virtual reality (VR). VR technology is used to provide a state-of-the-art training environment for key decision makers and people dealing with national emergencies, is capable of receiving data from a variety of different sources (including GPS, live news feed, and direct agent communications; see FP7 PANDORA project, Dastbaz and Cesta, 2011), and has been used to create surveillance and security systems.

Ott et al. (2006) stated that

> *Virtual Reality (VR) can become a key component of future surveillance and security systems, being used in a number of tasks such as: teleoperation of the actual*

data acquisition systems (cameras, vehicles, etc); providing multimodal interfaces for control rooms where information is analyzed; and empowering on-field agents with multimedia information to ease their tasks of localizing problematic zones, etc.

According to Ott et al. (2006), a general surveillance and security system typically has three key components: data acquisition, information analysis, and on-field operation.

Typically a VR device can be used to improve the ergonomics of existing systems. Today, visualization systems for video surveillance based on an augmented virtual environment (AVE) are also an important topic. AVE fuses dynamic imagery with three-dimensional models in a real-time display to help observers comprehend multiple streams of temporal data and imagery from arbitrary views of the scene.

NONGOVERNMENTAL ORGANIZATION POLICY RESEARCH: INTERVENTION AND ACCOUNTABILITY ON SURVEILLANCE

Over the last 30 years, surveillance studies has become a respectable academic field with a *Handbook of Surveillance* studies published by Routledge in 2012 (Ball et al., 2012) There is also a worldwide security and surveillance network run by academics to exchange views, sponsor publications, organize conferences, and build a critical community of academics to understand this burgeoning field and its associated industries (http://www.surveillance-studies.net/).

While academics ponder the adequacy of legally controlling the surveillance of mICT and associated proliferation, a number of nongovernmental organizations (NGOs) have systematically studied the changing state of the art and the need for intervention and if and when this goes beyond the limits of the law. Early work was accomplished by the British Society for Social Responsibility in Science, which controversially viewed much of what was being developed as a "technology of political control." This perspective was taken up by the Omega Foundation in a policy report titled An Appraisal of the Technology of Political Control for the European Parliament.

This Science and Technology Options Assessment (STOA) report called for a European Commission (EC)-wide oversight of interception procedures and activities after revealing that millions of e-mails, telephone calls, and faxes were routinely intercepted each hour by the secretive U.S. National Security Agency—often in direct contravention of privacy guarantees enshrined in individual Member State's national legislation. This exposure of a secretive global telecommunications system, known as Echelon, generated worldwide awareness of a new capability of mICT interception. The STOA report also revealed a new European Union (EU)–Federal Bureau of Investigation surveillance agreement researched by the NGO Statewatch. This plan was introduced to force service providers to make all of their traffic transparent via a document known as "the requirements," largely demanded because privatization had led to piecemeal evolution of systems that were opaque to the authorities. These measures were adopted by "written procedure"—literally 15 faxes sent to Member States—without any parliamentary scrutiny or debate and adopted "on the nod" by the EU Fisheries Council on December 20, 1996.

Since then, the EC has funded a number of research projects studying security and surveillance technologies. These include a number of projects concerned with taxonomic classification of surveillance systems such as the Stakeholders Platform For Supply Chain Mapping, Market Condition Analysis and Technologies Opportunities (STACCATO); Security technology Active Watch (STRAW); Supporting Fundamental Rights, Privacy and Ethics in Surveillance Technologies (SAPIENT); Security Impact Assessment Measures (SIAM); and Public Perception of Security and Privacy: Assessing Knowledge, Collecting Evidence, Translating Research into Action (PACT; see http://www.projectpact.eu/).

These substantial research projects are assembling a significant body of knowledge regarding the state of the art as well as policy impacts on areas such as privacy and human rights. However, the recent experience of the EU security research expenditure has suggested that there has been a process of bureaucratic capture by the military, industrial, security, and media entertainment industrial complex (Statewatch, 2009). In response to such perceptions, the EC set up a Societal Impacts of Security Panel that looked to create measures that would rebalance the mutual dependency of the triangle of research investments in security, freedom, and justice. This was a useful exercise since the sunset report from that panel made a recommendation, which has been largely accepted, that all future EU-funded research on security must have a core element devoted to a formal study of its anticipated societal impacts.

Some of these EC-funded security projects explore the different perceptions between experts, stakeholders, and citizens on such matters as the potential trade-offs of privacy and security (e.g., PACT). But is this merely an exercise in liberalism in which security is seen as one set of competing values that must be fairly balanced against others?

Are the checks and balances adequate? For example, there may be a formal body charged with ensuring surveillance activity operates within the rule of law, but if that body has a workload that massively exceeds its resource capacity, oversight becomes tokenistic rather than effective. How has this revolution in surveillance come about so quickly that the evolution of capacity appears to have outstripped controls?

Digitalization and Dataveillance

The need to increase bureaucratic efficiency necessitated by shrinking budgets proved a powerful imperative for improved identification and monitoring of individuals. Fingerprints, ID cards, data matching, and other privacy-invading techniques were originally fielded on populations with little political power such as immigrants, welfare recipients, criminals, and members of the military.

Older surveillance systems were slow, used film or tape, and were static. Now the content of the surveillance can be transmitted to other places by microwave links or through the Web. Many of the innovations owe their existence to the rapid increases in processing power now possible with digital technology. Modern systems can "piggyback" on other forms of telecommunications infrastructure such as the mobile phone network and associate satellites.

By the 1980s new forms of electronic surveillance were emerging, many of these directed toward the automation of telecommunications interception. The interconnection of visual and audio surveillance into networks of storage and data processing has enabled a new era of mass supervision and tracking, so-called dataveillance, initially pioneered in the UK.

We are only at the beginning of this era, but since it is happening within a specific sociopolitical context, that is, "the war on terror," we can anticipate that military communication command and control systems will become amalgamated with civilian systems of monitoring and management.

Surveillance Flows and Dataveillance Networks

Surveillance versus privacy is not some zero sum game, it is more complex than that. Modern surveillance no longer just stakes out individuals but looks at "flows" of information; in many senses we are always shadowed by a body of data that somehow "represents us," not all of which we can check for veracity.

Network is the operative term here since systems can be requested to record, "hunt," track, and alert. Emerging ID systems, for example, are networks reliant on more than one mode of tracking technology, such as facial recognition with an electronic card. This is already leading to a massive accumulation of personal data that cannot be kept secure. It is also leading to the evolution of a powerful architecture of surveillance that can sense, record, and identify specific individuals entering a designated surveillance zone. Privacy and surveillance now exist in a world of data flows, with modern surveillance offering increasing capacities to track mobility, whether it is physical or virtual.

Algorithmic Surveillance and Geolocation

Introna and Wood (2004) argued that when surveillance is digitized there is a step change in power; for example, so-called algorithmic surveillance, which has some intelligent reasoning and learning attributes. Algorithmic surveillance can be defined as the move toward smart semi-intelligent monitoring both at borders, on the Internet at strategic gateways and highways, and via mobile phones. There is a link between information gathering and assembly from information to intelligence, especially in times of national crisis such as the war against terror.

The software or mathematical instructions or "algorithm" enables the technology to scan unobtrusively without any need for cooperation from the target. Such algorithmic systems use neural networks to discover otherwise hidden patterns. Wikipedia defines a neural network as nonlinear statistical data modeling or decision-making tools. They can be used to model complex relationships between inputs and outputs in order to find patterns in data.

Graham and Wood (2003) argued that the silent nature of this technology makes it difficult for society to scrutinize it. Such lack of accountability enables the micro politics of surveillance to become pervasive. Although we have extensive community consultation and impact studies for motorway development, this is not done when CCTV systems are installed, and customers of mobile phones asked their permission to collect geolocation data. Every mobile phone routinely generates a host of data including its approximate geographical location. Mobile phone location data are typically based on the nearest mast from which the handset receives a network signal. Location data, together with other data about communications, are stored by mobile phone service providers for billing and legal purposes (Gorra, 2007). These data are used regularly in court cases and by the intelligence services, because they provide a rich picture about a mobile phone user's actions.

Accountability

The Surveillance Studies Network has raised concerns that the routine tracking and information-gathering mechanisms used in today's society are often not obvious to citizens. The complexity of the interconnections between surveillance devices and processing capacities makes it difficult to ask meaningful questions to the public about these capacities if only individual components of just some of the technologies are explored.

This makes it important to incorporate checks and safeguards when collecting data to ensure accountability. The retention of mobile phone communications data especially bears the potential for identifying patterns in the collected data. It is possible to analyze the behavior of particular groups of people or of mobile phone users located in a particular area without identifying specific individuals (Marx, 2002). A key question is the extent to which bodies charged with the responsibility to monitor the monitors have the access, staff, and resources to practically oversee such huge surveillance enterprises.

These capacities and their potential role and functions can only be truly comprehended as systems within an entire political and social context. For example, Western companies provided mobile surveillance technologies used to track down dissidents during the Arab Spring, including Syria (http://topics.bloomberg.com/wired-for-repression/).

President Obama recently introduced new export controls to prevent such proliferation of U.S. surveillance technologies to authoritarian regimes in the future, but EU countries have continued to do so. Companies such as ISS bring together suppliers and buyers at specialist conferences/expos such as that in Prague in June 2012. Are EU citizens comfortable with such prioritization of profit over principle? NGO's like Privacy International have made it a priority to gather evidence to link security company's mICT and related training with subsequent human rights abuse. They are currently saying companies that supply surveillance technologies to the likes of Syria must cease or face legal action.

NGO's such as Witness are now using satellite technology to monitor suspected human rights violations. Similarly, some protestors are beginning to adopt cheap unmanned aerial vehicle surveillance cameras to monitor community safety when riot squads are at work. What are the implications of such mICT-driven approaches to protect human rights defenders? What if they initiate response and counter-response arms races between the controllers and those who challenge that control? The next section explores some of the core issues in light of recent experience, especially during the so-called Arab Spring.

HUMAN RIGHTS AND SURVEILLANCE TECHNOLOGIES

Lannon and Halpin (2012) discussed the application of technology in the human rights world in *Human Rights and Information Communication Technologies: Trends and Consequences of Use* describing the innovation that has transformed the use of ICT as

> *...(having) helped enormously to move the promotion and protection of human rights forward...They (ICT) have transformed the capacity of the human rights community to highlight human rights abuse and to advocate for causes and victims of oppression. They have made it easier to access and share information, to facilitate human rights data aggregation and analysis, to offer new tactical approaches to campaigning, and to precipitate real-world activities ranging from local demonstration to intergovernmental agency lobbying.*

In 2000 there was no YouTube for video sharing. Social networking with Facebook or Twitter was still half a decade away, and blogging had not yet become mainstream. Web mashups were nonexistent, and wireless devices were still only emerging technologies. The World Wide Web was still relatively young at that time, and was all about read-only content and hyperlinked Web pages designed to be read by humans. The bursting of the dot com bubble in 2001 led many people to believe that this Web was overhyped, but organizations like O'Reilly Media (formerly O'Reilly & Associates) had a different view. They recognized that the Web was becoming more important than ever and exciting new applications and sites were popping up with surprising regularity (O'Reilly, 2005). To highlight these innovations they organized a conference in 2004 at which the term Web 2.0 was born.

Web 2.0 applications that facilitate participatory online information sharing and collaboration have transformed the human rights community. Blogging in particular has become a vitally important tool for individuals and organizations that want to keep the public or the human rights community informed about human rights issues. Very often the first people to present evidence of human rights violations publicly today are "frontline" bloggers who are either witnessing and documenting the violations themselves or posting someone else's information. Aggregation bloggers like Global Voices Online (http://globalvoicesonline.org) amplify this information so that it is more accessible. International human rights NGOs and libraries also publish and translate selected blogs, and sometimes editorialize what they consider to be the "good sources".

The value of Twitter, Facebook, YouTube, and other social media tools of the Web 2.0 era was demonstrated during the pro-democracy protests in Iran in June 2009 and in the Arab Spring uprisings in Tunisia and Egypt in January 2011. Real-time reports on what was happening on the streets went out on these social networks, as did calls to rally. Poignant images of suffering—a video recording of the death of Neda Agha-Soltan who was shot on her way to the election protests in Iran, or photographs of Mohamed Bouazizi, a street vendor who burned himself to death to protest harassment by the Tunisian authorities—were seen by millions of people around the world.

Today the Web is used by large groups of people to create collective works whose value far exceeds that provided by any of the individual participants. In 2009, Tim O'Reilly and John Battelle wrote that it

> *...is no longer a collection of static pages of HTML that describe something in the world. Increasingly, the Web is the world—everything and everyone in the world casts an "information shadow," an aura of data which, when captured and processed intelligently, offers extraordinary opportunity and mind-bending implications.*
>
> ***O'Reilly and Battelle, 2009***

We are now in the era of Web 3.0, which is all about personalization, intelligent searching, and the Semantic Web. The latter links up information on a global scale and has the potential to provide powerful data organization and query capabilities. These enable machines to understand the meaning of information on the Web through the addition of machine-readable metadata about pages and how they are related to each other. Resources can be aggregated, shared, and accessed from many different places, and users can choose the appropriate presentation for the tasks they need to accomplish (Hendler and Goldbeck, 2008). As a result we are crossing into a new learning paradigm, which offers a qualitative change in the way people think of interacting on the Web. With Web 2.0, interaction treats the Web as an information source and we learn by browsing, searching, and monitoring it. But with Web 3.0 the Web will be understood as an active human–computer system, and we will learn by telling it what we are interested in, asking it what we collectively know, and using it to apply our collective knowledge to address our collective needs (Gruber, 2008, p. 12).

An important factor in achieving this is to be able to draw on domain knowledge in areas where searches are difficult (Hendler and Goldbeck, 2008). One of the key challenges, therefore, for the human rights community is to bring human rights experts, information scientists, and technologists together to ensure that the necessary semantic linkages exist between the vast array of human rights-related information that is published online.

The human rights world has always been early adopters and adapters of emerging technology. In the days of the early Internet and Web organizations such as Amnesty International used bulletin boards to cut the time in transmission of their rapid response interventions for human rights defense, as reported to the European Parliament study on The Use of the Internet for the European Parliament's Activities for the Promotion and Protection of Human Rights (Halpin and Fisher, 1998). Throughout the period and since this trend has been continued with the use of GPS, crowd sourcing, blogs, and tweets, all playing a part in the repertoire of the modern human rights activist.

An examination of some evidence about current uses of mICT in human rights protection provides an interesting insight. If we look at the work of Ushahidi, who uses crowdsourcing, and the work by Douai on the Arab Spring, we can quickly view a small selection of these technologies in action (Lannon and Halpin 2012).

Douai, in Lannon and Halpin (2012), described research undertaken during the Arab Spring into the use of YouTube as a human rights advocacy resource. The "Arab Democracy Spring" has promised to end the entrenched history of human rights violations in Egypt, Libya, Syria, and Tunisia, among other Arab authoritarian states. However, the long fight against these abuses commenced years prior to the 2011 mass protests as an unprecedented era of virtual politics and activism took shape within Arab societies. At the forefront of these shifts, the Internet and other new communication technologies have been central forces for change. A few years after its inception, YouTube soon became an important tool for publicizing

Arab regimes' human rights abuses both locally and globally. Preliminary evidence suggests that YouTube has been effective in highlighting police abuse cases and prosecuting perpetrators. This work thus contributes to a growing body of research that underscores the vital role of communication and information technologies in promoting human rights worldwide.

Within the above framework, Arab citizens have similarly harnessed the site's video exchange capabilities to expose political corruption, police brutality, and demand political reform in the same way bloggers have countered official narratives and/or media blackouts on local events. YouTube and the new breed of social media have grown more effective as favorite political instruments for several reasons: high levels of anonymity, global reach, technical simplicity, absence of professional prerequisites, and local/global-organizing tool capabilities. Significantly, YouTube is hosting and abetting a new political discourse in which readers vent their frustrations and heap their scorn online before moving offline. The first tremors of this movement toward harnessing YouTube's social networking and video exchange capabilities appeared in 2007, as videos of police brutality and corruption in Egypt and Morocco were posted online.

Rosneau (2003, p. 149) argued that the twin forces of globalization of communication technologies mean that "the misdeeds of human rights violators no longer pass from human kind's conscience." For example, YouTube videos, in publicizing police abuse, corruption, and other human rights violations in Egypt and Morocco, have been a major factor in publicizing "misdeeds" by the abovementioned governments.

For activists, YouTube's repository and exchange capabilities provide audiovisual evidence for the excesses of the state. The more shocking the video evidence is, the louder the public and global outcry against those excesses will be. Also, the "permanent campaign" implies that the more "permanent" the record is, the more salient and constant the fight becomes. Permanent campaigning means constant surveillance of authoritarians' violations. Internationalization of human rights abuses builds on a well-proven record of transnational solidarity movements, similar to the movements behind publicizing human rights abuses in Latin America (Keck and Sikkink, 1998). In their campaign, these "activists without borders" are wielding cameras and low-tech skills whether they post amateurish videos showing police abuse or they construct highly edited videos.

Another interesting use of social media and emerging mICT is the use of technology to develop human rights organizations and campaigns. The Ushahidi organization developed in Kenya and describes itself as a "non-profit tech company that specializes in developing free and open source software for information collection, visualization and interactive mapping" (the word Ushahidi means testimony in Swahili). There work developed from engagement in the 2008 post-election period in Kenya, when they produced a Web site to map the violence that was occurring, working alongside Kenyan citizen journalists. In the short period since then Ushahidi the Web site mapped incidents of violence using reports submitted via mobile phones and the Web, accumulating approximately 45,000 users who provided evidence of the violence. Ushahidi stated that

> *Since early 2008 we have grown from an ad hoc group of volunteers to a focused organization. Our current team is comprised of individuals with a wide span of experience ranging from human rights work to software development. We have also built a strong team of volunteer developers primarily in Africa, but also Europe, South America and the US.*

The technology for communication is low tech, mobile phones are sufficient, particularly where Internet coverage or accessibility is difficult, and the free open source software used for creating the mapping provides an information and content management system that allows analysis of complex and dangerous events as they happen, which it is argued allows for early warning and visualization for response and recovery. These tools, along with others provided as open source resources by Ushahidi, help human rights activists hold perpetrators of human rights abuses accountable. Using these low-tech tools to achieve a high level of data gathering and analysis is vital in the very quickly changing world in which they work.

There are numerous other examples of the use of mICT, but there is also an alternative position, of which the human rights activist and citizens in general need to be aware. Reports indicate that governments do not stand idly by and watch their legitimacy challenged. In particular, Privacy International draws attention to the role that states take in repressive action of surveillance and monitoring of citizens and human rights activists. There is evidence of this from many sources, including the countries of the Arab Spring. They also report the exporting of these technologies by American and European companies to regimes known to be repressive and abusing human rights.

CONCLUSIONS

We started this chapter by stating that emerging technologies are creating fundamental changes to our daily lives and explored the role of these technologies in rapidly growing state surveillance capabilities. It is clearly obvious that the advent of the technology has had a tremendous impact in helping human society be better informed and hopefully better equipped to deal with natural and social ills. It is heartwarming to see how emerging technologies have helped human rights campaigners across the world to highlight state cruelty and suppression and how they have been able to mobilize public opinion in defense of peoples' rights in various countries. We are also very conscious and anxious that states across the world are using emerging surveillance technologies as an integrated part of their suppressive apparatus and citizens' right to privacy and their human rights are increasingly threatened and violated.

Borrowing from Charles Dickens (1859) it might be said that the world of human rights and the impact of mICT on it is a little like the introduction to his work, *A Tale of Two Cities*:

> *It was the best of times, it was the worst of times, it was the age of wisdom, it was the age of foolishness, it was the epoch of belief, it was the epoch of incredulity, it was the season of Light, it was the season of Darkness, it was the spring of hope, it was the winter of despair, we had everything before us, we had nothing before us, we were all going direct to Heaven, we were all going direct the other way—in short, the period was so far like the present period, that some of its noisiest authorities insisted on its being received, for good or for evil, in the superlative degree of comparison only.*

While technology marches on and surveillance capabilities are developing it is also important to recognize that the human rights of citizens should also be recognized as part and parcel of the national security of a country. It is the ethics of development and the ethics of practical application mICT that we need to balk at and rigorously, to ensure that they are justified within a democratic and free society. Everyone is responsible to ensure that they enhance human dignity rather than diminish it. The human rights community, which will always be an active stakeholder in mICT policies and possibilities, must also ensure that this is so.

SECTION 3

Technologies, Information, and Knowledge for National Security

CHAPTER

11

User Requirements and Training Needs within Security Applications: Methods for Capture and Communication

Alex W. Stedmon, Rose Saikayasit, Glyn Lawson, Peter Fussey

USER REQUIREMENTS ELICITATION

Understanding the valuable tacit knowledge and expertise of end users and other stakeholders is fundamental to the general acceptance, successful use, and overall effectiveness of new training programs, procedures, and technological solutions (Wilson, 1995). From a human factors perspective this translates into user-centered approaches in which a variety of methods have been applied in research areas as wide as healthcare, product design, human–computer interaction and, more recently, security and counterterrorism (Saikayasit, et al., 2012). A common aim across user-centered approaches is the capture of user requirements so that system requirements can then be designed to support users within specified contexts of use (Wilson, 1995). User requirements embody critical elements that end users and stakeholders want from a solution while system requirements express how that solution should be designed and used (Maiden, 2008). However, these two factors are not always balanced and resulting solutions may be developed that are not fully exploited, or used as intended, which may lead to serious incidents and potential disasters. A fundamental issue in developing successful solutions is the correct identification of user requirements that are then revisited in an iterative manner throughout the design process. Alongside this, it is important to clarify specific "contexts of use" for proposed solutions as this provides design boundaries as well as frames of reference for communicating issues back to end users and other stakeholders. In order to achieve this, it is often necessary to prioritize potential solutions to ensure that expectations are managed appropriately (Lohse, 2011).

Requirements elicitation is characterized by extensive communication activities between a wide range of people from different backgrounds and knowledge areas, including end users, stakeholders, project owners or champions, mediators (often the role of the human factors experts), and developers (Coughlan and Macredie, 2002). This is an interactive and participatory process that should allow users to express their knowledge and designers to display their understanding to ensure a common design base (McNeese et al., 1995). End users are often experts in their domain and possess deep levels of knowledge gained over time that is often difficult to communicate to others (Blandford and Rugg, 2002; Friedrich and van der Poll, 2007). Users often do not realize what information is valuable to the development of solutions or the extent to which their knowledge and expertise might influence the way they work (Nuseibeh and Easterbrook, 2000).

User requirements elicitation presents several challenges to researchers, not least in recruiting representative end users and other stakeholders upon which the whole process depends (Lawson and d'Cruz, 2011). Equally important, it is necessary to elicit and filter the relevant expertise and knowledge from end users and other stakeholders. Finally, having gained an insight into user requirements, it is essential to communicate these forward to designers and policy makers, as well as back to the end users and other stakeholders.

This chapter presents aspects of a user requirements elicitation exercise that was conducted as part of the "Shades of Grey" security consortium funded by the Engineering and Physical Sciences Research Council (EPSRC: EP/H02302X/1). This research aims to design and develop a suite of interventions that can be applied in crowded public spaces to amplify the signal-to-noise ratio of suspicious behaviors in order to improve the rate of real-time detection of terrorist activities. Specific attention is focused on defining requirements for security and counterterrorism within the context of a large public venue. The findings are mapped onto a macro-ergonomics framework and different communication tools are presented. Observations are then made in relation to current and future training needs within this context.

User Requirements Elicitation in Sensitive Domains

While user requirements elicitation could and should be conducted with users and stakeholders across many domains, some are more restricted and challenging than others in terms of confidentiality, anonymity, and privacy. These "sensitive domains" can include those involving children, elderly, or disabled users, healthcare systems, staff/patient environments, and other domains where information is often beyond public access (Gaver et al., 1999). In addition, some organizations restrict how much information employees can share with regard to their tasks, roles, strategies, technology use, and future visions with external parties to protect commercial or competitive standpoints; in others, staff is required to protect classified information through formal national security agreements. Such domains add further complications to ultimately reporting findings to support the wider understanding of user needs in these domains (Crabtree et al., 2003; Lawson et al., 2009).

In relation to privacy and confidentiality issues there are added ethical challenges to the process of user requirements elicitation, where researchers must ensure the following:

- End users and stakeholders are comfortable with the type of information they are sharing and how the information might be used.
- End users are not required to breach their agreements and obligations with their employers or their associated organizations (Kavakli et al., 2005).

Methods for identifying and gathering user needs in the security domain are relatively underdeveloped and little research exists on understanding the work of security personnel and systems, which leads to the lack of case studies or guidance on how methods can be adopted or have been used in different security settings (Hancock and Hart, 2002; Kraemer et al., 2009). As a result it is necessary to revisit the fundamental issues of conducting user requirements elicitation that can then be applied to security research.

CONDUCTING USER REQUIREMENTS ELICITATION

One of the first steps in conducting a user requirements elicitation is to understand that there can be different levels of end users or stakeholders.

While the terms end user and stakeholder are often used interchangeably, stakeholders are not always the end users of an intervention; they are more likely to be people who have a particular investment or interest in the process and outcome (Mitchell et al., 1997). The term end user or primary user is commonly defined as someone who will be using a particular solution (Eason, 1987). In many cases, users and stakeholders will have different needs and often their goals or expectations of the product can be conflicting (Nuseibeh and Easterbrook, 2000). This background information about the users, stakeholders, and the current situation allows designers and system developers to understand the contexts of use for new products (Maguire and Bevan, 2002).

For example, in the procurement of police body armor, the stakeholders could be the buyers who have control of the funding but are not the people who will wear the protective clothing on active duty. The end users in this case would the frontline police officers who would wear the protective clothing on a daily basis, so it is important to derive their views on what is acceptable at an operational level (Hooper, 2000). In addition, the general public could also be regarded as indirect stakeholders as perceptions of police spending, the necessity of protective clothing, and overall performance of the police force may be important issues to be explored.

A key challenge in the initial stages of user requirements elicitation is gaining access and selecting appropriate participants. In the security domain, snowball or chain referral sampling are particularly successful methods of recruiting participants through cumulative referrals made by those who usually share knowledge or interact with others at an operational level or share specific interests for the research (Biernacki and Waldorf, 1981). This sampling method is useful where security agencies and organizations might be reluctant to share confidential and sensitive information with outsiders. This method has been used in drug use and addiction research where information is limited and where the snowball approach can be initiated with a personal contact or through an informant (Biernacki and Waldorf, 1981). However, one of the problems with such a method of sampling is that the eligibility of participants can be difficult to verify as researchers rely on the referral process, and the sample includes only one subset of the relevant user population.

Human Factors Methods and Visualization Tools

There are a number of human factors methods such as questionnaires, surveys, interviews, focus groups, observations and ethnographic reviews, and formal task or link analyses that can be used as the foundations for user requirements elicitation (Crabtree et al., 2003; Preece et al., 2007). These methods provide different opportunities for interaction between the researcher and participant, and hence different types and levels of data (Saikayasit et al., 2012). A range of methods is often selected to complement each other and enhance the detail of issues explored. For example, interviews and focus groups can be used to provide further insights or highlight problems that have been initially identified in questionnaires or surveys. In comparison to direct interaction between the researcher and participant (e.g., interviews) indirect methods (e.g., questionnaires) can reach a larger number of respondents and are cheaper to administer, but are not efficient for probing complicated issues or tacit knowledge (Sinclair, 2005).

Focus groups can also be used, where the interviewer acts as a group organizer, facilitator, and prompter, to encourage discussion across several issues around predefined themes (Sinclair, 2005). However, focus groups can be difficult to arrange depending on the degree of anonymity required by each of the participants. In addition, they need effective management so that all participants have an opportunity to contribute without specific individuals dominating the interaction (Friedrich and van der Poll, 2007). When using interactive methods, it is important that opportunities are provided for participants to express their knowledge spontaneously, rather than only responding to directed questions from the researcher. This is because there is a danger that direct questions are biased by preconceptions that may prevent researchers exploring issues they have not already identified. On this basis, researchers should assume the role of "learners" rather than "hypothesis testers" (McNeese et al., 1995).

Observational and ethnographic methods can be used to allow researchers to gather information about how different individuals within an organization work with each other or interact with artifacts as well as how set tasks are conducted in various scenarios (Stanton et al., 2005). However, observation and ethnographic reviews can be intrusive, especially in sensitive domains where privacy and confidentially is crucial. In addition, the presence of the observers may prompt participants or end users to act/behave differently from the norm, either in the way they work or by providing socially acceptable responses to questions (Crabtree et al., 2003; Stanton et al., 2005). Furthermore, this method provides a large amount of rich data, which can be time-consuming to analyze. However, when used correctly, this method can provide rich qualitative and quantitative real-world data (Sinclair, 2005).

Researchers often focus on the tasks that users perform in order to elicit tacit information or to understand the context of work (Nuseibeh and Easterbrook, 2000). Thus the use of task analysis methods to identify problems and the influence of user interaction on system performance is a major approach within human factors (Kirwan and Ainsworth, 1992). A task analysis is defined as a study of what the user/system operation is required to do, including physical activities and cognitive processes, in order to achieve a specified goal (Kirwan and Ainsworth, 1992). Scenarios are often used to illustrate or describe typical tasks or roles in a particular context (Sutcliffe, 1998). There are generally two types of scenarios: those that represent and capture aspects of real work settings so that researchers and users can communicate their understanding of tasks to aid the development process, and those used to portray how users might envisage using a future system that is being developed (Sutcliffe, 1998). In the latter case, researchers often develop "user personas" that represent how different classes of user might interact with the future system and/or how the system will fit into an intended context of use.

While various methods assist researchers in eliciting user requirements, it is also important to communicate the findings to relevant users and stakeholders. Several techniques exist in user experience and user-centered design to communicate the vision between researchers and users. These generally include scenario-based modeling (e.g., tabular text narratives, user personas, sketches, and informal media) and concept mapping (e.g., scripts, sequences of events, link and task analyses) including actions and objects during the design stage of user requirements (Sutcliffe, 1998). Scenario-based modeling can be used to represent the tasks, roles, systems, and how they interact and influence task goals, as well as identify connections and dependencies between the user, system, and the environment (Sutcliffe, 1998). Concept mapping is a technique that represents the objects, actions, and events (or even emotions and feelings) so that both the researchers and users form a common understanding in order to identify gaps in knowledge (McNeese et al., 1995; Freeman and Jessup, 2004). The visual representations of connections between events and objects in a concept map or link analysis can help viewers recognize conflicting needs, create mutual understandings, and enhance recall and memory of critical events (Freeman and Jessup, 2004). Use cases can also be used to represent typical interactions, including profiles, interests, job descriptions, and skills as part of the user requirements representation (Lanfranchi and Ireson, 2009). Scenarios with personas can be used to describe how users might behave in specific situations in order to provide a richer understanding of the context of use.

SECURITY CASE STUDY

A series of user requirements elicitation exercises were conducted at a UK entertainment venue (referred to as "Venue X"). Venue X was recruited using a snowball sampling method, developed from an introduction by a local Counter-Terrorism Security Advisor (CTSA). To ensure anonymity and confidentiality, the full nature of the work, staffing, location, and other sensitive information are not fully disclosed. Venue X hosts conferences, teaching classes, and activities for a range of age groups. The venue is open to the general public throughout the day and late into the evenings when large-scale events take place. On these occasions, Venue X employs external security staff from a national agency in addition to their in-house employees.

In developing a strategy for the Shades of Grey project, the user requirements process began with a set of high-level issues designed to form an understanding of the overall work process (e.g., job design, roles, and requirements). Scenario-based interview questions were developed to compare responses between procedural training and actual work experiences. In addition, by asking participants to refer back to past events as well as hypothetical future events (such as a direct terrorist attack) it was possible to explore wider systemic issues and generate more detailed responses.

User Interviews

Informal semistructured interviews were conducted with venue managers to explore their day-to-day tasks and procedures in dealing with different aspects of security. This process enabled the managers to discuss their work while the researchers assumed the role of "learners" (McNeese et al., 1995). This approach also enabled the researchers to gather background information regarding the organizational culture, the work schedules, and the way different teams within the organization interact and communicate with one another. These data were used to create a profile of the organizational characteristics (Table 11.1).

TABLE 11.1 Organizational Profile for Venue X

Job	Team	Role	Acceptance of New Training
Customer facing	Bookings	Interact with customers on a daily basis Sometimes show customer venue facilities Priorities include: -Sales -Customer service	Medium
	Security staff	Internal staff and external staff employed by a national agency Only work during an event In direct contact with customers, checking for valid tickets, supervising ingress and egress Priorities include: -Security of customers, venue, performers and other staff -Ensure emergency routes and exits are kept clear -Customer service	Low–medium
	Catering staff	Employed by an external company Interact with customers directly but are not trained by venue Priorities include: -Sales of food and drinks -Catering for special guests in "VIP" hospitality areas	Low
	House keeping	An in-house team working for the venue Trained in hostile reconnaissance and other health and safety issues Staff are around during the day and on event nights, thus can have direct interaction with customers Priorities include: -Cleanliness of the venue	Medium–high

TABLE 11.1 Organizational Profile for Venue X—cont'd

Job	Team	Role	Acceptance of New Training
Control room	Control Room A	Works together with the decision marker during an event In charge of all house utility, safety, CCTV cameras, and access into the venue Also looks after all the keys used in the building and access rights to different areas of the venue Works closely with Control Room B on a large-scale event Priorities include: -Coordinate internal/ external activities of the venue -Safety and security of the venue and patrons	Medium–high
	Control Room B	Employed by the external agency and in charge of all communication for the security team in large-scale events Located above main area to gain a clear view of the event (e.g., audience and stage) Coordinates all house-keeping and security activities via radio communication Priorities include: -Coordinate all safety personnel -Safety and security of venue staff, patrons, performers, and the venue	Low
Management	Duty managers	Oversees all activities at the venue, including training, customer facing, sales, safety, and security as well as event management Typically two managers are on duty, where one assumes the role of the decision marker and the other as "front of house" in charge of customer service Both work closely with all teams to ensure the smooth running of an event and the highest level of customer service Priorities include: -Safety of the venue, patrons, staff, and performers -Generate revenue from ticket sales and good relationships with performers -Training and education for in-house staff to ensure customer service and a safety culture	High
	Head of security	Mainly responsible for supervising security staff and liaising with duty managers at venue	Low

This representation provides an overview of the characteristics of the control rooms and security teams; however, the management team coordinates all activities within the venue. The management operate a "hands-on" philosophy with the day-to-day running of the venue and are present at management and supervisor meetings leading up to a large-scale event where two managers are always on duty, although one will be assigned the task of "decision maker" and tends to remain in one of the control rooms. The decision maker has ultimate control in case of an emergency (e.g., whether to evacuate the premises, or call in emergency responders). The other manager on duty during an event will patrol the venue performing several safety checks (e.g., fire exits, car parking, customer queues) and dealing with localized incidents when requested by staff.

All teams work together and collaborate during an event to achieve the same shared goals. Each smaller team has their own goals that they need to accomplish, which contribute to the overall performance of Venue X as a total system.

Personas

In order to further explore the different groups of staff the main teams involved in security of Venue X (e.g., managers, control room, security staff) are illustrated using personas. Personas typically provide a profile of a specific role in an organization based on information from a number of sources. What is then communicated is a composite and synthesis of key features within a single profile that can then be used as a single point of reference. In some cases personas are given names and background information such as age, education, recent training courses attended, and even generic images/photos to make them more realistic or representative of the general worker in that role. In other cases, personas are used anonymously in order to communicate fundamental information about specific roles. Examples are provided in Table 11.2.

From these personas key attributes and goals are illustrated for each role, which can then be communicated back to staff in those roles or to other staff at Venue X to help illustrate different job demands across the organization. This profile provides a useful outline of the key roles within the overall security system and was used to assist researchers' understanding of the relationships between various end users and other stakeholders. It became apparent that Venue X relied on several teams to work closely with each other to ensure safety and customer satisfaction. These teams included various customer-facing staff as well as those working in two control rooms and a security management team responsible for the overall conduct of the venue. These early activities provided a basis for exploring user needs and future training needs in a field observation study that was conducted at Venue X from which a link analysis and tabulated communications network was developed.

Field Observation

An eight hour field observation took place at Venue X during an evening event where approximately 10,000 people attended a large-scale event. The purpose of this research was to allow researchers to gain a deeper understanding of the security activities in a live setting and explore emergent themes gathered from the informal interviews in a real context of work.

Prior to the event, several management and supervisor meetings took place so that the different teams involved in the event could prepare a joint strategy for the event. An event manager had been working for several months on arrangements leading up to the event and was thus able to pass on information and inform the supervisors of different teams about decisions that had been made prior to the day of the event.

The first pre-event meeting took place between managers and supervisors, who later held meetings within their respective teams to disseminate information. Smaller teams included security staff (including door supervisors, frontline personnel supervising the venue, CCTV operators, and supervisors patrolling the venue), housekeeping, and catering.

Based on past experience of similar events, and coupled with event intelligence from other venues, Venue X was able to predict the crowd demographics and expected behaviors at the event. This enabled the security of the event to implement coping strategies and safety precautions. This was also an important aspect of supporting the general public's experience of the event so that their user needs were also considered.

Informal interviews with staff on duty also took place during the observation on an *ad hoc* basis. Information gathered from these interviews, the field observation, and the initial information interviews were used to contribute to the framework of emerging themes (Saikayasit et al., 2012).

Link Analysis

From the communications between teams during the event, a link analysis was developed as

TABLE 11.2 Personas for Managers, Control Room, and Security Staff

Job	Attributes	Goals
Manager	Has been working in the service industry for over 15 years Has been working at the Venue X for 10 years Works full time Works closely with all departments within the organization and often knows people by names Works with other managers to plan for large-scale events (including security planning, strategies, execution of tasks, and external companies liaisons) Designs and conducts training for in-house staff including health and safety training, security, and hostile reconnaissance	Successfully host an event Successfully train staff on security procedures Generate revenue via sales
Control Room A	Works part-time (evening only) for the external security agency May work elsewhere during the day Has been working specifically at Venue X for six years SIA trained to supervise CCTV Experienced operator and knows Venue X well so is usually requested for large-scale events with potentially difficult crowds Tracks the rate of entry into the venue at the start of an event and reports to the decision marker Coordinates all radio communication between security staff Highly skilled and is very capable of multitasking Works under pressure in a loud and dark working environment	Effective team working of the security personnel Safety of the security team members, audience, performers, and venue
Control Room B	Works full time and employed by Venue X Has been at Venue X for 10 years Responsible for monitoring the CCTV around the venue during the day and often late into the night In charge of keys to all rooms on site In charge of secure car park and all alarmed entrances/exits around the venue around the venue (personnel use only) Monitors staff coming on and leaving shifts Works closely with Control Room A on large-scale events Been to in-house training and away day training on hostile reconnaissance organized by managers	Smooth running of the venue during the day Security of the venue
Security staff	Employed by the external agency as a part-time security night worker CRB checked for criminal records and has basic training Only works at Venue X on an event night Works within a team of other security personnel under a supervisor Prevent emergency routes and exits being blocked, keeps the stairways free SIA trained in crowd/conflict management Ensures that everything runs smoothly and helps the customers who suffer from heat exhaustion while preventing antisocial or drunken behavior Reports directly to supervisor and sometimes head of security	Security of customers, venue, performers, and other staff Ensures that only patrons with valid tickets can enter the venue Keeps emergency routes and exits free at all times

a communication tool to inform Venue X about the security network for a large-scale event (Figure 11.1). The analysis presents formal links between the different teams governing security at the event (as recorded by the researchers and with data taken from a formal incident log after the event). The line thickness represents the number of communications that occurred during the event. For added detail reference numbers are then tabulated to illustrate who communicated with whom, the information content, and the frequency of communications (Table 11.3). Interviews with management (before and after the large-scale event) suggested that they are keen on encouraging frontline staff to understand the thinking behind different sets of protocols and safety measures, so that they can assess situations for themselves. The managers empower the staff to take responsibility for security and report anything they feel is "out of the ordinary" directly to the control rooms. The control room operators are then able to decide whether to contact the duty managers for further advice on a situation. Regardless of whether this action is taken, all incidents communicated to the controls rooms are recorded in a formal incident log that provides a record of all raised issues for future reference or postincident analysis.

IDENTIFYING TRAINING NEEDS

The data collected throughout this case study provided evidence toward the emerging themes in terms of team interactions and security. This information was then presented in different formats back to Venue X to develop their understanding of security at the venue and elicit future training needs to support their current approaches in empowering the staff at all levels.

A number of themes were identified and are reported in the following sections. The comments are taken from researcher notes collected during informal interviews and therefore represent the views of the staff at the venue directly.

Empowering Staff

The work culture developed by the management team at Venue X is very proactive in encouraging staff to challenge decisions made by others

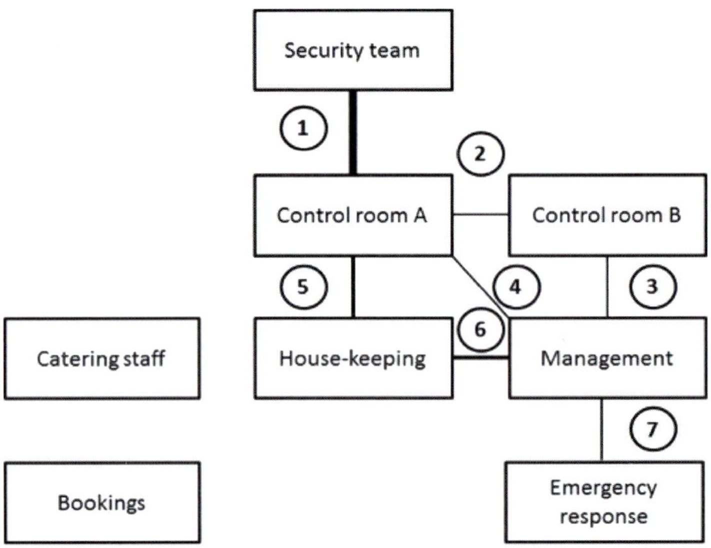

FIGURE 11.1 Security link analysis for a large-scale event at Venue X.

TABLE 11.3 Tabulated Patterns of Communications for Venue X

Link	Communications From/To		Information Content	Frequency
1	Security team	Control Room A	Security teams based in the main area were often the first to identify an incident and therefore tended to request backup from other teams via Control Room A	Constantly throughout the event
	Control Room A	Security team	Control Room A would coordinate and confirm backup or provide information for Control Room B to interrogate using CCTV	Constantly throughout the event
2	Control Room A	Control Room B	Any serious incidents reported to Control Room A were relayed to Control Room B	Infrequently
3	Control Room B	Duty managers	Requests for ground teams to be dispatched to control an incident that was identified by Control Room B	Infrequently
4	Control Room A	Duty managers	Serious incidents were reported to the duty managers	Infrequently
	Duty managers	Control Room A	Requests dispatched when any incident were identified directly by managers	Infrequently
5	Control Room A	Housekeeping	Requests for housekeeping (e.g., spillages, litter) when identified by the security team and communicated to Control Room A	Often throughout the event
	Housekeeping	Control Room A	Typically job confirmations upon completion	Often throughout the event
6	Duty managers	Housekeeping	Requests for housekeeping (e.g., spillages, litter) when identified by the duty managers	Not often
	Housekeeping	Duty managers	Typically job confirmations upon completion	Infrequently
7	Duty managers	Emergency response unit	Request for assistance	Infrequently

(including management) to ensure that all the decisions suit everyone and everyone is able to contribute. In this way training focuses on giving staff the responsibility to report things they see or feel are suspicious as well as developing the skills of specific teams to support security activities in the venue. For example, the in-house catering and housekeeping teams are regarded as a key resource in observing any suspicious activities. They are the staff members who have the potential to "see more" as they get access to different areas and can check toilets and bins around the premises. They can "spot things which are out of the ordinary" and are trained to call the control room. Sometimes the control room might receive too many calls regarding a random toolbox lying around (if there is construction work going on), but they would rather have this and are not going to stop their staff from reporting small incidents.

A major concern with this and similar venues is that pedophiles may target the area if there are young children attending an event. The staff are "very good at spotting things/people which are out of place." The staff take child protection training that allows them to relate to and protect children who may be lost or upset. "Anyone who is at the venue to watch children but didn't arrive with them will be noticed straight away" and approached by staff. In this way the person will realize they have been noticed and will feel exposed. The staff is not afraid to challenge people by asking what they are doing and why they are there. They will also check with the children as well to see if someone who looks suspicious has interacted with them. This training also has the benefit or "raising awareness around the venue for other activities (such as hostile reconnaissance) without explicitly placing this on the agenda." In a similar way, this kind of training has helped with staff noticing suspicious crowds at other events (e.g., religious events) as they did not appear to "fit in" with the target audience.

Access to Training Resources

It is "very difficult to send a whole team of staff on dedicated training courses, although staff enjoy attending them," and it can also be expensive to take people out of their shifts or pay for extra training time. It would be "very beneficial to have training material or techniques that staff can use at their own pace," "something short and easy to understand in about 15 minutes during a routine briefing." "Simple bullet points and headings of important facts/procedures to prompt people on different subjects/situations would be useful." "A toolkit that covers generic procedures that venues can personalize and adapt in the way that works best for them and their staff would be very beneficial." Online materials that staff can "access at different times" could serve as a reminder of different strategies or procedures "when staff have time to review them rather than a specific one-day training course that they may not be able to attend." Having materials within the venue would allow for new recruits to begin their training whenever they start rather than perhaps missing a key training session that will not be repeated for a while. At a counterterrorism training day conducted 2 years ago, only 60% of the attendees are still working at Venue X, and the new 40% of the staff have not taken that specific course (although they have attended others). This means that the training profile of the staff is variable when management should be looking for more consistent training methods.

In-house Training

The management team conducts a lot of training with staff. Some of the training is preplanned (i.e., staff are informed that there will be a drill "to assess how staff prepare and respond to an incident") and others are unexpected (i.e., announced through the control room as it would in a real incident) to simulate realistic contexts and "follow training under pressure." Conducting the training in-house has certain advantages as it can be "fitted around normal routines" and also embeds learning within the real context (rather than at a detached location). Staff feedback has been very positive. They enjoy live drills as they are "a change from normal work" and they value the opportunity to learn new procedures or practice existing protocols. Venue X recently carried out a mass decontamination training exercise with staff from a specific department. The department was chosen as if the staff there felt they would not be a particular target for such an attack. The training exercise highlighted the vulnerability of the department as well as educated the staff about what to do, where to go, and what would happen to them (including how they will be contacted by the management team). This was a particularly useful training event as "the end product is to know what will happen to real people" and therefore underlines the human factors approach of user-centered approaches in the real context of work.

Generalizing from Specifics

As Venue X hosts many different events it is difficult to model a single scenario for training purposes. They develop specific scenarios for different incidents, but these often promote different strategies for dealing with each one that can then be difficult to generalize. In terms of streamlining training, they would like to have "a set of generic strategies and teach staff members to learn to adapt their approach and response to different situations." "There will always be an incident that you haven't specifically trained for." For example, at a recent religious event a man entered the premises carrying a knife; however, there was no guidance for what to do once they had contacted the police (although Control room B used CCTV to track the suspect). There are many different threats and "being able to assess threats quickly would really help" so that staff know what to do (in terms of their own safety as well as others). "It would be nice to have something to help them define the threat" and relate this back to the generalized strategies for dealing with it.

Integrated and Tailored Training

Venue X outsources several of their security activities to a national agency who typically oversees crowd control at large-scale events. The agency is responsible for the training of its own staff, but they are also invited to attend training days at the venue. This is to help develop an integrated approach to security so that during a live event, there is a common understanding for procedures, and so forth. Venue X is also aware that training needs to be done at different levels. For example, "senior managerial staff will receive a certain type of training that is different to those working in the control room or on the frontline." However, there is no specific training for how management might pass on their knowledge to staff at a lower level. There is "value in training, but there isn't much point if they cannot pass their knowledge to other people in their team because they need everyone to deliver."

Raised Awareness and Skill Fade

After each training session, there is a tendency for an "increased number of reports for similar incidents" illustrating that the training might act as a stimulus to raise awareness (or perception) of such activities. It could also be that staff are more motivated or feel empowered to report such incidents when they previously might not have known the procedures or felt it was their responsibility to do so. However, this fades over time and "typically the level of awareness drops after six months," so refresher training is a necessary and valuable activity. For example, First Aid training requires retraining every three years. From an operational perspective, there needs to be a way of reminding staff of what they should be on the lookout for and to stay sharp throughout their work, not just after training. "A key question is how do we keep them engaged?"

Certificates

An interesting factor that emerged from the training needs exercise was the value of a simple certificate to the end users. Certificates offered after training "can be used as a reminder of what people have done"; it is also a record of formal training undertaken. Venue X retains a photocopy of all certificates while allowing their staff to take pride in their achievement and keep their original copies (many of whom either take them home to show family members or stick them up in their work lockers). Certificates could also be designed as training aids with key points of information on them to remind staff of useful information. "Staff really like to have a certificate after they have been to a training course, for their pride and also for a reminder that they have actually been trained in that area."

Knowledge Elicitation

Another key area that was identified during the training needs exercise was the importance of tacit knowledge. It is "difficult to try and understand the knowledge of an experienced control room operator or manager so that they can package it in a certain way and use the information to train new staff." For example, "if we bring one of the best control room operators up for an interview, all their answers will be perfectly in accordance to the strategies/procedure, but in a real-life situation, these reported answers might not be the same as their actual actions." "The operator is very experienced and is able to adapt their reaction to different situations, but this decision process is very hard to capture." Furthermore, the managers highlighted that they "are not trying to teach common sense, but we are trying to teach someone the thought process that they have to go through to make a decision, and that's very hard." It is also hard to interview people and ask them why they reacted in a certain way after an incident; "it might have been the right reaction, but they would say they don't know why." Knowing what prompts those to react correctly can help train the less experienced members of the team. In this way the responses from the training needs exercise returned to the fundamental principles of user requirements elicitation and the problem of capturing tacit knowledge.

DISCUSSION

Using a variety of human factors methods provided the researchers with a clearer understanding of how security, as a process, operates within the current protocols that have been designed for different situations. Without the use of interviews and field observation, indirect methods such as questionnaires would have been insufficient to gather the data presented. In addition, without using the tools for communicating the findings the process would have been incomplete and end users and other stakeholders would have missed an opportunity to learn about their security system as well as contribute further insights into their roles.

The implemented methods included interviews, which provided an opportunity to build a rapport and trust for future activities in this sensitive domain. Informal interviews were used to ensure that data were collected in an appropriate format and gave the researcher exposure to the context of interest. In practice interviews could only reach a limited number of participants; however, these were coupled with field observations to enable a rich understanding of the tasks being conducted by different teams.

Observation data gathered in the field study allowed insights into real work practices. This data collection method also gave the researchers an opportunity to talk to staff *in situ* and discuss issues that arose during their shift, which may have been missed during interviews. However, as the observation session lasted eight hours, it was necessary that the researchers defined the focus of their observation prior to the start to ensure key events were captured and irrelevant data were avoided. Furthermore, gaining access into an organization to conduct a field observation is not always possible as they sometimes have more conservative outlooks on allowing external bodies to observe their work practices. This is often to protect the privacy of their staff or avoid disruption to highly demanding work where public safety is paramount. Therefore researchers should work closely with the target organization when designing an observation session to manage expectations and establish a mutual understanding of the process (e.g., duration, access issues) to ensure that the activity is conducted without breaching ethical issues or causing delays and disruptions in the workplace. A further requirement is for the target organization to have a clear view on the purpose of the research, as well as knowledge of the intended usage of data collected during the observation session.

User requirements elicitation with users working in sensitive domains also presents issues of personal anonymity and data confidentiality (Kavali et al., 2005). In order to safeguard these, anonymity and pseudonymity can be used to disguise individuals, roles, and relationships between roles (Pfitzmann and Hansen, 2005). In this way, identifying features of participants should not be associated with the data or approaches should be used that specifically use fictitious personas to illustrate and integrate observations across a number of participants. If done correctly, these personas can then be used as an effective communication tool without compromising the trust that has been built during the elicitation process. From the training needs exercise the following issues were identified:

- Empowering staff
- Access to training resources
- In-house training
- Generalizing from specifics
- Integrated and tailored training
- Raised awareness and skill fade
- Certificates and knowledge elicitation

These were developed from the initial interviews and feedback after the observation session and highlighted a range of operational issues as well as an awareness of key human factors issues regarding the understanding and communication of expert knowledge. Without the range of user requirements elicitation activities it would not have been possible to gain such an insight into the training at Venue X.

This combination of user requirements and training needs analysis work has led to the development of the next stage of development and a key focus for the Shades of Grey consortium: how future interventions might be incorporated into established practices. At a fundamental level, interventions should not be perceived as extra work for the staff. For example, within the control rooms at Venue X, the staff can have an extremely high workload during a large-scale event. Introducing additional steps in procedures or extra work might delay the coordination and communication of activities in an emergency. It is therefore vital and integral to the whole organization, particularly during a large-scale event, that such potential barriers to effective organization do not compromise the overall system integrity. In order for a new system of interventions to be effectively used and successfully implemented, the various needs of different end user groups as well as those of other stakeholders (i.e., management and policy makers) need to be understood and addressed in the overall design and implementation of solutions. Failure to address the needs of the end users may lead to the lack of acceptability and lack of use of the system and compromise its overall effectiveness.

CONCLUSION

A user-centered approach is essential to understanding security from a human factors perspective, as evidenced by the case study at Venue X. This case study has also shown the importance of understanding the context of work and related factors contributing to the overall performance of a security system through user requirements elicitation activities. These have been illustrated in various ways (e.g., organizational profiling, personas, link analysis, and tabulated patterns of communications) that each provide valuable tools for communicating issues back to users and stakeholders. From this, key training issues were identified across a range of end users and other stakeholders, which will provide a basis for developing future interventions designed as part of the Shades of Grey project.

ACKNOWLEDGMENTS

The Engineering and Physical Sciences Research Council funded the research in this chapter as part of the Shades of Grey project (EP/H02302X/1). The authors would also like to thank all of the end users and stakeholders from Venue X who took part in the user requirements elicitation.

CHAPTER 12

Exploring the Crisis Management/Knowledge Management Nexus*

Mats Koraeus, Eric Stern

INTRODUCTION

During August 2012, the Gulf Coast of the United States was threatened by Hurricane Isaac. Isaac's impending arrival necessitated intensive consultations between political–administrative leaders and a wide variety of technical experts—engineers, meteorologists, hydrologists, emergency managers, and so forth—providing input into numerous critical decisions about how to prepare for, respond to, and recover from the storm. Furthermore, Isaac did not come alone. Isaac was shadowed by the specter of Hurricane Katrina, which devastated the same region seven years to the day before. A question on everyone's mind was whether the lessons of Katrina had been adequately learned and remembered in ways that would protect the people and the livelihoods of the region.

Crises are becoming a ubiquitous phenomenon and are, like the example of Hurricane Isaac above, typically surrounded by complex social and technical factors beyond the competence of generalist leaders. By their very nature, crises are associated with considerable uncertainty. Knowing this, crisis managers frequently call upon experts to provide relevant information on specific subject matter. During the past decade and half, a theoretical and methodological field of study has emerged addressing this issue: how to find missing knowledge, transfer it to where it is most needed, and institutionalize it for future use. This subject is called "knowledge management" (KM) and is rooted in older theories on organizational knowledge and organizational learning. In fact, these very same organizational learning theories form the foundation for the post-crisis learning perspectives, which are increasingly prominent in the field of crisis management (CM). Hence, one may question why these two management subdisciplines have not been more systematically combined and integrated. In CM, learning is often seen as a process that takes place after a crisis has been resolved, in preparation for the next crisis. Yet, KM considers knowledge creation and learning to be a constant process. Ideally, combining these two perspectives could stimulate some kind of "instant learning" during an actual crisis, so that relevant lessons are learned and implemented for the current crisis as well as for future crises.

A potential problem with combining CM and KM is the difference in typical time frames

*The argument in this chapter is developed at greater length in Koraeus (2008) (http://www.fhs.se/Documents/Externwebben/forskning/centrumbildningar/Crismart/Publikationer/Publikationsserier/Crismart36-WhoKnows.pdf).

associated with the two management subfields. By definition, crises involve a strong element of urgency and thus require immediate action. KM, however, was born out of attempts to improve consumer product innovation cycles, which can run several months or even years. The possibility of tapping into the KM techniques for finding and accessing new or unfamiliar knowledge (e.g., calling in experts) is very alluring, but the question is if such techniques can fit into the tight time frame associated with crises.

The purpose of this chapter is to explore the conceptual nexus between CM and KM with an eye toward exploring the potential to enrich the study and practice of CM with "tools" from the KM "tool box." The first section will provide a brief and selective overview of some central notions from the field of CM. The second section will explore a parallel selection of the KM literature.

CRISIS MANAGEMENT

As a field of study, CM is a multidisciplinary mix of theories covering a vast number of issues associated with the field of CM. Among the contributions to this amalgamation, we find political psychology, organizational theory, communications theory, public policy, management theory, and many other interconnected fields of study. Together, they try to explain how crises emerge; are identified and framed; how decisions are made and communicated under pressure; how individuals and organizations learn and, hopefully, improve their CM skills from one crisis to the next; and how crises can be avoided or mitigated.

CM does not only deal with different subsets of problems, but also considers the myriad actors involved in each step and the vast number of arenas where crises may occur. It is easy to fall into the trap of only looking at policy makers when studying the decision-making process in public administration, but these policies and decisions do not jump out of thin air; they are influenced by the media, by public sentiment, by future policy implementers, and by public and private interest groups. There are numerous informal and even "invisible" decisions that are affected. Anyone's opinion has the potential of affecting the decision-making process and ultimately the outcome: anyone from the a decision-maker's spouse, to a private and largely unofficial advisory committee, to an *ad hoc* group of experts, or even a journalist's controversial article in a major newspaper. These groups and individuals all act as framing advocates; they work to promote their particular interpretation of the situation. Unlike the classical "threat entrepreneur," these advocates do not necessarily promote any given decision or course of action to solve the problem at hand, but rather offer an alternative view for the decision makers. CM has received a fair amount of attention not only in the foreign policy/international relations literature, but also in the sociological (e.g., disaster studies), public administration/public management, and business/management literature. All of these fields of research have contributed to the conceptualization and evolving practice of CM.

The term "crisis" can easily become overused and pointless lacking a firm analytical framework. While there are many ways to define crisis, a significant strand of the literature has defined crisis as constituted by the challenging combination of perceptions of threat to core values, high degree of uncertainty, and necessity for urgent decision making. Let us take a closer look at each of these in turn.

Threat to Core Values

A crisis, by definition, is an event that invokes a threat to core values. The exact nature of these values at stake can vary wildly depending on the actors involved. A national government could consider national security, financial stability, public confidence, and the lives and well-being of citizens as threatened values. A business establishment could consider delays in production, disruptions in cash flow, supply or distribution chains, or risk to brand reputation to be threatened values. Threatened values can encompass anything from physical resources to symbolic commitments or assets.

Due to the interconnectedness of actors in the modern world, scope and extent of values in play in a crisis is not always clear to the actors themselves at the outset (and sometimes not until the dust has settled afterwards), making the identification and prevention of crises even harder. Core values distinguish crises from mere inconveniences. An inconvenience can be endured, worked around, or bargained away, even if one would prefer not to. Crises are ignored at the peril of crisis managers and their constituents. Crisis situations commonly involve competing values and may force decision makers into making "tragic choices" and sacrificing one value in favor of another.

These kinds of value conflicts also run the risk of causing secondary crises. In the aftermath of an otherwise well-handled situation, the decisions regarding which values were upheld and which were sacrificed may come into question. If these choices were not made explicitly clear, and sometimes, even if they were, an accountability crisis can evolve and it might turn out to be even more threatening and difficult to manage than the original crisis.

Uncertainty

The second defining characteristic of a crisis is a high degree of felt uncertainty. This should not be confused with a simple lack of information, which in our modern information society is not very often the case since we are often hit by too much information. Instead, the uncertainty comes in the form of not being able to make sense of the situation. What has caused it? What is going on? How to solve the problem? What will the consequences be if a certain course of action is taken?

This confusion intensifies the problem of value conflicts. At times, the uncertainty is such that the decision makers cannot even see that a problem exists. The scope of what is at stake is lost in the chaos of the moment and mistakes easily occur. For example, the decision is made to save one core value at the cost of another, only to find out that the value sacrificed was the one that was actually more important. It can even come to the point where the decision makers fail to even see that there is a crisis, since only one core value seems to be threatened and they feel it can easily be saved. Only in hindsight does it show that it was saved at the cost of something far more precious and that there actually was a value conflict. In short, the real crisis might occur long after the other "dramatic" events have occurred, as an outcome of the accountability process that will inevitably follow.

In general, decision makers are often painfully aware of this very problem. They may have the available data and the best of plans in mind, but they may still be uncertain about what will happen if those plans are activated. There just might be some tiny, but crucial, detail everyone is missing, which might cast the entire problem in a completely different light and turn that "best plan" into a complete disaster.

What ultimately feeds the uncertainty of a crisis is its dynamic nature. Even if they are predicted, correctly identified, and, at least initially, addressed, they can evolve at a moment's notice into something different and completely unforeseen. At every point along the way, a crisis needs to be re-analyzed and re-evaluated in order to detect any subtle changes. This process increases the demand for information as well as the amount of information that needs to be filtered for relevance. Consequently, the very act of trying to reduce the amount of uncertainty can ironically result in even more confusion.

Urgency

The last distinguishing characteristic of a crisis is the sense of urgency, real or imagined. Decisions and/or actions are being demanded and there is no luxury of waiting for all the facts. This time pressure might stem from the concern that the crisis will cascade and start affecting new parts of the system, thereby causing secondary, spin-off crises. A sense of urgency may be felt when attempts are being made to save the remaining resources, which are rapidly being consumed by the crisis. The sense of urgency can even be instigated by the desire or need to make a symbolic gesture to show that you are still in control and capable of action.

The urgency felt can be completely artificial; for instance, when the media is pushing for immediate answers when there are no answers or for immediate action when there is actually very little to do. It can also be systemic, caused by lags and delays in the decision-making process. If the feedback loop is slow, the decision makers can be forced into a "hurry up and wait" pattern, where they feel the need to make quick decisions because they know it takes a while for their actions to take effect. The sense of urgency can of course also be entirely functional. Some, if not all, crises present windows of opportunities where the right decision/action at the right time can stop the crisis dead in its tracks, such as the quick containment of an epidemic or the apprehension of a hijacker when he suddenly dozes off.

CM Tasks

The main contribution of Boin et al. (2005) is the unpacking of CM in the form of a recurring set of critical leadership tasks: sense making, decision making, meaning making, termination, and learning and reform. Sense making is the task of detecting and interpreting the policy implications of a developing crisis, either before it strikes (which can enable preventive intervention), or as it unfolds and mutates. At the center of the decision-making task is the challenge of determining "what to do now," but it also involves coordination, division of responsibility and labor, and establishing processes to provide both differentiated and integrated views of crisis problems and courses of action. Note that effective decision-making processes vary greatly according to context and time frame. The purpose of the meaning-making task is to provide communicative leadership, boost morale, and secure/maintain public and elite support. The termination task deals with managing the transition back to "normalcy," handling the demands for accountability and attempting to ensure the situation does not spin off into a "crisis after the crisis." Both of these latter tasks revolve around deliberate, focused, and effective crisis issue framing and communicative action.

According to Boin et al. (2005), the final task in CM is learning and reforming—a theme that, as we will see, runs parallel with a key thrust of the KM literature. According to Boin et al. (2005), the core problem with learning after a crisis is that it is fundamentally opposed to actually managing a crisis. The fundamental task in managing a crisis is often to resume things to the way they were, whereas reform introduces change. By successfully dealing with the meaning making and termination tasks, the decision maker has convinced his or her audience that drastic measures are needed to maintain the status quo, which makes it rather difficult to then turn around and argue that the status quo needs to be changed. As a result, reformists might not actually want to end the crisis, just keep it sufficiently under control without causing too much damage, so that the implemented measures will, over time, be accepted as the norm. Additionally, there is the problem with sweeping post-crisis reforms. As already mentioned, crises are ripe with equifinality. Hence, it may be very difficult to determine whether the outcome was caused by poor management or by the fact that the crisis was truly unmanageable. Without knowing this, the genuine need for reform cannot be determined, and any measures put into place might actually reduce the crisis-handling capabilities. The trick, then, is to manage the crisis while it happens and be able to argue for change once it is over if change is really needed at all. Boin et al. (2005) claim that the best way to combine these two is to have a system with continual evaluation (before, during, and after a crisis), long-term learning, and gradual change, not dramatically sweeping reforms.

KNOWLEDGE MANAGEMENT

KM sits quite apart from these kinds of political considerations and instead attacks the core problem of how to identify what change is needed, if any, and where to find the inspiration for new solutions. As a concept, it was by most accounts invented by Ikujiro Nonaka and Hirotaka Takeuchi in 1995 as a refinement and extension of the existing theories of organizational learning. At that point in time, most of the literature on the subject was concerned with the existence, nature, and implications of "the learning organization."

Nonaka and Takeuchi (1995) took an interest in the mechanisms that made learning at all possible. Like most literature in the genre, Nonaka and Takeuchi took as their point of departure Polanyi's (1966) landmark work on the nature of tacit knowledge and further described the role it plays in the evolution of an organization's internal practices via learning or other mechanisms.

Nonaka and Takeuchi (1995) set apart from early theoretical works, and helped establish KM as a field of its own, by speaking of organizational learning strictly as a byproduct of the much larger process of knowledge transfer. Apart from learning, knowledge transfer also deals with the active search for, and the creation of, completely new and previously unknown knowledge. This includes the different ways in which various types of knowledge are transferred between individuals and between organizations, and how one form of knowledge is converted into another form. KM deals with the ways in which the various exchange processes can be controlled, actively and purposefully, to yield better results.

What is Knowledge?

At the heart of KM lies, of course, knowledge, and the complexity of this word alone makes it abundantly clear that it is something that could benefit from being managed. The first hurdle in trying to understand "knowledge" is to differentiate it from the legion of closely related, but slightly dissimilar, concepts. The KM literature is ripe with such distinctions, but each author offers a slightly different explanation of the various concepts depending on where their different foci lie. There is some common ground, however, at least when it comes to listing and describing the nature of related topics.

Generally, a distinction is made between three main components: data, information, and knowledge. A common trait in defining these concepts is to use a system of progressive embedding, where each step adds a layer of interpretation and contextualization, but where each such addition also increases the level of cognitive complexity. Most definitions of knowledge also strongly emphasize the subjective nature of knowledge, which is an attribute that is lacking in the other concepts. Alvesson (1993), Alvesson and Kärreman (2001), and Blackler (1995) specifically noted how the concept of "knowledge" is inherently contested and ambiguous, not just between different knowledge holders, but often even within a single subject. Thus, they argue that it is a fiendishly difficult concept to work with.

A final commonality between the different definitions is that knowledge tends to be described in terms of a cognitive process as opposed to data and information, which are simply passive meaning-bearing objects of varying complexity. At first glance, the other two may seem just as procedural in their nature. They are defined in terms of "measurements" and "observations" and they are "organized" and "processed," but these should be interpreted as products of a process rather than a process in and of itself. On the other hand, knowledge, as originally defined by Polanyi (1966), is the act of knowing; the process of believing, interpreting, and understanding; and the act of making use of information and data. In short, knowledge is what the organizational learning school of thought calls "actionable"—it is the art of making the right decision and implementing the right action in order to move forward.

Tacit and Explicit Knowledge

Perhaps the biggest contribution made by Polanyi was not this definition, but rather his division of knowledge into two separate categories: tacit and explicit knowledge, which roughly equates to "knowing how" and "knowing that." The basic idea behind this distinction is older than Polanyi's work and can be found in a number of different disciplines under an array of different names. The names have varied according to the discipline and have been made to reflect the focus of field of study. In computer science, for instance, the key distinction is assumed to be between data that is already given—"declarative" knowledge—and data that comes as a result of a process—"imperative" knowledge. The field has even adopted a naming convention that reflects this distinction. The realization here is that the act of knowing can range from knowing simple facts by heart, to knowing how to perform very complex tasks in order to achieve a specific final result. Polanyi focused on the problem of knowledge transfer, which has also been adopted by the organizational learning field. Here, too, the focus explains why the terms "tacit" and "explicit" were used to describe the two kinds of knowledge. The problem with transferring knowledge is to properly articulate exactly what it is you know, so the two names hint at how one class of knowledge is more difficult to verbalize and explain than the other.

Tacit, or silent knowledge, generally comes in the form of some kind of "know-how." It can be anything from a craft or art (such as creative writing or painting) to a mechanical skill (such as hearing or feeling what is wrong with some piece of machinery). Another example is filling out a request form in a slightly quirky way (by skipping relevant parts or using shorthand for certain things) simply because you know that the people who will read it prefer it that way and will get back to you faster if you do. It is often described as the kind of knowledge you only get with experience—a kind of "gut feeling" for what is right and what will work. For this reason, tacit knowledge is not something that can easily be transferred between people, or, when attempts to transfer it are made, generally manifests itself as horribly complex manuals or point-by-point procedure checklists. This same kind of knowledge can be found within organizations in their standard operational procedures, workflow, hiring and personnel schemes, or the lack of any of the above. Each of these reveal how work tasks are organized and represent one piece of tacit knowledge; that is, one lesson learned about what seems to work best. Just as with individuals, these lessons are not always easy to transplant to other organizations, because the right person with the right skills for the right task is in the wrong place or maybe there is some internal friction that did not exist within the prototype organization. Either way, organizations can exhibit different types of intangible knowledge that set them apart from one another, even if they presumably look alike on the surface.

At the other end of the scale we find explicit knowledge, which, as the name suggests, is more open to being transformed into unambiguous statements of facts and thus easily shared with other people. Explicit knowledge is generally described in terms of "knowing that" something is true—the kind of knowledge that can commonly be acquired just by reading and memorizing it. This can be anything from knowing that 1+1=2 to knowing that the storming of the Bastille took place on July 14, 1789. The distinction between explicit and tacit knowledge, at its simplest, can be described as the difference between knowing the recipe for a cake and actually being able to bake it. In an organizational context, such knowledge can be found in organizational charts, mission statements, and so on. Here, the difference between explicit and tacit organizational knowledge can be expressed as the difference between having an accounting department and actually being able to keep the company books in order.

Nonetheless, the distinction between these two classes of knowledge is not always as clear as it may seem. "Articulability" might be an important factor for determining how easily transferred a given piece of knowledge is, but it is not necessarily an inherent or given part of that knowledge. The problem here is that the "act of knowing," itself, is something completely different from the source of that knowledge, and that two different sources can result in knowing the exact same thing. For instance knowing that $5 \times 3 = 15$ because you have memorized the multiplication table, is a good example of explicit knowledge. However, knowing that $5 \times 3 = 15$ because you understand what multiplication is, and can quickly run the calculation process $5 \times 3 = 5 + 5 + 5 = 15$ in your head, is more akin to tacit knowledge. Regardless of how you know it, you still know the same thing: that $5 \times 3 = 15$. Ultimately, then, before discussing the transfer of some particular piece of knowledge, the first questions to ask include the following: Exactly what is the knowledge that is being transferred? Is it the end result, or one of the numerous potential means to arriving to that result?

Transfer and Conversion of Knowledge

Strictly speaking, knowledge transfer between people can only occur within a given knowledge class; that is, tacit knowledge can only be transferred as tacit knowledge from one person to another and, likewise, explicit knowledge can only be transferred as explicit knowledge. In order to turn explicit knowledge into tacit knowledge, or vice versa, it needs to be converted, which is an internal process rather than an actual transfer between persons. Reading a cake recipe is a transfer of explicit knowledge (from the author of the cookbook to the reader), but the actual process of preparing the recipe in order to make a real cake requires converting the text to the process of baking, which is something the reader has to do on his or her own. In order to be able to write the cookbook, the author had to examine his or her own actions and experiences and convert those into words, which could then be transferred to the reader. Alternatively, the two could have skipped the book and shared the experience by baking the cake together, thereby transferring the tacit knowledge through demonstration and repetition or some form of apprenticeship.

A second and more troublesome limitation is that tacit knowledge can never be completely converted since it is always based on a subjective viewpoint. Therefore, trying to transform tacit knowledge into explicit or vice versa always results in the loss of some knowledge, either because the tacit knowledge cannot be completely verbalized or because the explicit knowledge must be interpreted by the recipient in order to be fully comprehended. Ultimately, the purpose of KM is to manage, control, and improve these transfers and conversions of knowledge in order to minimize losses and enable the accumulation of knowledge as it passes from one person to another, rather than letting the losses escalate with each transfer. The goal is innovation—the creation of new knowledge—without significant losses to what is already known.

The SECI Process

Describing innovation was what motivated Nonaka and Takeuchi (1995) to write their seminal work on KM, and their conclusion was that innovation comes as a result of sharing and comparing knowledge in a continuous cycle of knowledge conversions and transfers. This process was named the SECI process after the names of the different phases in the cycle: socialization, externalization, combination, and internalization. Each of these phases describes one way of creating new knowledge, and the process as a whole deals with the problem of ensuring that this knowledge is relevant to the problem at hand. There is no predetermined beginning or end for this process; everything depends on where the realization of the problem starts and what kind of knowledge is needed. While illustrated as a circular process in Figure 12.1, Nonaka and Takeuchi originally imagined the transfer and conversion of knowledge in the form of a spiral of ever-increasing refinement of the knowledge exchanged.

The model in Figure 12.1 also describes the interplay between the group and the individual in making sense of any given situation, where each revolution of the cycle increases the common understanding until a consensus on a solution is reached. This interplay is particularly important since it relies on the basic assumption of KM that the more knowledge applied to solving a problem and the greater number of perspectives being considered, the better the solution will be. The same assumption can be found in CM in the form of "multiple advocacy." CM sees multiple advocacies as something of an unreachable ideal—something that immensely improves decision making, but which often succumbs to the stress of a crisis. KM tends to treat multiple advocacies as an everyday, and at times even annoying, occurrence. This is one of many instances where the business-oriented roots of KM are clearly evident. The evaluation of multiple solutions will take place as a natural part of a product innovation cycle, but each such attempt increases the risk of introducing time lags and information leaks in production. Therefore, when discussed from a KM perspective, multiple advocacies are at times seen as a necessary evil, rather than the near-utopian solution described in the CM literature.

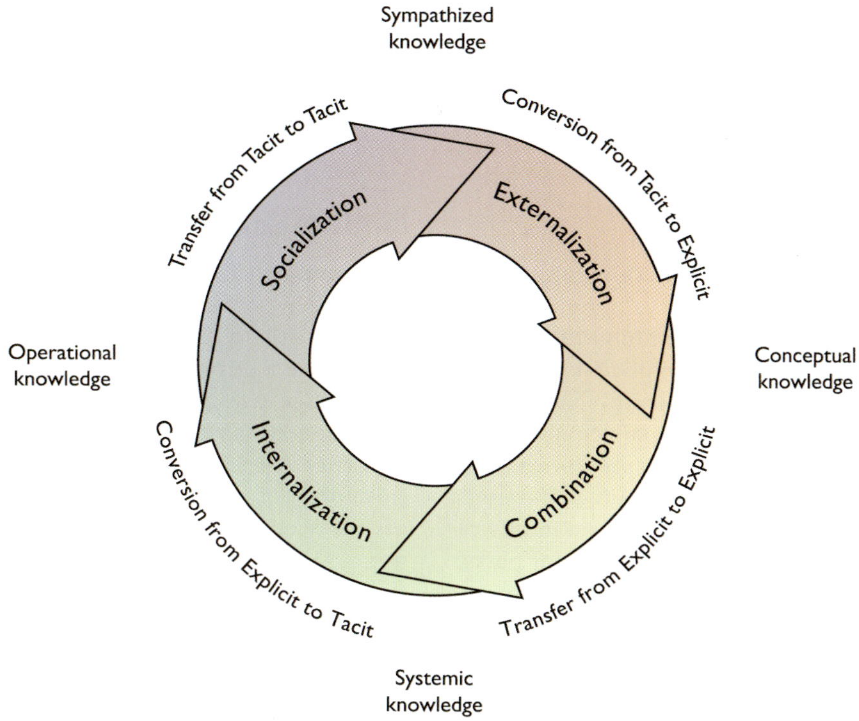

FIGURE 12.1 The SECI process.

As a process, then, SECI is more complex than might be immediately apparent. Each phase of the cycle represents a different combination of three key variables: (1) process, whether it is a knowledge transfer or knowledge conversion; (2) output, whether it produces tacit or explicit knowledge; and (3) participation, whether it is an individual task or a group task. This is further complicated by the fact that the exact nature of how these variables work and interact is largely unknown.

Socialization. Socialization is a group exercise in transferring tacit knowledge between group participants about current management practices. The aim of this process is getting each person to at least experience, and perhaps even adopt, the different solutions discussed in the group. The result of this phase is a shared "sympathized knowledge" of each perspective in the group. This has more to do with seeing how others try to solve similar problems rather than understanding exactly why those particular solutions were implemented. This phase is best exemplified by various forms of excursions, participating observations, and on-site studies. While the classic example of a tacit-to-tacit transfer is an apprenticeship, it does not have to be that intense or exhaustive to qualify as socialization. In fact, the only requirement is that people participate, if only passively, in a new and unfamiliar aspect of the problem-solving process.

Externalization. The externalization phase is where each individual takes this new sympathized knowledge and expresses his or her interpretation of it, including previous experiences of similar situations and potential solutions, thereby converting the tacit understanding of the situation to an explicit statement of facts. This creates a large repository of conceptual knowledge about the current situation for the group to use in the following combination phase. This is where the "why" that was missing from the socialization phase comes into play. Each person gets to explain why they chose some particular way of doing things or, for that matter, why they chose not to go down some given path. A simple example of externalization is any kind of after-action report or other form of evaluation of past or recent experiences. This kind of knowledge distinguishes itself from the systemic knowledge derived in the combination phase in that conceptual knowledge represents the thoughts and ideas of an individual, rather than a compilation of a larger set of ideas.

Combination. During the combination phase, all of the different pieces of explicit knowledge from the externalization phase is shared, i.e., transferred, among the group participants, and different combinations of interpretations and solutions are discussed. These group discussions result in systemic knowledge, an understanding of the larger picture as well as of its different parts. This phase generally takes the form of seminars, panel debates, and similar intellectual forums. Similar to the socialization phase, the end goal is a common goal or vision of what the problem is and how it should be resolved. The difference is socialization creates a common feeling of what is going on, and combination aims at producing something akin to a mission statement or a "best practices" list, which everyone involved can get behind and support.

Internalization. The last in the list of SECI phases is internalization: the individual process of taking the systemic knowledge of the combination phase, interpreting it, and converting it into action. The result is an operational knowledge about how to deal with a problem. The subjective aspect of this phase is particularly important to note: while the combination phase is meant to ensure an explicit common goal, this goal will be interpreted differently by each individual. Throughout the entire SECI process, there is a constant movement between consensus and subjective (re)interpretation. The socialization phase creates a common set of tacit knowledge, which is then interpreted to mean something slightly different for each person. These different interpretations are then combined to create a common set of external knowledge, which in turn is interpreted to create a different work task for each person. The final link in the cycle is where this personal internalized knowledge is once again shared in a new socialization phase, where the idiosyncrasies that came out of the last round form the basis for the next cycle.

Implications of KM

As previously mentioned, CM uses a slightly older view of knowledge and learning, and deals mostly with the interplay between tacit and explicit knowledge whenever learning and knowledge use are discussed. On the surface, the same holds true for KM—the SECI process that is the most obvious example—but in this interplay, it also demonstrates a new set of knowledge types that turn out to be quite important in describing how to "learn on the fly" and to find and incorporate new knowledge as the need arises. While the SECI process lies at the very heart of KM, it is only a way of transporting knowledge; it does not explain where that knowledge comes from. Expanding on the ideas of learning, KM has also uncovered a number of different ways to find knowledge, and thus adds a number of options to the organizational learning toolbox. In more general terms it exposes a number of otherwise hidden details about the nature of knowledge and learning, many of which are lacking in the existing CM framework.

The three key concepts here are meta-knowledge, empathic knowledge, and second-order knowledge. Just as tacit and explicit knowledge can be expressed in terms of "knowing how" and "knowing that," respectively, these three new knowledge types can be described as "knowing why," "knowing who," and "knowing how to know more." KM also describes how these knowledge types in practice come into play through the use of communities of practice and through knowledge brokerage.

Communities of Practice. Often in KM literature, there is a great fascination with modern information communication technology systems combined with so called "expert systems": computer programs that can access, cross-correlate, and interpret huge amounts of data from a large number of sources and that can even draw conclusions from this data. In many ways, this is the ideal image of KM at work. A vast bulk of information is "mined" for knowledge and, through repeated processing, is distilled down to form new, nonobvious, bits of knowledge. In reality, however, technology has not advanced so far yet that these expert systems can draw conclusions beyond a very narrow and predefined set of problems. When more complex issues are at hand, it is still down to ordinary people to get together, compare notes, and come up with a solution. Expert systems are not as versatile as their human counterparts, and this fact has reconfirmed what we already knew: large, loosely coupled groups of people will have more knowledge embedded in them than immediately meets the eye.

This idea goes back to the foundation of organizational knowledge, where the knowledge processes of individuals were extrapolated onto organizations. In its more recent form, however, these ideas have been expanded to also describe larger communities with similar interests and activities, so-called communities of practice (COP). These are groups of both people and organizations who have chosen to share and exchange ideas since they have come to the conclusion that many of their knowledge needs cannot be properly fulfilled any other way. Historically, when KM was still only focused on describing firms and similarly closed organizations, COPs were primarily seen as cross-division setups within the same organizations. As KM ideas spread and communications and information-sharing technology improved, these communities expanded outside of organizational boundaries. Now they span over large sets of institutions that can contain everything from individuals and smaller interest groups to large companies or even entire states and governments.

Having grown this large, the problem with new knowledge within a community is no longer about inventing knowledge (most answers already exist somewhere within the community), but rather how to actually find that knowledge in the vast network and how to transfer it to where it is needed. These processes are called "sourcing" and "grafting" knowledge. It is in solving these problems that the three new knowledge types are exposed. It is worth mentioning that the terminology surrounding these processes can be quite confusing. Sourcing, on its own, means finding (the source of) some particular knowledge, and should not be mixed up with the concepts of "in-sourcing" or "out-sourcing," which actually are both forms of grafting: in-sourcing and out-sourcing describe the different ways of incorporating new-found knowledge or, more accurately, ways of incorporating knowledge holders into an existing organization.

Meta-knowledge

The first step in finding new knowledge is understanding exactly what it is you need to know or what is missing from your current base of knowledge. In order to do so, you need to have a firm grasp about what your "current base of knowledge" actually contains; perhaps the right information is already available, only you forgot that it exists. Sometimes knowledge is overlooked by large organizations, either because it is a fairly daunting task that requires constant updating or because it may intuitively seem like something you should automatically know. Yet as organizations grow, so do their interests and their contact networks, and before long, all of the different ways of obtaining outside knowledge become anything but intuitive. In order to make a community of practice more effective, this knowledge needs to extend beyond an individual organization and include the other community participants' knowledge and needs, which can only be obtained by actively searching for it.

Therefore, "knowledge audits" can be performed in an attempt to detect knowledge gaps and to assess the community's current and future knowledge needs. In short, meta-knowledge is knowledge about knowledge. It represents a knowing why, analogous with the knowing how of tacit knowledge and the knowing that of explicit knowledge, and it is the key to working within a community of practice. Meta-knowledge is the facilitator of knowledge exchange in a COP, since participants become aware of the kind of knowledge that is available and that they are likely to need.

Empathic Knowledge

Understanding why different knowledge needs exist is a good start, but does not actually solve the problem of satisfying those needs. In order to do that, actual contact with relevant knowledge holders is needed. This means interacting either with the experts directly or with someone who is capable of pointing you in the right direction. In short, accessing knowledge is not so much knowing how, but rather of knowing who to ask.

Empathic knowledge is the skill in knowing who to ask for advice, knowing who will understand your request, knowing with whom you get along and with whom you can work, and so on. This knowledge goes beyond looking at an organizational chart and assuming that the title of a person will reveal who can and, more important, is willing to provide some useful knowledge. Empathic knowledge can be seen as the knowledge that comes from having a large network of contacts and having personal connections with most of the other members in that network. As the label empathic suggests, it is centered on having a feeling for other people: knowing who and how they are as individuals and understanding their personal interests, needs, and feelings.

Second-order Knowledge

The final link in using KM to access and use external knowledge is second-order knowledge, which is the art of knowing how to know more. In comparison to empathic knowledge, it is less personal and more structured.

As previously mentioned, empathic knowledge deals with knowing what person has access to what knowledge, directly or indirectly, rather than where that person is in the organizational structure. But the question still remains: "Where do you turn to if no one seems to be able to provide the kind of knowledge that is currently needed?" While KM in general is all about "getting to know more," second-order knowledge deals with the complete unknown; when you do not know what knowledge is needed or who might be able to provide it. Since going directly to the right person is no longer available, a good place to start might be to take a look at the organizational chart and use the titles of the different positions to navigate you. However, given the fact that the exchange of knowledge often occurs within a larger community of practice with a large array of different actors, no such chart may exist.

The solution to this problem lies in understanding the common patterns of functions that exist in most knowledge-oriented organizations. This kind of knowledge largely consists of understanding the different bottleneck and gatekeeper roles that are more or less inherent in any sizeable knowledge network: where they are generally situated in the network; how they control the flow of knowledge; and, if needed, how to work around these bottlenecks.

Knowledge Brokerage

Being born from the same, rather narrowly focused, realm of theory, it should come as no surprise that all these three new types of knowledge are aimed at more or less the same thing: finding new knowledge. They are all part of a larger whole and fit together like the pieces of a puzzle—all are required to get the most out of a knowledge network. Meta-knowledge provides an understanding of what knowledge needs exist and also provides a means of predicting possible future needs. Empathic knowledge provides an understanding of who can fulfill these needs, and what they might want in return. Second-order knowledge provides an understanding of where to look if something pops up that was not foreseen by the meta-knowledge or if there is no one in the personal network that can deal with it.

When all of these skills work together and are connected to one or more COP, they give rise to what could be considered the Holy Grail of KM: knowledge brokerage. Knowledge brokerage is a group exercise where the different participants in a COP continuously and proactively trade knowledge with each other. This is commonly done through specialized "knowledge workers" who keep tabs on the information that flows in and out of their own organization and who deliver it to where it might be best served. The term "knowledge worker," in its most general sense, means any person who in his or her line of work adds value to data, information, or knowledge through interpretation, cross-referencing, or other forms of knowledge processing. In KM, the term refers to the people who actual interact with the other COP members.

By understanding the web of knowledge needs and requests within and outside their own organizations, knowledge workers are better able to identify the value of new incoming information. They can share this information with people, who will benefit most from it, even if these recipients sit in vastly different organizations. Depending on where it is used, this may seem overly naive and idealistic, but this arrangement relies heavily upon the other COP participants who have agreed to freely share knowledge for the common good.

Aside from the previous discussions on optimal knowledge transfer techniques and so on, the key ingredient in getting KM to work is nurturing an environment of mutual trust. Without this trust, no matter how well intentioned, elaborate. and thought out the knowledge sharing network is, it will simply not be used. In addition to creating a basic level of trust and well-functioning knowledge network, the COP must be accessible and encourage its members to take advantage of the network. Knowing that it exists is not enough. Knowledge workers must also be able to identify what knowledge is relevant to their own organization and, ideally, to the other organizations in the network. Without this ability, the knowledge network is, in practical terms, void of any useful content.

Earlier knowledge workers tended to treat their role as merely editorial: to sort and categorize knowledge or, more accurately, information, in vast information technology-based repositories from where it could easily be restored. As the importance of tacit knowledge became more apparent and when the distinction between information and knowledge was more clearly established, the knowledge worker role was transformed into something more akin to a switchboard, from information storage to knowledge brokerage. This meant you know what you need to know, but also what others around you might find interesting so that you know where to direct their inquiries. It is important to note that these knowledge brokers are generally not the people who actually "own" the knowledge in question. Their role is to provide it to others (most commonly, decision makers) and on behalf of others (such as experts in some given field), or to arrange for the two ends of the problem-solving spectrum to meet each other face to face.

This is essentially the nexus of empathic and meta-knowledge. The explicit and tacit dimensions of knowledge might still be the common elements of KM, but the brokerage of knowledge—the actual use of "knowing who" and "knowing why"—is the quintessential mechanism behind finding, extracting, and using external expertise.

CONCLUDING REFLECTIONS: IMPLICATIONS FOR CM

What implications do the SECI process, this expanded view of knowledge, and KM have for CM? The more elaborate categorization of knowledge provides CM with a more precise and up-to-date view of organizational learning in general terms, but nothing that is particularly CM specific. The transfer and conversion modes modeled by the SECI process provide a new perspective for the different CM tasks and further specify some of the key challenges inherent in these tasks.

Likewise, the use of expertise has a long history of coverage in CM-related literature, but is usually only discussed in terms of how experts influence (or do not influence) the decision-making process. Little or no comments are given on how experts should, or even could, be selected and recruited. Adding KM to the mix of CM tools thus highlights the issues of sourcing and grafting knowledge in a way that is often overlooked, and it also puts a somewhat different spin on a number of central CM issues.

SECI and CM

Since KM traces its roots back to the theories on organizational learning, it should come as no surprise that it almost completely revolves around adding new knowledge to an organization. While this can be done in a number of ways, a promising way of using it in CM is to improve learning during crises. Post-crisis learning has recently received increased attention in the existing CM literature, which already discusses the special problems and benefits inherent in learning from crises. During crises, there is, by definition, little time available for self-reflection and reform craft that is usually inferred by post-crisis learning, so relying on external expertise for quick infusions of "instant learning" can be worthwhile.

Hence, the SECI process immediately comes to mind. It can play a useful role in creating a better understanding between decision makers and their advisors, and assist in transferring relevant expert knowledge. Essentially, there are five uses for these kinds of advisory teams, each of which entails a different set of SECI transfer modes:

- **Analysis**: Combination of known facts and the creation of new knowledge that must then be internalized by the decision makers.
- **Framing**: Internalization of a certain point of view (a particular framing of a situation), which must then be shared through socialization among the advisors, decision makers, and anyone the decision makers want/need to convince.
- **Implementation**: Either codifying existing practices through externalization, or internalization of new best practices and operational procedures, which are then shared through socialization.
- **Building credibility**: Depending on the medium, the internalization of trust using explicit statements or by sharing feelings, which are then spread through socialization.
- **Evaluation**: Scrutinizing how something was managed (externalization in SECI terms), comparing these results with some specific model (combination), and then reintegrating the findings into the system (internalization).

Interestingly, none of these functions represent a full SECI cycle; instead they follow a classis query response interaction pattern. The one that comes closest to completing a full cycle is the evaluation function, and at a stretch, it can be expanded to also include a socialization aspect in the form of participation in and observation of the process to be evaluated. The reason for this is that these functions are generally only a part in finding a solution, whereas the SECI process is meant to describe the entire arc of solving a problem. In addition, these functions only really represent the work done by the experts in finding and delivering their part of the solution, leaving out the precursory step when the decision makers formulate and communicate the questions they want answered. This is also why the evaluation function is the closest to making a full SECI circle since tacit knowledge is transferred, interpreted, discussed, dissected, and converted into action. For example, some "problem" (a particular course of events) is compared to the ideal "best-case scenario" (i.e., where the expert advisor(s) can participate in every step along the way to finding a "solution") and then usually some kind of report is made on how well the events were handled. This example also illustrates the generality of what kind of problems and solutions can be discussed using the SECI process. For the crisis manager, "the problem" could be the crisis in its entirety. For the people in the field, it could be the implementation of a decision. And as shown here, "the problem" for the advisor could be defined as having to provide a good analysis of the situation.

All in all, it is clear that by applying KM perspectives such as the SECI process to CM can provide a more specified, integrated, and comprehensive approach, both when trying to analyze various CM efforts and designing CM systems.

Expertise and CM

While the crucial role of expertise will not surprise CM scholars (or reflective practitioners), the KM literature provides a good catalog of the kinds of knowledge various experts can provide—something that is often only mentioned implicitly in the CM literature. The most apparent is the casual way in which it treats practical expertise. Even if many instances can be identified proving where it might be useful, it is discussed in rather passive terms and is still fairly theoretical in nature: the planning of a communications strategy, the preparation of decision implementation, and the evaluation of lessons learned. Very little is said about those tasked with communicating, the implementers in the field, or the teachers of the lessons learned, especially if one compares them to the job of advising and making decisions. A clear example of this is the list of the experts' roles. Only once is the implementation of expertise mentioned; that is, before the crisis actually breaks out and as a part of the decision-making task.

A better picture of how experts can be used in crises can be achieved by mapping out their potential contribution in relation to the five CM tasks discussed earlier. Here, it is important to make a clear distinction between the role of the decision maker and the role of the expert advisor. As already mentioned, Boin et al. (2005) focused heavily on the leader's part of the equation and did not take on the related question of what experts can contribute to the various task. So while part of the sense-making task entails implementing early-warning and analysis functions, experts cannot really contribute to the implementation process. Rather, they are the implementation process: they provide the analyses and influence the frames of reference that help identify a problem looming on the horizon. Figure 12.2 illustrates how the other CM tasks can be viewed in terms of the different expert functions.

Knowledge Workers in Crises

Just as KM adds granularity to the CM tasks by introducing the SECI process and its perspective on knowledge workers and on the different types of knowledge, they provide additional insights into a number of common CM issues. Perhaps the clearest example of this is how KM deals with the concept of "gatekeepers." In the CM literature, gatekeepers are seen as something of a mixed blessing. On the one hand, they provide a valuable filtering service that helps the decision makers avoid information overload. On the other hand, they can quite easily become so-called mind guards—persons who ensure that no dissenting views are allowed into the decision-making process. In KM, gatekeepers have a rather different role. Here, gatekeepers are not just passive filters (or devious manipulators), but fully-fledged knowledge workers who actively seek out much needed information. It is worth pointing out, however, that these knowledge workers are not necessarily the ones holding this knowledge. Most commonly these gatekeepers merely act as a switchboard providing knowledge for others, most commonly decision makers, and on behalf of others, most commonly external expertise.

As various organizations and individuals start to assemble into COPs, some people will, almost inherently, become more involved in these communities than others and become the intuitive interfaces between the actors within those communities. As a result, they will be perceived as virtual, if not actual, authorities on certain topics, although in fact acting more as knowledge brokers—as they often speak for and represent the knowledge base of the

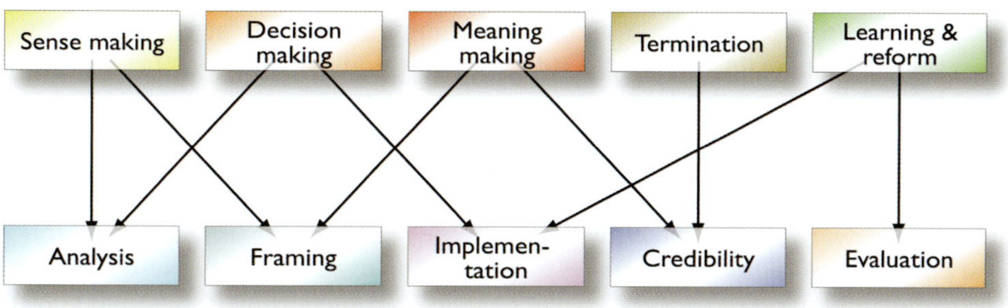

FIGURE 12.2 Types of expertise used in crises.

organization they represent as a whole. A common cause for the collapse of knowledge diffusion is that these gatekeepers fail to forward the requests for knowledge or the answers, either because the position gives them a high degree of control over the problem-solving agenda or because they are simply not up to the task of identifying and sorting the incoming knowledge.

By the time the KM literature noticed the importance of this role, gatekeepers as mediators and negotiators of information had already been closely studied in foreign policy decision making, especially in the context of crisis. In this literature, it is generally seen as a way to avoid information overflow, which at first might seem to be the opposite of their intended role in the KM context. In KM, the role has even been adopted for noncrisis situations, since similar skills are needed for finding, evaluating, and sorting knowledge from various knowledge communities and networks.

This does not mean that KM disregards the problems of groupthink-related mind guarding and manipulation; instead it puts a new spin on the issue. In imbuing the knowledge workers with this much control over what comes in and out of the decision-making group, KM acknowledges that the impact of any bias could be much larger than it would be otherwise. If the knowledge available in the COP is skewed in some particular direction or if the knowledge workers pick out a biased set of experts (knowingly or unknowingly), then the result will be an even more biased decision-making process than if the old-fashioned mind guards had been at work. In short, by using KM, what was once a problem and could be considered a character flaw or just a quirk in group psychology, suddenly runs the risk of becoming a systemic flaw—one that cannot be detected from within the system, no matter how observant the participants are. A second and somewhat related problem with KM is that, almost by definition, external experts will not be part of the established CM team. As a result, when the two must interact, they run the risk of succumbing to the "newgroup syndrome" where the unfamiliar composition of the group makes it harder to air divergent opinions.

Yet by using the KM style for building an advisory system, the problem of bias and groupthink is moved outside the decision-making group and onto the community of practice as a whole. Since everyone involved in the community will proactively provide everyone else with the knowledge they need, the entire COP will indirectly take part in the decisions made within the participating groups. This means that far more people must succumb to these cognitive breakdowns for the decisions to be affected by them, which means that there is a much larger chance of multiple advocacy taking place than if the decision-making groups was cut off from the COP.

Secondly, the use of knowledge workers should somewhat mitigate the problem of newgroup syndrome. One of the side effects of having knowledge workers as gatekeepers is that they effectively create a layer of separation between the CM group and the experts they choose to employ. As mentioned, knowledge workers provide information for someone on behalf of someone else, and these two parties do not necessarily have to meet with each other face-to-face. This mediation between experts and decision makers is most likely to introduce additional delays in the advisory process, but the advantage is that since the knowledge workers are intimately familiar with both parties, there is never a problem with voicing concerns, dissenting views, and objections.

Knowledge Under Pressure

The delay in the decision-making process is still cause for some concern when trying to connect the use of expertise with the special circumstances of a crisis. Looking back at the three defining characteristics of a crisis discussed in the beginning of the chapter (urgency, threat to core values, and uncertainty), the problem of uncertainty is most obviously connected to the use of external expertise, since experts are generally brought in to solve a specific problem.

Nevertheless, the other two aspects still have some relevance for the purpose of this study. The issue of urgency raises the question of whether there is enough time to find and consult external experts, when this potential unfamiliarity could result in further uncertainty. Consequently, the aspect of uncertainty ultimately determines the relevance of KM theory for CM. Lastly, the conflict between several threatened core values often forces decision makers into making "tragic choices" where one non-negotiable value must, in fact, be sacrificed to save another. The benefit of using external experts is that they can offer different views of a problem by casting new light on the different values and can potentially help resolve the apparent conflict.

Regardless of how they are connected to the use of expertise, it is the combination of these three characteristics that makes crises special. In general, it is sufficient to resolve, avoid, or eliminate just one of these aspects in order to improve managing a crisis. Without a threat to core values, the issues of urgency and uncertainty do not matter much, because nothing of greater value is being sacrificed. If there is no uncertainty, value conflicts and time pressure are not major problems—the best solution is already obvious. Without urgency, there is enough time to work through uncertainty and conflict.

(Re)integrating KM

Perhaps the most obvious intersection between CM and knowledge management discourses centers around the learning and reform task. When CM discusses the concept of learning, it generally falls back on a long-established discourse on organizational learning. This concept was first popularized in the 1960s through the research of Cyert and March (1963), Hermann (1963), and Cangelosi and Dill (1965), and has since gone through several stages of refinement. The basic and, at the time, revolutionary idea was that organizations contain knowledge in their structures, policies, practices, and standards (which are independent from the individuals actually inside these organizations) and that by changing these elements in response to a changing environment, the organization itself could "learn" and improve itself. Over the years, organizational learning theories have divided into four distinct, but still interconnected, topics: organizational knowledge, organizational learning, "the learning organization," and KM. It is worth mentioning that the field as a whole is still commonly referred to as "organizational learning," which can sometimes be a source of confusion. This chapter, however, has focused on the last of these topics—knowledge management—so there is little need to refer to the subfield of organizational learning.

The division into four different subfields occurred fairly recently. While parts of CM share the same ancestry as KM, this common ground was established prior to the division of the field. Since then, the field of crisis studies has had its eye on rather different matters and has not kept entirely up to date on the developments within organizational learning. Missing some of these developments might not be a big loss to the CM field, since they have little to do with CM. The concept of the learning organization, for instance, is an idealized institutional design with an enhanced capability of learning as its main focus. While this is certainly a good intention in preparing for a crisis, it is mostly a tool for analyzing organizational design and, as such, does not really fall within the scope of CM studies. Likewise, organizational knowledge is mostly the theoretical foundation behind the more interesting subfield, from a CM perspective, of organizational learning.

Thus, of the four topics, two fall outside the main focus of CM studies, and one is already integrated in the form of one of the five critical CM tasks. This leaves the fourth topic, KM, as a potential candidate for inclusion, or perhaps rather reintegration, into the CM framework.

Contemporary CM is a knowledge-intensive endeavor. As has long been recognized by shrewd leaders as well as philosophers, knowledge and power are closely entwined; some would say two sides of the same coin. The effectiveness of the knowledge work done before, during, and after crises heavily impacts preparedness, CM capacity, and the prospects of success in a very difficult domain. Scholars and practitioners of CM can benefit significantly by drawing upon the growing literature and evolving set of practices associated with the field of KM.

CHAPTER

13

A Semantic Approach to Security Policy Reasoning

Ivan Launders, Simon Polovina

INTRODUCTION

The automation of business transactions across the Internet has brought about many benefits. As such, enterprises and individuals enjoy the provision of many goods and services that hitherto were inaccessible to them. This success has relied on the use of payment cards instead of cash, as these cards enable instant electronic transfer directly between supplier and customer including remotely across the Internet. However, as Schreft (2007) identified, these systems are information-dependent transactions and extend across checks, credit cards, and debit cards. As a consequence, this information, that is, the identity details of the participants in the transactions are known. Consequently, they can be misappropriated by unauthorized parties (e.g., organized criminal gangs and terrorist groups) who in turn can use this information for illegal purposes. This leads to fraudulent transactions that fuel the criminal misuse of identity details (WenJie et al., 2006) and pose a serious threat to national security.

The growing sophistication of business transactions using payment cards continues to heighten the issue, causing its policing to be increasingly difficult. Nonetheless both public and private sector organizations are required by law to comply with industry standards for information technology (IT) security, and these standards mandate the creation of security policies and audit trails for these enterprises, especially from key systems holding sensitive information. Enterprise architects, who have to holistically capture the enterprise as a business rather than just its technical systems, face ongoing design challenges securing enterprise systems with safeguards to protect against an ever-increasing number of these cyber threats and misuse. The enterprise architect needs to be able to secure each architecture component providing risk assessments and countermeasures for these threats.

Almost every hardware and software component within enterprise architecture has the capability of producing a detailed record of its activity. Records are produced through audit logs often used for problem diagnostics and problem identification. Audit logs are a vital resource used to improve information security and compliance with regulatory standards (ISO2702, 2005). Security architects have traditionally relied upon technical controls, for example, firewalls and disabling all but the necessary technical services. Both public and private sector organizations are required by law to comply with industry standards for IT security and these standards normally mandate the creation of audit trails, especially from key systems holding sensitive information. However, within ever more complex transactional enterprise architecture it is difficult to achieve compliance without the ability to capture and model the complexity of security policy across the enterprise.

CURRENT APPROACHES

Organizations are looking to industry best-practice standards for guidance on how best to manage these new enterprise security challenges (Parkin et al., 2009). Industry best practice has tended to rely on technical controls, assessing identifiable vulnerability and then assessing the associated risk of a threat on that vulnerability. Industry-best practice is provided through organizations such as the Information Security Forum (ISF, 2012a). The ISF provide a good practice for information security, in association with leading organizations. The ISF provides good practice through working with tools such as Information Risk Analysis Methodologies (ISF, 2012b). IRAM provide practical guidance on how to conduct effective threat and vulnerability assessments, analyzing threats and vulnerabilities, and making assessments as a method for determining the likelihood of incidents occurring in an enterprise.

IRAM works at different levels with a business impact assessment, allowing an enterprise to determine (1) whether a possible business impact could arise as a result of an incident that compromises information in a system, (2) the business security requirements of the system being assessed, and (3) the steps that should be taken to protect information in the system. This analysis results in the production of design artifacts such as a threat matrix from which residual risks and countermeasures can be taken. Figure 13.1 shows an example threat matrix, which identifies that although threat assessments may be high the impact of those threats is medium; therefore, it has less impact if the threat is carried out. This approach is, however, a subjective measure taken by an enterprise architect's knowledge and judgment.

Threats are identified and the knowledge and judgments are captured about the risk of those threats on each architecture component. Threats are then assessed through the use of visual inspection tools such as the color-coded risk radar shown in Figure 13.2. A red color indicates that there is a high risk of external attack and internal misuse and abuse. An indication can then be used by an enterprise architect to show where to focus the security design effort.

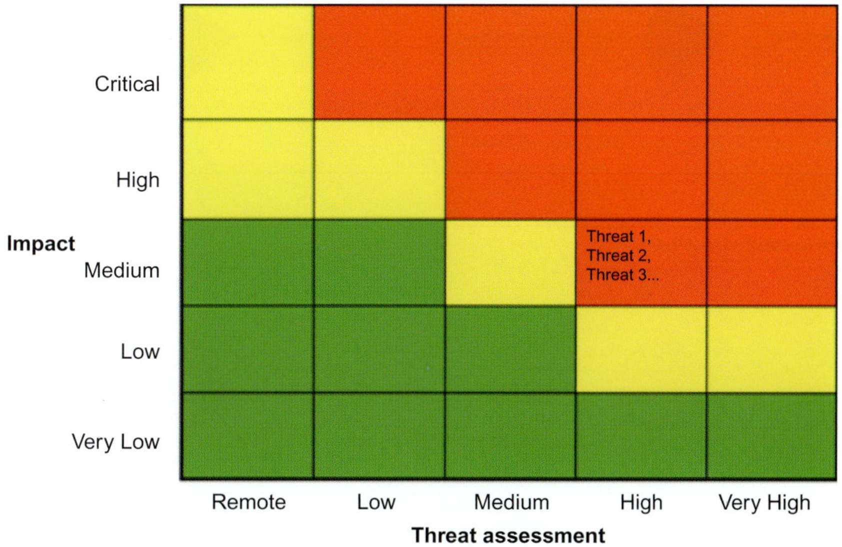

FIGURE 13.1 Example impact and threat assessment matrix *(ISF, 2012b)*.

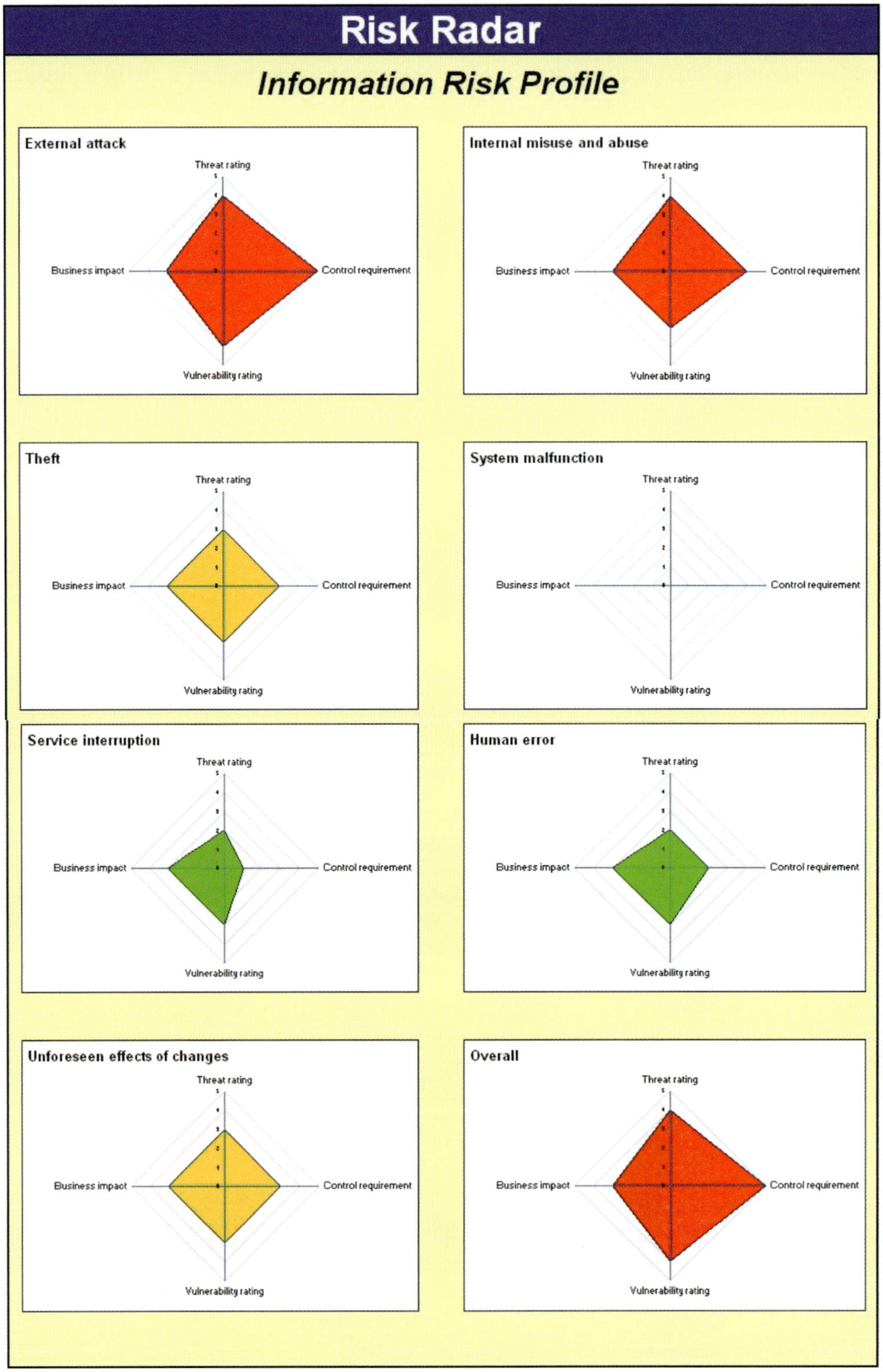

FIGURE 13.2 Risk radar categorization and severity *(ISF, 2012b)*.

Figure 13.2 provides an example visual indication of risk categorization and severity. The risk radar shows the perceived inter-relationship between threat rating and vulnerability rating balanced between business impact and control requirement. Working with visual information in this way allows the enterprise architect to be able to weigh up a level of security risk identifying those areas of greatest vulnerability. One of the design roles of an enterprise architect is to identify vulnerabilities and then to carry out a risk assessment of those vulnerabilities assessing the probability of a threat manifesting and therefore what controls and countermeasures are required to mitigate the risk of a vulnerability being exploited (ISF, 2012b).

BEST PRACTICE

Should industry best practice be extended to focus greater attention on understanding the semantics of the security threats and in turn their threat to the enterprise? If we accept that security threats and vulnerabilities to architecture hardware and software components are continuous and that current approaches are partially subjective, and that there is an ever-growing stream of new viruses and malware, then not all threats will be totally eliminated. However, we still need to capture high level business threats and vulnerabilities that we do know about in ways that are less constrained by deficiencies in the models used.

In cloud computing, for example, employees are selected by the cloud service provider to administer cloud infrastructure. These employees may be able, via their privileged levels, to have access to the tenant's resources (including data). They may not be screened (security cleared) to ensure that their security skills or trustworthiness meet the standards promised to the tenant. Trustworthiness or trust is a complex concept and has a lot of shades of meaning (Schineier, 2012). When we trust employees, we can trust their intentions and their actions. It can be argued that we need security when we do not have trust (Schineier, 2012). Do we therefore trust systems more than employees? Regardless of screening, employees can pose an insider threat (internal misuse and abuse as shown in Figure 13.2) in that they may abuse their administrative rights to compromise the tenant's security. Outside threats shown in Figure 13.2 may also occur as a party breaches a tenant via cloud-level infrastructure components that the tenant is forbidden from monitoring. It may be impossible for the tenant to identify the cause of the breach and prevent it from happening again. Shared infrastructure with other mutually suspicious tenants means that tenants may be expressly forbidden to investigate infrastructure components because doing so might violate the security guarantees made to other tenants. These scenarios and their associated security policies are complex and can benefit from modeling to fully explore their semantics.

Identity management and authentication is an essential part of enterprise security. Identity management involves defining which users exist and what roles they have, and which have access and to what degree (Woods and Matters, 2006). Authentication is then a process by which the software or user attempts to confirm identity. Securing more than the first service encountered by a user extending security across the enterprise architectures involved in the business process. For example, what a user orders must be verified upon receipt of delivery.

BUSINESS RULES

Roles are used with identity management to permit access to what they need at a business level. The definition of what a role should have access to, including down to the level each role can access, forms a security policy. Each service must be secured through business rules if we intend to place trust in the enterprise system. That is a compact statement describing what must be the case in terms of business process flow (Morgan, 2002). The less secure the business rules the less trust we place in the system (Schineier, 2012). For example, I trust my cash machine to dispense money from my account and to debit it by the correct amount. Enterprises do, however, have individuals who are not trustworthy and who will test business rules to their limits; for example, the financial sector often employs individuals who have a willingness to take risks to exploit business opportunities. This level of behavior can exploit enterprise vulnerabilities, resulting in security incidents and at worst case serious economic crime. For example, the 2008 collapse of the real estate industry in the United States was partially caused by employees manipulating financial systems for their own gains—loaning mortgages to homeowners who could not afford them and then repackaging and reselling those debts in ways that hid the real risk. Financial analysis made further money rating the debts to ensure repeat business (Schineier, 2012). These serious economic crimes happen over and over again because financial systems allow people or groups of people to misuse the system and commit fraud. Financial networks allow people to transact and do business around the world. Hacked financial accounts mean illegal or fraudulent transactions can take place; these can be complex or as simple as stolen credit card numbers.

The Semantic of Business Vocabulary and Business Rules community (SBVR, 2011) provides an approach that enables people and organizations to treat business in a consistent legal way. Business rules and policy are usually presented in natural language, which is then designed into software as part of the implementation solution. The misinterpretation of policy rules from natural language by designers and developers making their own assumptions means rules and policies may not provide an explicit representation of rules and reasoning. Inconsistency and ambiguity in policy and rules are hard to identify since they may be implemented by various people with different views (Croitoru et al., 2008). Inconsistency and ambiguity in the implementation of enterprise architecture can therefore lead to potential security vulnerabilities. The interrelationship among security policy and business rules must be established (Croitoru et al., 2008). A graph that interconnects the policy and rules to check their relationship while ensuring validation would support the design of policy in enterprise architecture. This is also due to the fact that when some rules are applied, extra knowledge is obtained to be reasoned with. This could be extended to include security therefore combining aspects of logic and natural language, business rules, security, and conceptual modeling (Launders, 2012).

The use of business rules translated from natural language into conceptual graphs (CGs) with ontology provides control in order to perform reasoning (Sarraf and Ellis, 2006; Croitoru et al., 2008; Launders, 2011). Business rules are compact statements that specify an aspect of a business declaration of policies or conditions (Morgan, 2002; Croitoru et al., 2008). Croitoru et al. (2008) proposed the use of CGs to represent more appropriately the structure of information and to employ reasoning for consistency checking and analyzing the interdependencies among policies. We will explicate CGs shortly; meanwhile, a further real-world example of a CG rule-driven Web-based enterprise architecture that builds upon concepts used by Tesco.com is Sonetto (Sarraf and Ellis, 2006). Sonetto is used to reason with product information for its electrical and wine categories. Sonetto's solution allowed business users to provide online customers with content-rich customer-focused information demonstrating a business-to-customer application for business rules. Sonetto is also based on ripple-down rules (RDRs) for rule acquisition using CGs logic as its basis for representing product documentation, rules, and categories therefore reasoning with CGs rules. RDRs build systems incrementally while they are already in use; the expert adds a new rule based on the context of an individual case (Compton and Jansen, 1990). RDRs are similar to other methods in that they are a rule-based method. In traditional rules systems rules are a long list of individual pieces of knowledge, whereas in the RDR knowledge base the rules are stored in a tree where each branch of the tree is considered a rule.

ENTERPRISE ARCHITECTURE FRAMEWORKS

Frameworks for enterprise architectures have been adopted to help organize design thinking about business scenarios and business goals including their security-specific architectures (Zachman, 1987, 2012; TOGAF, 2009). The Open Group Architecture Framework (TOGAF), for example, recognizes the importance of security design early on in a project life cycle and has provided a significant move toward including security in the design of enterprise architecture (SABSA, 2011). Architecture frameworks such as TOGAF and Zachman can be extended to include the capture of semantics in the business architecture (Sowa, 1984; Sowa and Zachman, 1992; Launders and Polovina, 2010). The TOGAF Architecture Development Method starts with a preliminary phase to determine the architecture capability desired by the organization as well as establishing current architecture capability. Defining the security capability and security policy requirements is a key part of this phase.

As part of an increased focus on security The Open Group has released new guidelines developed in collaboration with the SABSA Institute to enable enterprise and enterprise architects to integrate security and risk management approaches into enterprise-level architectures with the aim of aligning IT security decisions with business goals across the enterprise, therefore adopting a business-driven approach to IT security decisions. The SABSA methodology was designed for integration with TOGAF based on its objective of developing security architectures that facilitate the business (SABSA, 2011).

Security is considered in the Information System Architecture phase (phase C) in TOGAF (TOGAF, 2009). This phase involves assessing the baseline for the current security-specific architecture elements. The task involves identifying safe default actions and failure states in order to evaluate applicable guidelines and standards. This allows a security architect to determine the relationship of a system with business architecture and determine potential avenues of attack and what can go wrong. Reference models can therefore be used to capture patterns of security activity that are reusable assets in enterprise architecture.

Enterprise architecture frameworks provide a starting point asking the analysis questions What are the vulnerable components? How will behavior exploit those vulnerabilities? Who are the agents likely to provide the threat to security? Why are agents likely to be a threat? Are they under pressures that may change their behavior? TOGAF security considerations would start by determining the following:

1. Who are the legitimate actors who interact with an enterprises product service and process?
2. What is the baseline for those current security-specific business processes?
3. How much is it acceptable to inconvenience in using security measures?
4. How to determine the assets at risk if something goes wrong, that is, what is being protected
5. How to determine the costs both in terms of qualitative and quantities of asset loss in case of failure

For example, will a trader in a financial trading company in personal debt behave the same as a trader with no debt? Is it likely that the trader in debt will take greater risks than the trader with no debt? These types of questions bring focus to the security design identifying and assessing the vulnerability and control concepts for an asset and therefore the value at risk.

THREATS, VULNERABILITIES, AND SECURITY CONCEPTS IN CGs

The purpose of a best practices threat and vulnerabilities assessment is to help the enterprise architect understand the likelihood of incidents occurring in the organization (ISF, 2012b). Threats and vulnerabilities can be assessed from an internal, external, or combined overall threat rating position. The factors are combined to assess the likelihood based on the probability of a risk occurrence becoming an issue; this, however, is subjective judgment based on domain knowledge. CGs (Sowa, 1984) can provide a means to model and explore the semantics of that domain knowledge. They are useful forms of semantic networks, as they include a generalization hierarchy of types of relations and complete graphs (De Moor, 2004).

Conceptual Graphs

CGs are a system of logic based on Peirce's existential graphs (Roberts, 1973; Peirce, 2012). They provide a powerful knowledge representation and inference environment (Polovina, 2007) and can represent model and high-order logic, which can be extended to model security concepts. An example of a CG representing security concepts in "linear" text based form:

[Vulnerability] -> (target) -> [Threat].

The example graph reads as "The target of vulnerability is the threat".

Vulnerability. Adjective; Exposed to damage by a computer threat in an enterprise.

Threat. Noun; Declaration of an intention to cause harm to a computer in an enterprise.

Target. Links [Vulnerability] to [Threat].

Example: A computer vulnerability is the target of a threat.

[computer_vulnerability] -> (target) -> [threat].

[Asset] -> (characteristic) -> [Vulnerability].

This example graph reads as "The characteristic of Asset is Vulnerability".

Asset. Noun; Regarded as having value to an enterprise.

Characteristic. Links [Asset] to [Vulnerability].

Example: An enterprise asset has the characteristic of being vulnerable. [enterprise_asset] -> (characteristic) ->[vulnerable].

CGs capture the nuances in natural language in an automated way, aiding the process of inference (Polovina, 2007; Launders, 2012), therefore automating the capture of the lexicon (vocabulary) in a security model. The intention of the model is to relate security words to their grammatical categories and their underlying concepts in order to build a security-specific conceptual catalog using as its general basis Sowa's conceptual catalog (Sowa, 1984). In this general (and not security-specific) catalog the relations are not definitions, but constraints on the use of the relation in a context that would be tailored for security scenarios. This will be illustrated shortly. Automation of the security model provides ontology, which describes meaning as a formal specification of the terms in a domain and the relationship between them (Gruber, 1993). It therefore follows that security ontology can define vocabulary for enterprise security architects who need to share security information in a domain (Launders, 2012).

Business Rules for Vulnerabilities

Industry best practice would determine whether a threats and vulnerabilities risk assessment needs to be conducted (ISF, 2012b). A Business Impact Assessment in the case of IRAM would then focus attention on the areas of confidentiality, integrity, and availability in relation to the business process under assessment. Stakeholders are typically asked questions pertaining to four areas of their business processes including financial, operational, customer related, and employee related.

Business rule design provides the opportunity to examine roles and identify enterprise security vulnerabilities. Different kinds of roles will have different perspectives and different values on the business transaction that need capturing (Launders, 2011). For example, in the case of the financial trading case study discussed shortly the role of the trader and portfolio manager are modeled. The business rule for these roles states that "An order must not be placed both by the trader and by the portfolio manager." This is a banned combination of roles shown in Figure 13.4 that if not complied with poses as a security threat to the financial trading enterprise. If orders were to be placed by the same individual acting in both roles, then the potential conflict of interest could potentially expose the enterprise to errors of judgment and risk taking. In this case a trader has a conflict of interest between carrying out a trade that impacts the value of a portfolio. The trader needs to act only as a trader in terms of the financial transaction. Acting outside a role as stated in the business rules creates business vulnerability, which has a potential risk.

Business Threats in Enterprise Architecture

A simplified theory of transactional analysis defines that people expect some sort of return for their action and that the exchange of goods for services is honorable and trustworthy. For example, eBay escrow services facilitate trust by contracting the buyer and seller into a commitment that they both agree on. Without trust a business or individual has a potential risk of transactional failure and at worst case a security threat (Schineier, 2012).

A central element of a business transaction is the exchange of resources between agents. A Transaction Agent Model framework (TrAM) exploits the formal language of CGs, economic accounting, and a transactions-oriented architecture that is based upon Geerts and McCarthy's Resource-Events-Agents (REA) model (Geerts and McCarthy, 1991; Polovina, 1993, Polovina and Hill 2009; Hill, 2012). REA enables models to be built that reflect business activities, which may include economic transactions using the following core concepts:

Resource: Any resource that is the subject of an exchange or transaction.
Event: The activities that are required for a transaction to take place.
Agent: A person, system, or organization that participates in the transaction.

TrAM, based on REA, captures the qualitative dimensions of business transactions and business processes developing the semantics and testing the subjective human judgment that otherwise leads to errors of omission or miscommunication in transactions (Polovina and Hill, 2009).

The automated TrAM involves building ontology to describe the form and meaning of concepts used, expressing enterprise transactions in the form of a transaction graph (TG) (Launders, 2012). A TG can be extended to examine security concepts and policy in the context of a business transaction. For example, in the financial trading case study that we will illustrate shortly, "The firm buys and sells numbers of shares of securities and manages its clients' assets" including three different categories: real estate, cash, and securities. Each type of category has different properties and therefore different vulnerabilities and would be potentially targeted with different security threats. By identifying assets and the controls that can be employed (process or technology), it is possible to inform the development of internal policies for their protection (Parkin et al., 2009). Producing the TG involves analyzing high-level business goals, modeling, visualizing, automating business improvements, and identifying business assets and the potential security threats to those business assets.

Security Threats

Security threats take many forms, some of which are within the control of an enterprise architect and some that are not (external threats). As previously discussed, a visual tool such as the risk radar shown in Figure 13.2 provides a perception of external and internal attacks. These may typically include the following:

- **Undertaking malicious probes and scans:** Detecting network connections that are likely probes originating from malicious sources, and detecting scanning activity
- **Carrying out denial of service attacks:** An attempt to make a machine or network resource unavailable to its intended users
- **Cracking passwords:** Recovering passwords from data that have been stored in or transmitted by a computer system
- **Eavesdropping:** Secretly listening to the private software communications of others
- **Cracking keys:** Decryption of security software keys

- **Spoofing user IDs:** A person or program successfully masquerades as another by falsifying data
- **Modifying network traffic:** Capture network packets of data, modify them, and then resend them on the network
- **Distributing computer viruses:** A computer program that can replicate itself and spread from one computer to another causing harm
- **Introducing Trojan horses:** Malware computer programs presented as useful or harmless to introduce the user to install and run them
- **Introducing malicious code:** Software used or created by hackers to disrupt computer operation, gather sensitive information, or gain access to private computer systems
- **Carrying out social engineering:** Manipulating people into performing actions or divulging confidential information

Examples of internal misuses and abuses could include the following:

- **Gaining unauthorized access to systems or networks:** Nonpermitted access either by hacking or unpermitted use of account and password details
- **Changing system privileges without authorization:** Changing a user access to systems and networks either granting more access that is allowed or reducing access to deny services
- **Misusing systems to cause disruption:** Helping a person gain access to protected files with the intent of causing confusion or disorder

The internal misuse and abuse threats are those threats that an enterprise architect has the greatest control over. For example, employees for a financial trading company could assume roles and privilege levels higher than they are allowed and may then act with such privilege level in a way that is potentially harmful to an enterprise asset. Parkin et al. (2009) explained that an asset may be perceived as vulnerable to some form of exploitation but may not be necessarily exploited. The vulnerability becomes a higher level of risk or threat when there is a means of exploiting it. We can now turn to illustrating the previous and further arguments by referring to the following financial trading case study.

FINANCIAL TRADING CASE STUDY

To illustrate our arguments, we present a simple financial trading (FT) case study. This case study is a fictitious asset management firm called TechRules Advisors (TRA Inc.). The firm buys and sells numbers of shares securities and manages its clients' assets. Portfolio managers create and manage accounts.

Evaluating the TG through a FT case study leads to a better understanding of semantics within the business transaction (Launders and Polovina, 2010; Launders, 2011, 2012). Semantics are captured in those concepts and relations specifying the FT business transaction, therefore, enhancing understanding in the business transaction through the exchange of meaning. CoGui (2012) allows for further model automation and reasoning extending the FT model to include security concepts such as threat and vulnerability. Figure 13.3 shows a rich FT TG of the case study but before any security reasoning.

The Automated FT TG in CoGui

CoGui (2012) is a visual tool for building our ontology, rules, and reasoning in CGs. It allows for the creation of both a concept- and relation-type hierarchy, allowing the structure and content of the vocabulary (ontology) to be edited and expanded. The structure of CoGui is such that it supports facts, rules, queries, and constraints. CoGui allows the TG to reason with security rules, selecting facts and step testing rules against those facts to project and join, therefore, discovering greater security knowledge for a transaction in a given context. The output ontology model is saved in an XML format called COGXML, and supports RDF for the Semantic Web (CoGui, 2012).

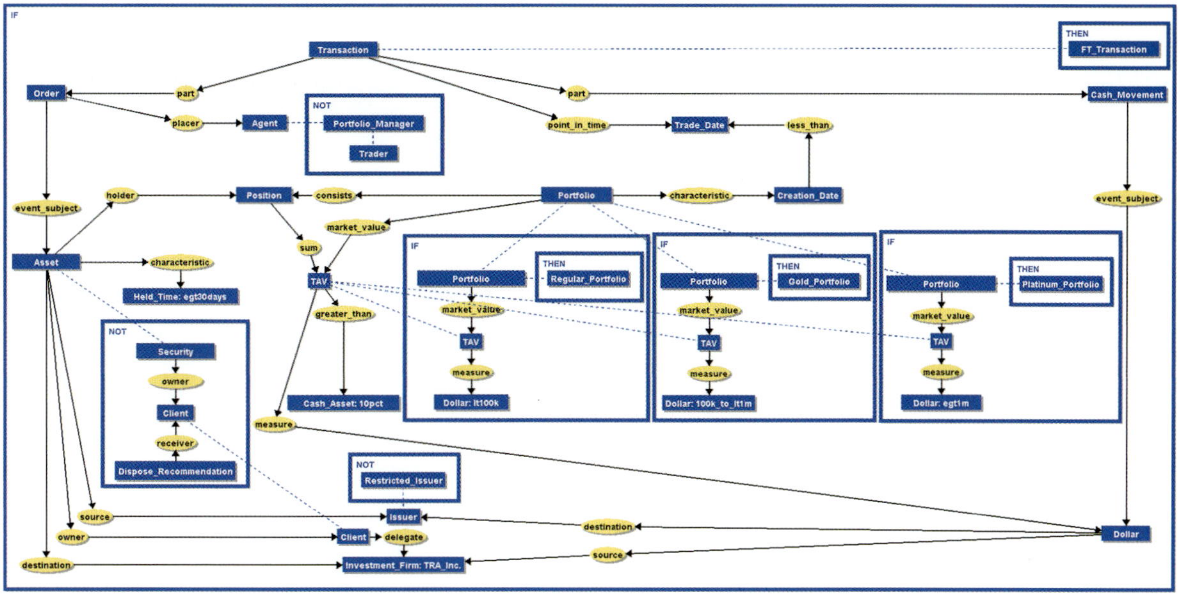

FIGURE 13.3 The TG for TRA Inc.

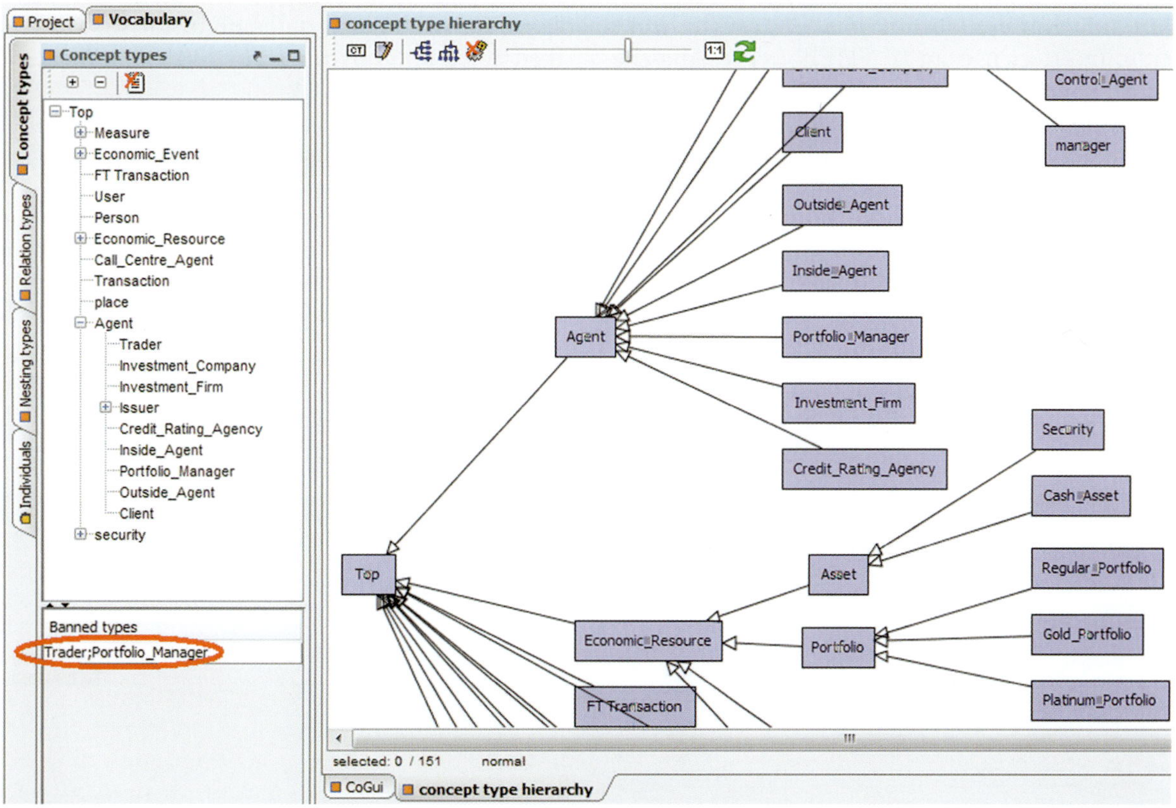

FIGURE 13.4 Banned types: trader and portfolio manager.

Figure 13.4 shows the implementation of ontology for the FT case study (Launders, 2012) highlighting the banned type specifying that "an order must not be placed both by the trader and by the portfolio manager." This is a banned combination of roles, so a portfolio manager cannot be a trader as there is a conflict of business interests. Banned concept types mean that the concept types cannot have a common subtype. For example, a person cannot be a boy or a girl at the same time. This is useful for building restrictions with security concepts within the ontology. CoGui refers to banned types as a means of extending the vocabulary.

The vocabulary in CoGui shown in Figure 13.4 can be viewed as a basic ontology. Rules are then applied to reason with and to enhance the expressiveness of that ontology. Figure 13.5 shows a business rule for a regular portfolio defining the hypothesis as a CG on the left side and conclusion as a CG on the right side. In effect it is an IF THEN expressing the rule "It is rated regular, if TAV is less than 100.000$."

Rule reasoning is carried out by first projecting the CG in the hypothesis side of the rule into a fact CG. If the hypothesis projects into a fact CG then the conclusion part of the CG is applied to the fact. So in the case of Figure 13.6 if the condition of a total asset value (TAV) has a measure of less than100K (a regular portfolio) the portfolio type is set to be Regular_Portfolio. In CoGui rules can be applied individually or in a sequence so that the outcome of the first rule is then acted upon by the second rule in a forward chaining scheme.

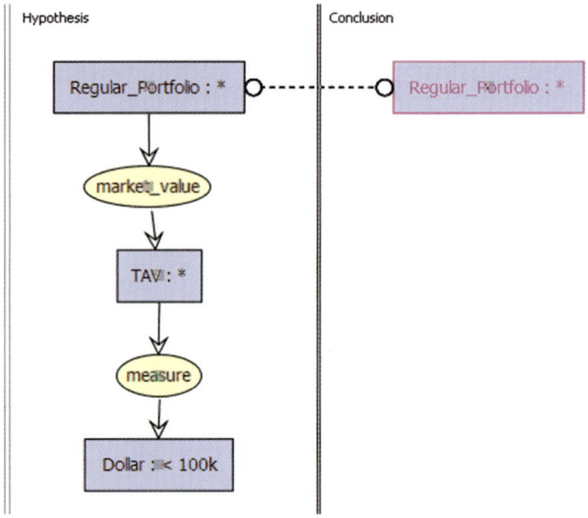

FIGURE 13.5 IF THEN rule in CoGui

Reasoning with Business Rules

Business rules can be selected individually or as a sequence and then applied to a fact in ontology. The resulting conclusion of applying the rule is then added to the fact CG as

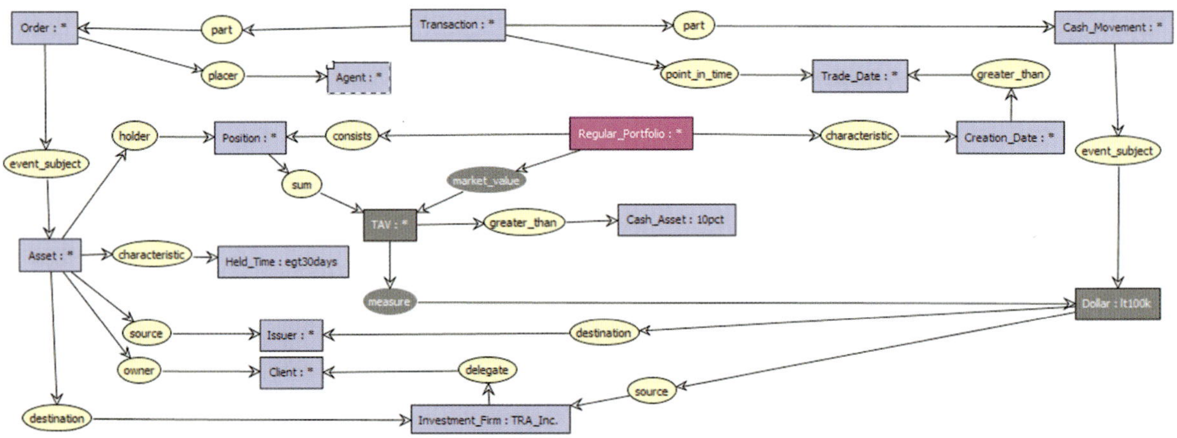

FIGURE 13.6 Reasoning with the Regular_Portfolio rule.

shown in Figure 13.6; it shows an automated TG for TRA Inc. sell (and buy) assets on behalf of their clients. The assets come from an issuer. Figure 13.6 shows an initial automated TG reasoning with the rule "regular_potfolio" as identified in Figure 13.5. For simplicity we have not reproduced the full TG of Figure 13.3, and we will re-illustrate the "if–then" aspects of that figure according to its potential security semantics.

Figure 13.7 refines a fragment of the FT TG shown in Figure 13.6 showing the reason hypothesis and conclusion on the left side and the result of applying this reasoning on the right side. Figure 13.7 shows the role of the concept agent through the use of a business rule, which states that "The placer of a transaction is a trader and that trader is the trader_of an asset." We also know from the rule that there is a banned combination of roles and that a Portfolio manager cannot also be a trader, as there is a conflict of business interests as previously discussed. This conflict of interest between a portfolio manager and a trader is a potential security vulnerability and therefore security threat if acted upon.

SECURITY REASONING WITH THE FT TRANSACTION GRAPH

The further application of security reasoning on the FT TG shown in Figure 13.8 allows the inter-relationships among business rules and security to be explored more deeply. The fragment of graph shown in Figure 13.8 expands upon the TG shown in Figure 13.7 to include the interconnection between the concept asset and the concept vulnerability, therefore, expressing knowledge that assets in the FT case study are vulnerable. The rule expressed as hypothesis and conclusion on the left side of Figure 13.8 shows the reasoning that if the model contains the concept asset (a value at risk) then it also follows that it has a characteristic of the concept vulnerability.

Figure 13.9 builds upon the knowledge applied by the rule in Figure 13.8 to identify that the asset is also the target of the vulnerability. While this may seem an obvious piece of reasoning it does not always follow that because an asset is vulnerable it is automatically a target. For example, the asset may be so small in value it is not a security target.

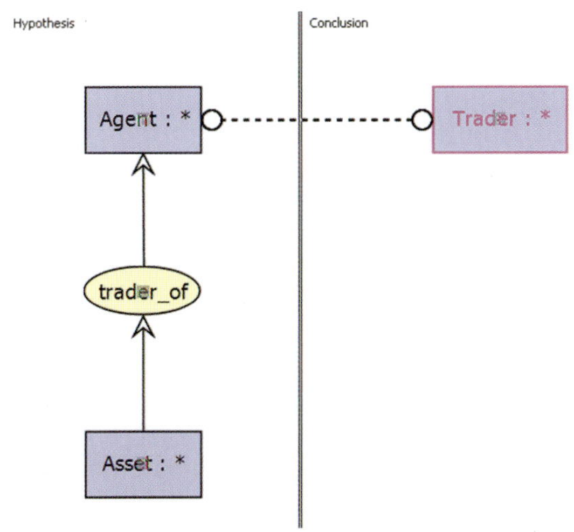

FIGURE 13.7 Reasoning on the trader placing an order.

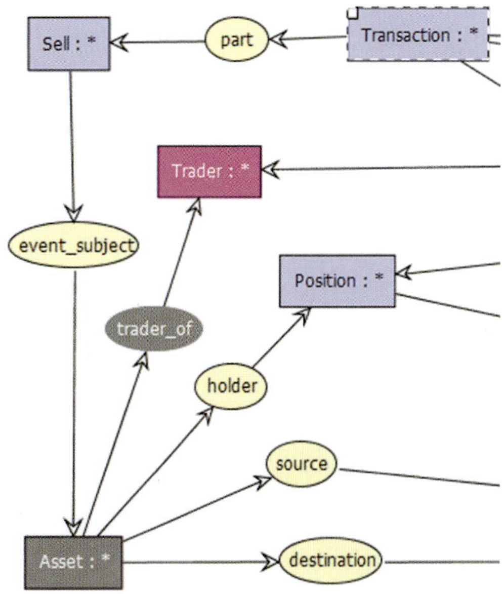

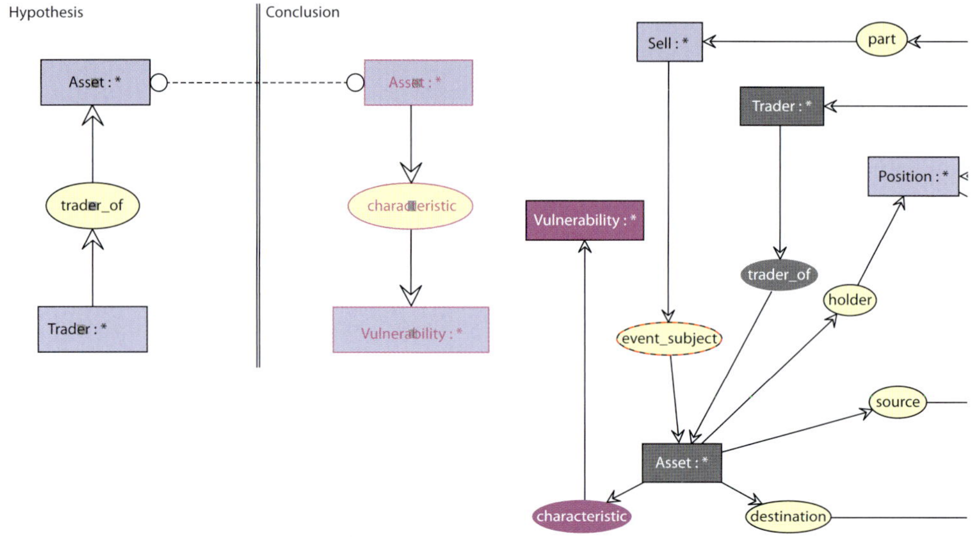

FIGURE 13.8 Asset rule reasoning.

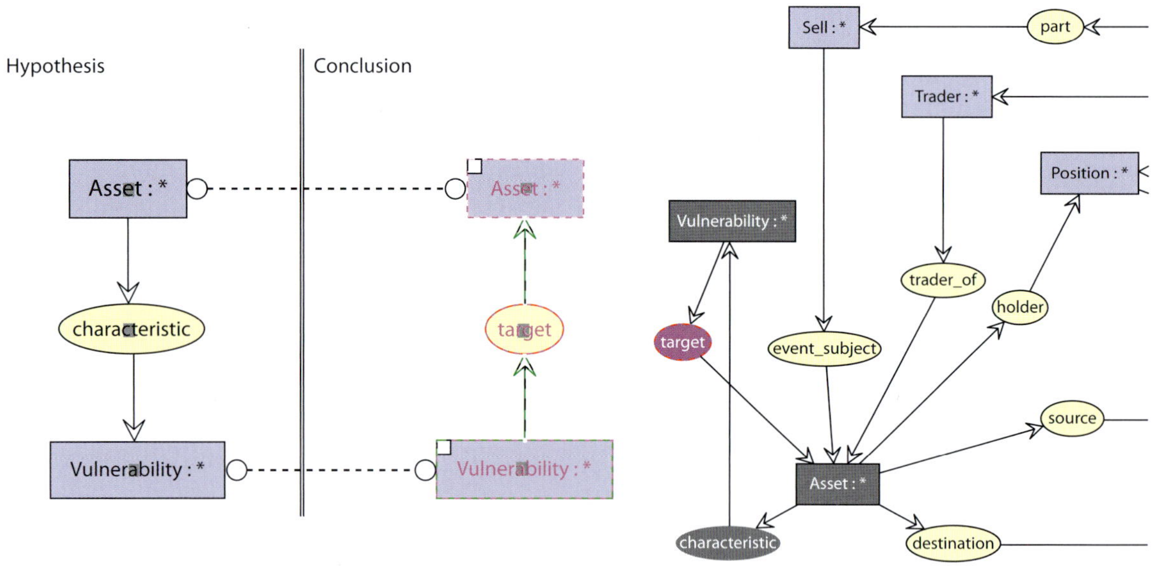

FIGURE 13.9 Expanded asset rule reasoning.

Figure 13.10 shows a further new rule refining upon the results of previous rule shown in Figure 13.9. This new rule extends the security reasoning to identify the target of the threat as being Investment_Company. Again, while this may not be a significant piece of reasoning it highlights that although the Investment_Company is not the owner of an asset it is by virtue of the fact trading with the asset, which is also the target of a security threat. In this case the security threat may be an internal misuse threat. For example, the internal misuse and abuse of a trader's system privileges may lead to the adjustment of a real estate value on paper to leverage a share trade above that allowed by a portfolio limit.

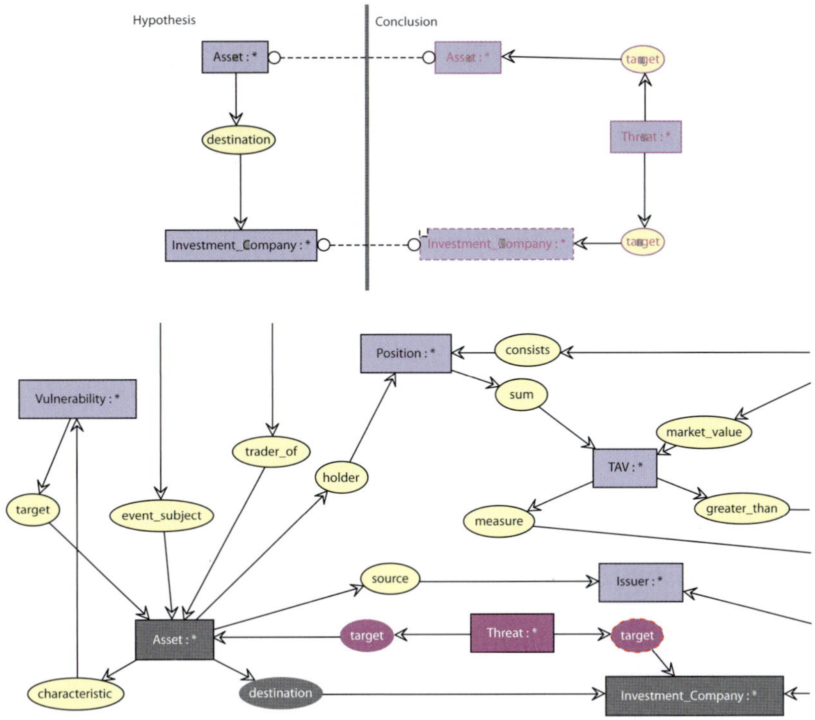

FIGURE 13.10 Refining asset rule reasoning.

BUSINESS RULE AND EVOLVING SECURITY POLICY

Through the reasoning of concepts and relationships in a TG it is possible to explore potential connections between business rules and evolving security policy, which enterprise architects can then use to counteract potential threats. As previously discussed, model automation with CG provides the mechanism for building up rules and security policy enabling the enterprise architect to explore and test the semantics of concepts used within the business transaction (Launders, 2012). Security model reasoning extends the inter-relationship between business rules and security policy in order to infer more knowledge through evolving security policy.

For example, if you own a bank account or use a credit card you may be aware that there is a security policy governing the use of transactions with those card payments (PCI, 2006). Criminals are using credit card data to make purchases and withdraw money from accounts of unknowing and unsuspecting victims. It is estimated that more than 80% of data stolen in breaches are payment card data (Verizon, 2009). Enterprises are being formed based on technology working to the Payment Card Industry (PCI) standard (Semafone, 2012). A lack of trust in the way merchants are able to handle and protect customers data means that merchants have to be PCI compliant in order to be able to process a card transaction. These security standards are also continuously evolving; for example, an early credit card transaction would have involved a signature and then a PIN code was introduced, and today authentication is via chip and PIN. The PCI data security standard is a set of 12 specific requirements that cover 6 different goals (PCI, 2006), including the following documents:

- **Payment Application Data Security Standard (PA-DSS)**: This document is to be used by payment application-qualified security assessors conducting payment application reviews so that software vendors can validate that a payment application complies with the Payment Application Data Security Standard.
- **PIN Transaction Security (PCI PTS)**: These documents include the PIN security requirement, the hardware security, and the point of interaction security requirements.
- **Point-to-Point Encryption (PCI P2PE)**: This program provides a comprehensive set of security requirements for solution providers to validate their hardware-based solutions.

These specific requirements detail how the card payment needs to be secure and how to become secure. The goals contain information on how transactions can be complaint from, for example, P2PE through to how the PIN number is entered and used (PTS).

Extending the FT case study still further it is possible to see an example of how a PCI security policy would need to be reasoned within the FT TG. For example, if a customer wished to increase their trading position through the use of a payment card it is possible to reason that if the card payment instruction is taken over the phone with, for example, a call center agent, then a part or all of the card number and the security code should not be visible to the call center agent taking a payment.

Figure 13.11 shows the hypothesis that there is a relationship between the theft of Card_Number_Details and a Call_centre_Agent. Figure 13.11 also shows the conclusion that there should be no visibility of the Card_Number_Details in the transaction to increase a trader's position in order to avoid the theft of card details. This is a valuable piece of reasoning to explore in that it identifies to that enterprise architect that a solution must be found so that the call center agent handling the transaction is unable to steal or pass on card details. The solution may then be operational, technical, or a combination of both; however, the reasoning reflecting the business activity is centered on the following core concepts (REA):

- **Resource**: customer credit on an account
- **Event**: payment increasing a trader's position on an account
- **Agent**: Customer making a card payment and call center agent taking a card payment

Interestingly it is the agent that creates the potential vulnerability and if exploited threaten part of the transaction, and it is the agent that it is already assumed to be untrustworthy by the PCI. The PCI starts from a point of not trusting any person and/or any merchant taking a card payment. It is the event, that is, "a payment increasing a trader's position on an account," which can be managed in order to secure the transaction. For example, taking the payment can be automated removing access to the data for the call center agent both visually and audibly.

CONCLUDING REMARKS

While industry best practice relies on technical controls, assessing identifiable vulnerabilities and then assessing the associated risk of a threat on that vulnerability suggests that it is possible to extend the focus to examine the associated concepts and to represent the semantics of security in terms of business transactions using the TG. Automating the FT case study using CoGui demonstrated in a basic form that it is possible to extend the inter-relationship between business rules and security policy and then to reason with those inter-relationships. The effect of this level of reasoning serves to support the overall enterprise decision-making process through modeling an understanding of the concepts and relations used in a representation of the business transaction.

Sonetto's solution using RDR for rule acquisition with CG logic as its basis for representing product documentation, rules, categories, and schema had previously demonstrated the power of developing CG systems incrementally

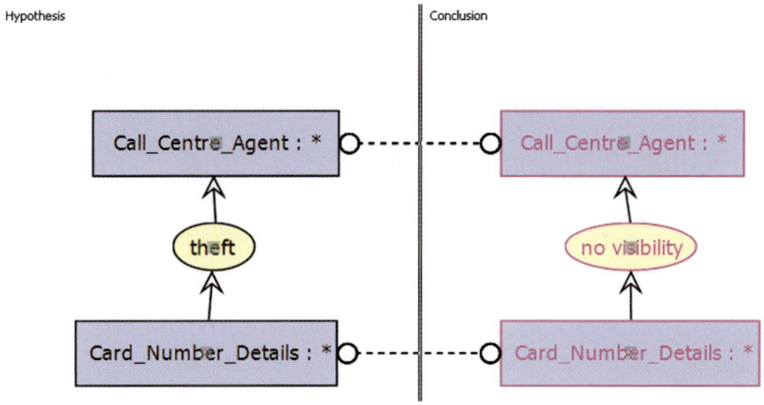

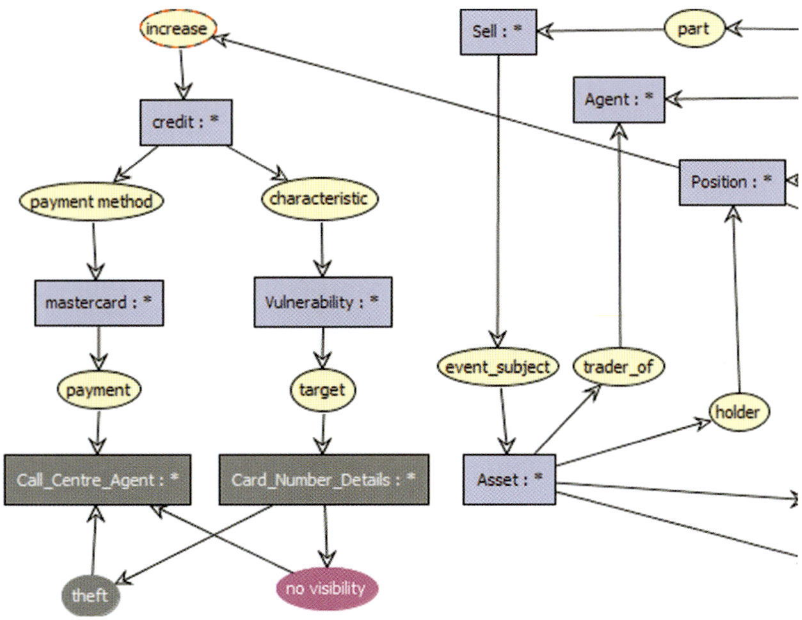

FIGURE 13.11 Increasing the credit of a trading position.

while they are already in use (Sarraf and Ellis, 2006). In the Sonetto example a domain expert was able to add new rules based on the context of an individual case (Compton and Jansen, 1990). This same approach can be used to represent security policy in the context of business transactions. The use of CoGui demonstrated that it is possible to build up a model and layer the results of applying one rule on top of the next rule, therefore, inferring more knowledge, albeit in a basic form and not in the productions sense as deployed by Sonneto.

The FT case study illustratively highlights how business rules can capture the elements of fraudulent transactions. We envisage that we can deepen this capture through the encoding of PCI policies as ontology, using emergent semantic technologies such as CGs (as we have illustrated using CoGui), the Semantic Web, and

Common Logic (Delugach, 2009; Burners-Lee et al., 2001). From this basis, the enterprise architect can analyze the threat vulnerabilities and security concepts that epitomize fraudulent transactions. We have to resort to security when we do not have trust (Schineier, 2012). Trustworthiness is inherently a complex concept and has many shades of meaning; therefore, in order to trust enterprise architecture it is necessary to be able to analyze and represent not only the business transactions but also the underlying vulnerabilities and potential security threats, therefore, going beyond the use of technical controls currently deployed today. Once we have addressed these issues then we begin to address the current limitations in card payment transactions that ultimately threaten national security.

CHAPTER

14

The ATHENA Project: Using Formal Concept Analysis to Facilitate the Actions of Responders in a Crisis Situation

Simon Andrews, Simeon Yates, Babak Akhgar, David Fortune

INTRODUCTION

In recent years, a series of security events have occurred across the globe that demonstrate the broad spectrum of crisis situations in which today's ordinary citizens are effectively utilizing their mobile communication devices. The prolific use of mobile communication devices and so-called smartphones during public disorder, terrorist attacks, extreme weather events, and public health scares are providing new challenges and opportunities for first responders responsible for securing public safety (Crowe, 2010; Lindsay, 2011). Hence, according to Edwards (2009):

> *Government, its agencies and emergency services should leverage the potential of social media for disaster management and emergency planning. Social media is important for building resilience due to its reach, usability, accessibility, and quick response time. It also opens channels of communication and allows individuals and communities to share and cooperate with one another outside the framework of traditional institutions and organisations.*

Thus the goal of the ATHENA project is to enable and encourage users of new media to contribute to the security of citizens in crisis situations and for search and rescue actions. ATHENA will deliver a suite of software tools to enhance the ability of law enforcement agencies (LEAs), police, first responders, and citizens in their use of new media in crisis situations. This project will explore how the huge popularity of new communication media, particularly Web-based social media such as Twitter and Facebook, and the prolific use of high-tech mobile devices, can be harnessed to provide efficient and effective communication and enhanced situational awareness during a crisis. The project aims to produce guidelines for enabling and encouraging the users of social media and high-tech mobile devices to contribute to the security of citizens in crisis situations as well as other contingencies demanding search and rescue activities.

By making use of high-tech mobile devices and social media for sourcing information, the project will develop the means of obtaining and delivering crisis information efficiently and effectively to and from crisis stakeholders: the emergency services command and control, emergency services first responders, and the public as pre-first responders.

THE ATHENA VISION

The public are underutilized crisis responders; they are often first on the scene, vastly outnumber the emergency first responders, and are creative and resourceful. In a crisis, the public self-organize into voluntary groups, adapt quickly to changing circumstances, emerge as leaders and experts, perform countless life-saving actions, and they are increasingly reliant upon the use of new communications media to do it. ATHENA will help them by joining their conversations and adding an enabling voice. ATHENA will give them the information they ask for in a way they can understand. ATHENA will assist them in targeting their actions by directing them to the places they need to be and away from danger. ATHENA will identify emergent behavior that is beneficial and provide support with agency resources to develop that behavior. The public has an extraordinary ability to adapt in a crisis, because it can mean the difference between life and death. ATHENA will help the public help themselves by empowering them with their own collective intelligence and the means by which they can exploit that intelligence. ATHENA will provide the emergency services with new real-time intelligence from crowd-sourced information, greatly assisting in their decision-making processes and making search and rescue more efficient. ATHENA will create a fundamental and permanent shift in the way crisis situations are managed by helping the public as victim to turn into the public as part of the crisis team. ATHENA will utilize social media and smart mobile devices as part of a shared and interoperable two-way communication platform. By developing an orchestrated cycle of data, information, and knowledge, ATHENA will empower both the public and emergency services with the intelligence they need in dealing with a crisis.

ARCHITECTURE NARRATIVE

ATHENA System Overview

The ATHENA system is a crisis communication and management system that encourages and enables the public to participate ethically in the process of emergency communication to contribute to the security of the citizen in crisis situations and for search and rescue actions. ATHENA makes use of new social media and high-tech mobile devices to efficiently and effectively acquire, analyze, and disseminate crisis information and intelligence that is appropriate and useful to LEAs/police/first responders and the public.

In the event of a crisis, such as a terrorist attack, extreme weather event, or disease outbreak, the public will be encouraged to use an ATHENA Crisis Mobile "app" to report on crisis events and activities and to participate with crisis-related activities in the social media. These reports and social media activity will feed into a Command and Control Center via an information processing center. In the information processing center, feeds from the Crisis Mobile app and information scanned from social media will be analyzed and aggregated in various ways to provide relevant, real-time information to LEAs in Command and Control. In particular, an ATHENA crisis map will be created and populated from the information processing center to visualize in real time such things as danger zones, safe routes, crisis events and activities, LEA resources, and distress signals. Via the ATHENA Crisis Mobile app the public will be able to access the crisis map and important crisis information provided by Command and Control. The app will also provide the public with a means of requesting help in an emergency situation, even when the normal routes of communication have failed.

ATHENA Components

Figure 14.1 gives a snapshot view of the system. The system consists of six main components (some with subcomponents) as follows:

1. Crisis Mobile
   a. Sending tools
   b. Receiving tools
2. Crisis Information Processing Center (CIPC)
   a. Acquisition and preprocessing tools
   b. Aggregation and analysis tools
3. Crisis Command and Control Intelligence Dashboard (CCCID)
   a. Crisis map
   b. Mobile communications center
   c. Social media content management tool
   d. Crisis Summary and Query Tools
4. Social Media
5. Interoperability
   a. Crisis Management Language (CML)
   b. Decentralized intelligence processing framework (DIPF)
6. ATHENA Cloud Secure Information Center

The following narrative describes the proposed operation of each of the components/subcomponents in terms of the tools/systems they comprise.

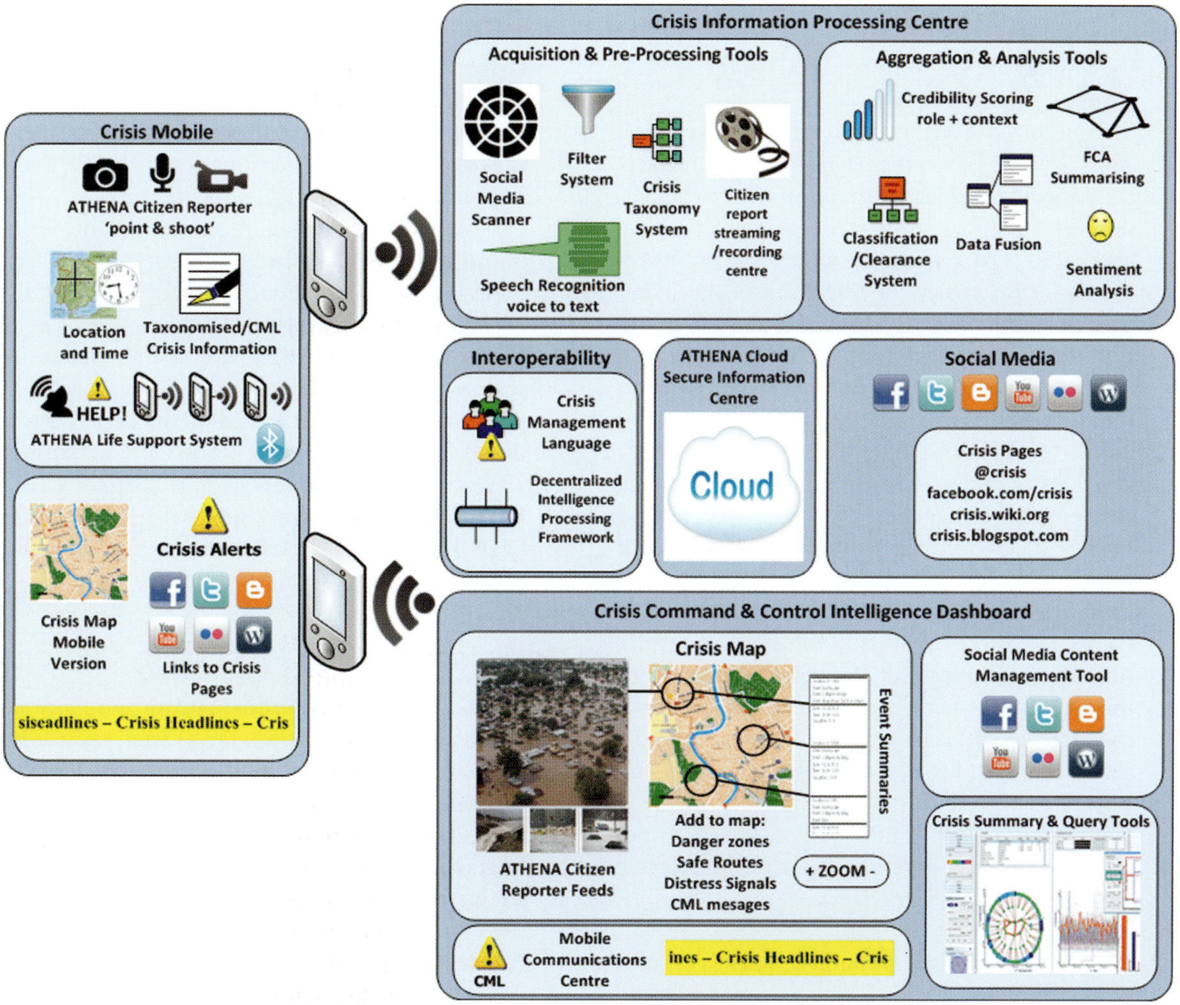

FIGURE 14.1 Snapshot view of ATHENA.

Crisis Mobile

This is an application/Web service for crisis pre-first responders comprising of a number of information-sending tools and information-receiving tools/systems.

Sending Tools

ATHENA Citizen Reporter "point and shoot" system: A "super simple" button interface to stream live photos or video as well as sound from a suitably equipped mobile device to the citizen report streaming/recording center of the ATHENA CIPC. GPS and temporal information will be included in the data stream.

Location and time: If permitted by the user, this will send GPS and temporal information with any information/messages sent to either the CIPC Filter System or the CCCID Mobile Communications Center.

Taxonomized/CML crisis information tool: This is a text-based message sending interface, combined with a crisis taxonomy and/or CML selection/menu system for the user to compose and send crisis information/event report/requests for help/offers of help to the CIPC Filter System and/or CCCID Mobile Communications Center. The destination of the information will depend on the options chosen by the user and the protocols of the DIPF.

The "ATHENA Life Support System": Provides a new means of communication that enables emergency messaging when land lines and cell communications are disabled. If land lines or emergency numbers are disabled (e.g., overloaded), the cell/Internet-enabled communications will still allow geolocated, time-stamped requests for help to be sent to the CCCID Mobile Communications Center. If the Internet and cell systems are also disabled a mesh-connected communication system, using phone-to-phone Bluetooth and/or infrared transmission, will route a geolocated, time-stamped distress signal to the CCCID Mobile Communications Center.

Receiving Tools

Crisis Map (mobile version): This is a simplified mobile version of the crowd-sourced ATHENA crisis map (the full version is part of the CCCID). Crowd-sourced crisis information, combined with geospatial and temporal information, is used to display event reports, automated summaries, danger zones, safe routes, and ATHENA Citizen Reporter feeds. A symbology will depict the level of a report's credibility, cries for help, offers of help, severity of event, type of event, type of report, and so forth, or aggregated versions where appropriate. A zoom facility in the interface will allow drill down to specific reports/areas when information has been aggregated in the display. The information available will depend on the ethics protocols, the clearance level of the user (generally the public will have a default "public" level of clearance), and the protocols of the DIPF.

Crisis alerts: An automated text message receiving system that will alert the user with targeted (by user/location/area) CML messages/warnings/instructions from the CCCID Mobile Communications Center.

Crisis headlines display: A continuous moving-banner display of current crisis news and crisis progress information from the CCCID Mobile Communications Center.

Links to crisis pages in the social media: A set of one-button links to the dedicated LEA/police/first responder social media crisis pages: Crisis Facebook page, YouTube channel, Crisis Wiki, follow Crisis on Twitter, Crisis Blog, and so forth. Users will be encouraged to join the conversations and provide information.

Crisis Information Processing Center

This is a collection of information acquisition, preprocessing, aggregation, and analysis tools. All crisis information is stored in the ATHENA Cloud Secure Information Center.

Information Acquisition and Preprocessing Tools

Social media scanner: A powerful, real-time social media scanning/crawling system that uses the ATHENA crisis taxonomies and ATHENA crisis hash-tag syntax to detect and collect crisis information sources from the general social media. Citizen contributions are also collected by the scanner from the dedicated LEA/police/first responder social media crisis pages: Crisis Facebook page, Crisis Wiki, re-tweets of Crisis on Twitter, and so forth. Photographs, video, and sound recordings are collected along with text-based sources. The acquired information sources are fed into the CIPC filter system to remove irrelevant/unwanted sources.

Citizen report streaming/recording center: This system will receive the streamed photo/video/sound output of ATHENA Citizen Reporter Systems of mobile devices and either stream them onward as available feeds to the Crisis Map ATHENA Citizen Reporter System or record them for preprocessing if bandwidth is not available for streaming.

Speech recognition system: This tool will convert into text, streamed voice recordings from the ATHENA Citizen Reporter System, and voice recordings from scanned social media sources. The text information is then fed into the CIPC filter system to remove irrelevant/unwanted sources. The system will support multilingual conversion. The proposed demonstration will be dual language: English and German.

Filter system: Acquired information sources will be parsed by the filter system using sophisticated NLP (in the case of text) and image recognition software (in the case of photo/video) to remove irrelevant/unwanted material before information is stored in the ATHENA Cloud.

Crisis taxonomy system: A set of dynamic/tailorable crisis taxonomies and Twitter hash-tag syntax with a separate interface for population, management, and so forth. The crisis taxonomy system is used extensively in the acquisition and recognition of crisis information and messages, and is also used by the aggregation and analysis tools of the CIPC.

Aggregation and Analysis Tools

In all cases below, results of aggregation and results of analysis are stored in the ATHENA cloud and may be available to the Crisis Map System if geolocated and/or the CCCID Crisis Summary and Query Tools in accordance with clearance, the ethics protocols, and the protocols of the DIPF.

Classification/clearance system: This system will use the crisis taxonomies/Twitter hash-tag syntax/CML, and other NLP and recognition techniques to classify information in terms of its type (e.g., offer of help, request for help, event information) and the clearance level required to use/obtain/process the information as determined by the DIPF.

Formal Concept Analysis (FCA) summarizing system: This system will use FCA to aggregate and summarize textual crisis information sources when they contain similar information. This will cope with large volumes of information to assist in giving a clear view of the crisis.

Data fusion system: Outputs of the various classifiers of different types, text mining, and clustering processes will be fused in order to identify and supply combined intelligence of critical situations by using a modular inference engine, supporting a combination of weakly coupled distributed fusion processes to determine aspects of the crisis such as levels of danger, crisis time line and progress, status of resources/hospitals, emerging volunteer groups and their actions, and so forth.

Credibility scoring system: This combines role- and context-based systems of reliability assessment to produce credibility scores. Information from users with recognized roles will be scored according to role (e.g., no known role, trusted pre-first responder, social media "brand leader," etc.). Information will be context scored using NLP content assessment (e.g., writing age, sexual/profanity content) and sentiment-type analysis to gauge the user's intent and background. Thus a combined credibility score can be calculated to weight the merit of the information and determine its future use in accordance with the DIPF.

Sentiment analysis tool: This NLP/text mining analysis tool, based on the development of crisis-sentiment taxonomies, will be used to determine public opinion and emotion regarding the progress of the crisis, the perceived likely outcomes of the crisis/crisis events, the performance of the LEAs/police/first responders, and feelings concerning fellow citizens in the crisis situation and their actions.

Crisis Command and Control Intelligence Dashboard

This represents the main interface for the LEAs/police/first responders to the ATHENA system. It is assumed that the CCCID is an addition to, not a replacement of, current information systems. The subcomponents are as follows.

Crisis Map (CCCID Version). This is a full PC-based version of the crowd-sourced ATHENA crisis map. The key difference will be in the size and configuration of the displays, particularly the ability to have Citizen Reporter feeds, event summary information, and the crisis map side by side.

Crowd-sourced crisis information, combined with geospatial and temporal information, is used to display event reports, automated summaries, danger zones, safe routes, and ATHENA Citizen Reporter feeds. A symbology will depict the level of a report's credibility, cries for help, offers of help, severity of event, type of event, type of report, and so forth, or aggregated versions where appropriate. A zoom facility in the interface will allow drill down to specific reports/areas when information has been aggregated in the display. The information available will depend on the ethics protocols, the clearance level of the user (generally the CCCID will have a default "all information" level of clearance), and the protocols of the DIPF.

Mobile Communications Center

CML messaging Tool: This is an interface for sending and receiving CML messages and communicating with other LEAs/police/first responders and with citizens who are using Crisis Mobile. CML requests for help and distress signals from the ATHENA Life Support System are received and alerted here.

Crisis headlines tool: This is a tool for populating and managing the Crisis Headlines Display.

Social Media Content Management Tool. This is a set of social media templates, Web-interfaces, and links to social media sites for the creation, population, management, and maintenance of dedicated LEA/police/first responder social media pages, such as a Crisis Facebook page, YouTube crisis channel, Crisis Wiki, Twitter profile, Crisis Blog, and so forth.

Crisis Summary and Query Tools. This is a set of visual analytics tools for accessing the ATHENA cloud and displaying crisis information such as casualty statistics, first responder resources, communications details, crisis progress and time line information, sentiment analysis results, and so forth. The tools will have corresponding query functionality for further, focused situation analysis and drill down. All crisis information that is used for annotating the crisis map is also available here in alternative views, for example, summary/temporal/or other facets of interest. Nongeolocated information, not suitable for the crisis map, is also accessed and included in analyses via these tools.

Social Media

Although not strictly a component of the system, the social media, such as Facebook, Twitter, BlogSpot, Word press, flickr, YouTube, and so forth, are main sources of crisis information for the ATHENA system. The social media will also host a set of dedicated LEA/police/first responder social media pages, such as a Crisis Facebook page, YouTube crisis channel, Crisis Wiki, Twitter profile, Crisis Blog, and so forth.

Interoperability

The ATHENA system has heterogeneous and distributed tools, information, and users. Interoperability is provided for these diverse human and technical elements via the CML and the DIPF.

Crisis Management Language

CML is a reduced, structured language for crisis management communications utilizing the underlying data models, ontologies, and taxonomies of LEAs/police/first responders. CML will provide a means to communicate directives (assignments and requests) and reports in a formalized and unambiguous way with other LEAs/police/first responders and with the public via Crisis Mobile. In crisis operations, it can be assumed that each responder resource as well as the Command and Control Center and the participating hospitals and the public via Crisis Mobile are all part of a network. CML alerts and warnings can be directed to individual users, locations, areas, and so on. The connected systems should all contain a CML interface. The types of CML message and types of CML communication available to users will depend on their role, clearance level, and the protocols of the DIPF.

Decentralized Intelligence Processing Framework. This is a framework supporting secure information flows and decentralized processing of data and intelligence. The framework will establish secure information/communication flows between providers of information and analysts, and between LEAs/police/first responders and the public in CML-based communication; channels are automatically established, such that the right information is delivered to the right analyst/process/user at the right moment in time; and information sources and analysis resources are incorporated into distributed analysis processes on the fly, as they become available and on an as-needed basis. This framework will be adapted to the operation in the ATHENA Cloud Secure Information Center. The DIPF will control the access to different information sources and provide an effective protection of the confidentiality and integrity of information over its whole lifetime, combining users and automated processing units from different roles and clearance levels and data with different classifications.

ATHENA Cloud Secure Information Center

Athena services will be made available in the ATHENA Cloud Secure Information Center. This will be a robust environment in which the DIPF will be embedded. It will support robust operation by distributing (1) the CIPC aggregation and analysis processes and (2) the secure information storage including a knowledge repository for European LEAs. ATHENA will integrate with many existing cloud information channels through an extensible, flexible environment supporting diverse inputs and monitoring diverse channels. The environment will be configurable so that it will route specific types of information to a variety of community channels—some public, some protected. and potentially some encrypted—depending on the requirements of the particular service or community channel as defined by the protocols of the DIPF. Based on community/channel requirements criteria, the framework will aggregate and filter the incoming streams to provide relevant information to the community or service. The ATHENA client services will be available on a range of fixed and mobile devices.

FORMAL CONCEPT ANALYSIS

At the heart of ATHENA is a new data analysis method based on FCA to categorize, process, and deliver relevant key information to crisis stakeholders. The method is based on semantics—capturing the meaning of information as it develops during a crisis situation, facilitating credibility assessment of such information, and computing associations between key crisis facts in the form of formal concepts. Related information on crisis events and actions can then be usefully presented to selected stakeholders in an appropriate manner.

FCA was introduced in the 1990s by Bernhard Ganter and Rudolf Wille (1998), building on applied lattice and order theory as developed by Birkhoff and others in the 1930s. It was initially developed as a subsection of applied mathematics based on the mathematization of concepts and concepts hierarchy, where a concept is constituted by its extension, comprised of all objects that belong to the concept, and its intension, comprised of all attributes (properties, meanings) that apply to all objects of the extension (Wille, 2005). The set of objects and attributes, together with their relation to each other, form a formal context, which can be represented by a cross table (Figure 14.2)

The cross-table in Figure 14.2 is a formal context representing destinations for five airlines. The elements on the left are formal objects; the elements at the top are formal attributes. If an object has a specific property (formal attribute), it is indicated by placing a cross in the corresponding cell of the table. An empty cell indicates that the corresponding object does not have the corresponding attribute. In the airlines context, Air Canada flies to Latin America (since the corresponding cell contains a cross) but does not fly to Africa (since the corresponding cell is empty).

In mathematical terms, a formal context is defined as a triple $K{:=}(G,M,I)$, with G being a set of objects, M a set of attributes, and I a relation defined between G and M. The relation I is understood to be a subset of the cross-product between the sets it relates, so $I \subseteq G \times M$. If an object g has an attribute m, then $g \in G$ relates to m by I, so we write $(g,m) \in I$, or gIm. For a subset of objects $A \subseteq G$, a derivation operator $^{\wedge\prime}$ is defined

Airlines	Latin America	Europe	Canada	Asia Pacific	Middle east	Africa	Mexico	Caribbean	USA
Air Canada	×	×	×	×	×		×	×	×
Air New Zealand		×		×					×
Nippon Airways		×		×					×
Ansett Australia				×					
Austrian Airlines		×	×	×	×	×			×

FIGURE 14.2 Airlines formal context.

to obtain the set of attributes, common to the objects in A, as follows:

$$A^{\wedge\prime} = \{m \in M \mid \forall g \in A : gIm\}$$

Similarly, for a subset of attributes $B \subseteq M$, the derivation operator $^{\wedge\prime}$ is defined to obtain the set of objects, common to the attributes in B, as follows:

$$B^{\wedge\prime} = \{g \in G \mid \forall m \in B : gIm\}$$

Now, a pair (A, B) is a formal concept in a given formal context (G,M,I) only if $A \subseteq G$, $B \subseteq G$, $A' = B$ and $B' = A$. The set A is the extent of the concept and the set B is the intent of the concept. A formal concept is, therefore, a closed set of object/attribute relations, in that its extension contains all objects that have the attributes in its intension, and the intension contains all attributes shared by the objects in its extension. In the airlines example, it can be seen from the cross-table that Air Canada and Austrian Airlines fly to both the United States and Europe. However, this does not constitute a formal concept because both airlines also fly to Asia Pacific, Canada, and the Middle East. Adding these destinations completes (closes) the formal concept:

$$(\{\text{Air Canada, Austrian Airlines}\}, \{\text{Europe, USA, Asia Pacific, Canada, Middle East}\}).$$

Another central notion of FCA is a duality called a "Galois connection," which is often observed between items that relate to each other in a given domain, such as objects and attributes. A Galois connection implies that if one makes the sets of one type larger, they correspond to smaller sets of the other type, and vice versa. Using the formal concept in the previous equation as an example, if Africa is added to the list of destinations, the set of airlines reduces to {Austrian Airlines}.

The Galois connections between the formal concepts of a formal context can be visualized in a concept lattice (Figure 14.3), which is an intuitive way of discovering hitherto undiscovered information in data and portraying the natural hierarchy of concepts that exist in a formal context.

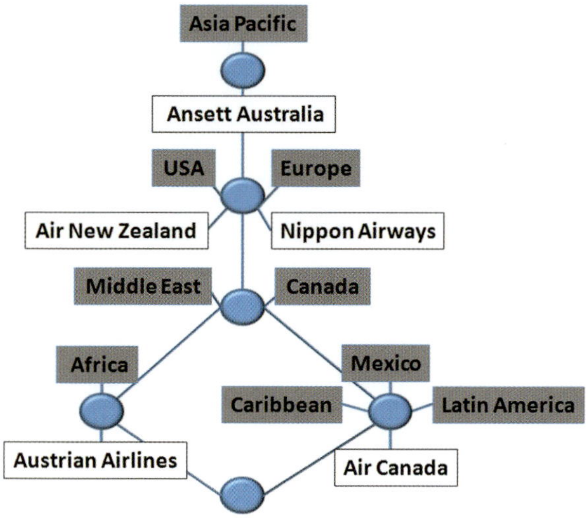

FIGURE 14.3 Airlines concept lattice.

A concept lattice consists of the set of concepts of a formal context and the subconcept–superconcept relation between the concepts. The nodes in Figure 14.3 represent formal concepts. Formal objects are noted slightly below and formal attributes slightly above the nodes they label.

A concept lattice can provide valuable information when one knows how to read it. As an example, the node labeled with the formal attribute "Asia Pacific" will be referred to as Concept A. To retrieve the extension of Concept A (the objects that feature the attribute Asia Pacific), one begins from the node where the attribute is labeled and traces all paths leading down from the node. Any objects one meets along the way are the objects with that particular attribute. Looking at the lattice in Figure 14.3, if one takes the attribute Asia Pacific and traces all paths leading down from the node, one will collect all the objects. Thus Concept A can be interpreted as "All airlines fly to Asia Pacific."

Similarly, the node labeled with the formal object "Air New Zealand" will be referred to as Concept B. To retrieve the intension of Concept B (the attributes of Air New Zealand), one begins by the node where the object is labeled and traces all paths leading up from the node. Any attributes one meets along the way are the attributes of that particular object. Looking at the lattice once again, if one takes the object Air New Zealand and traces all paths leading up from the node, one will collect the attributes USA, Europe, and Asia Pacific. This can be interpreted as "The Air New Zealand airline flies to USA, Europe, and Asia Pacific." As a further example, the formal concept involving Air Canada and Austrian Airlines, from Figure 14.3, can be clearly seen in the concept lattice as the third node down from the top of the lattice.

Although the Airline context is a small example of FCA, visualizing the formal context clearly shows that concept lattices provide richer information than by looking at the cross-table alone. However, concept lattices are not suitable visualizations for pre- and first responders in crisis situations; they require some expertise and time analysis. Thus ATHENA will develop new FCA visualizations appropriate to the Crisis Summary and Query Tools in the Command Control Center and use the FCA-derived crisis summary information to populate the ATHENA Crisis Map.

FORMAL CONCEPT ANALYSIS FOR DERIVING CRISIS INFORMATION

Central to ATHENA is the use of FCA to aggregate and summarize textual crisis information sources when they contain similar information (Figure 14.4). High performance FCA software, based on fast concept mining algorithms (Krajca et al, 2008; Andrews, 2011) will cope with large volumes of information to assist in giving a clear view of the crisis. This high-performance data capability will enable FCA to be applied in a dynamic fashion. ATHENA will develop a new facet to FCA: Dynamic FCA. Parallel processing and bit-wise optimization (formal contexts are implemented as bit-arrays enabling 64-way computation) will be combined with FCA-based aggregation, noise filtering, and simplification techniques to supply crisis stakeholders with real-time information and analytics. "Super-simple" mobile visualizations and interfaces will be provided to present important information in a clear manner via the ATHENA crisis map and to allow pre- and first responders to scan and

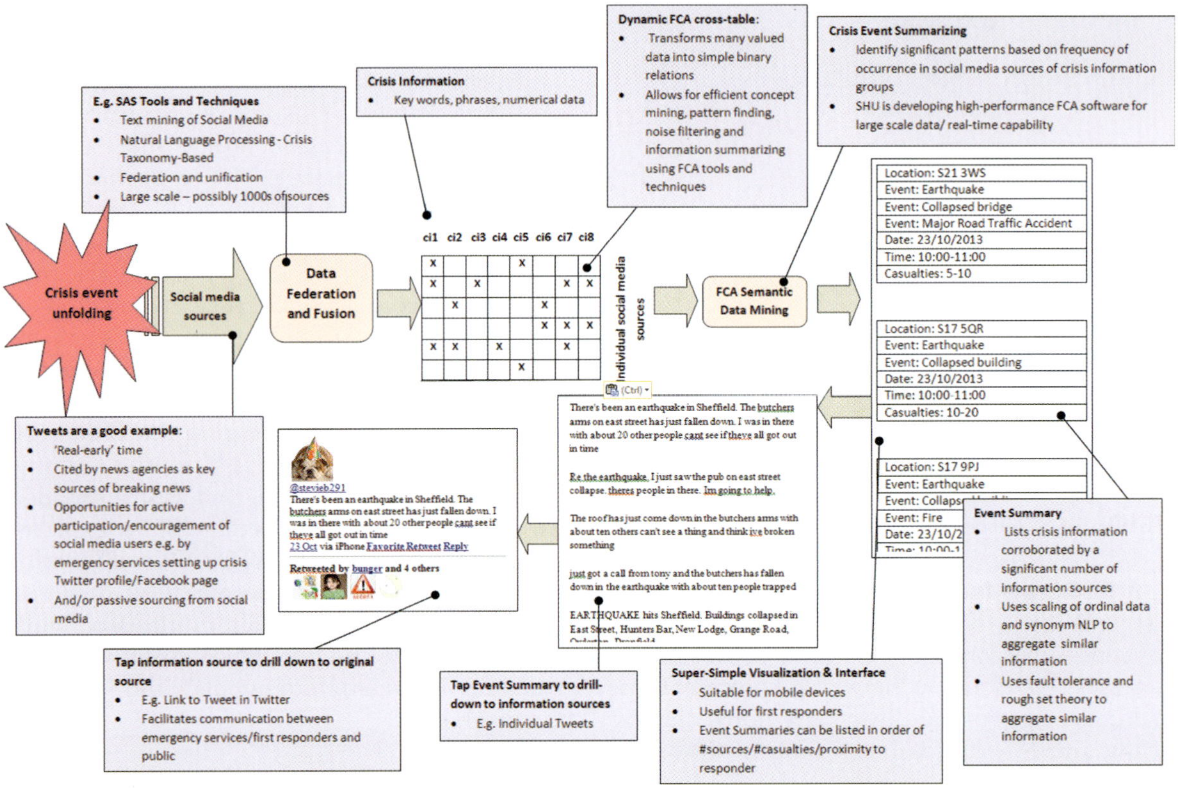

FIGURE 14.4 Using FCA to derive crisis information.

drill down through FCA-derived crisis summary information, from social media such as Tweets and Facebook posts.

London Bombing Example

To provide a demonstration of use, a simple example is taken from the July 2005 London bombings. As social media was in its infancy eyewitness accounts are taken from online media available at the time. A total of 66 accounts were used in the analysis. ATHENA will adapt and develop new crisis taxonomy-based text mining techniques and tools, but for the purposes of this example simple well-known techniques were used. The text of all the accounts was aggregated into a single document. This was fed into tag-cloud software to extract keywords to be formal attributes. The 66 accounts were then used as the formal objects thus creating a formal context. FCA software was then used to extract the largest (most significant) concepts (Andrews, 2011) and visualize them as a concept lattice (Figure 14.5).

The lattice in Figure 14.5 appears to indicate an incident on a train and another on a bus. A high proportion (48%) of the eyewitness accounts includes both "smoke" and "train" (Figure 14.6). When the analysis is repeated using this 48% of accounts, a finer level of granularity appears; the single train-related incident appears actually to be two incidents on the tube: one near Edgware Road and another near Russell Square (Figure 14.7). The drill down also reveals how serious the incidents appear to be with the common use of words such as "explosion" and "bodies."

One account in particular (no. 052) links the tube and bus incidents by the words "smoke," "people," and "screaming" (Figure 14.8). It may be worth examining the account to see if any insight can be gained on a possible connection.

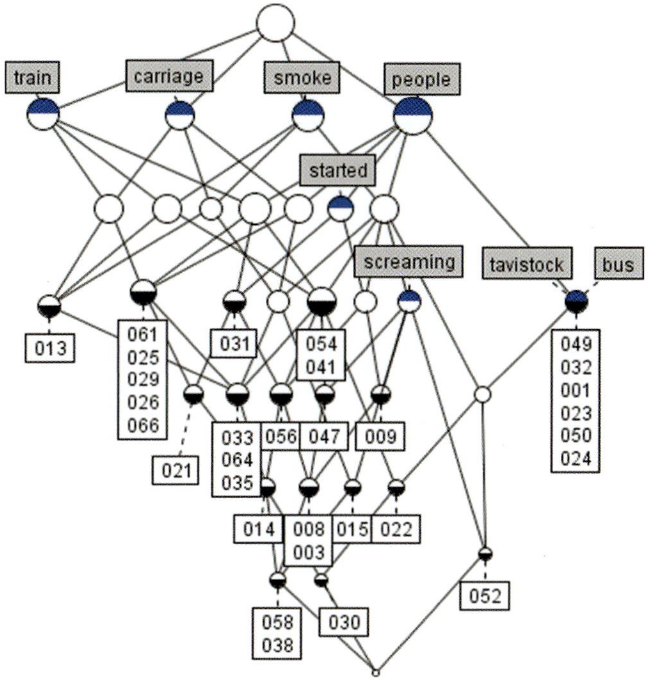

FIGURE 14.5 Concept lattice of London bombing eyewitness accounts.

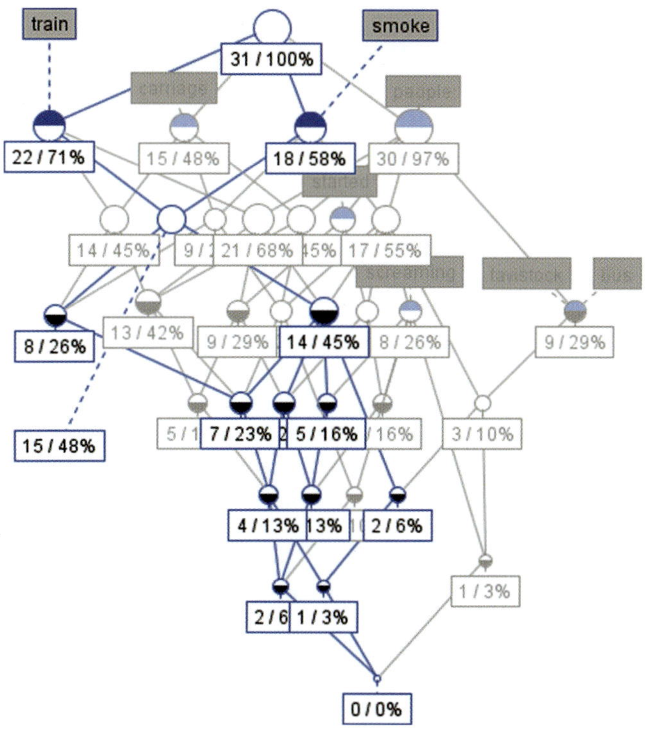

FIGURE 14.6 Highlighting a key piece of information using FCA.

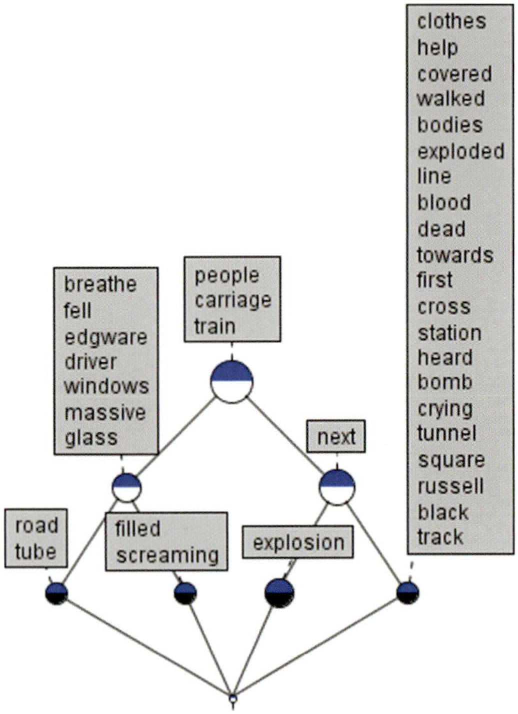

FIGURE 14.7 Drill down to reveal finer detail.

Account 052: "I'm in Tavistock Square. There was a big bang. After the smoke went away I realised there was a double decker bus exploded. People were running towards me screaming and crying. I saw at least five people jump from the top deck of the bus. Half of it was blown away. They were jumping onto the street to escape. It was such a big explosion and the bus was packed because the tube was closed. People were covered with dust and debris. I didn't see any blood."

Account 052 raises the question that if the tube was not closed, could this have been another tube incident? It is possible that someone with a bomb was unable to deliver it to its tube target and perhaps improvised an alternative public transport target. The analysis is beginning to suggest that this might be a coordinated multitargeted terrorist attack on London's tube system.

BUILDING ON PRIOR PROJECTS

To maximize the benefit from prior European Union (EU) investment Project ATHENA will build upon prior work under PASR and FP7. Key partner projects whose outcomes will be used in the ATHENA project include the FP7 projects Odyssey, CUBIST, DIADEM, and INDIGO, and other security agency funded projects such as C-BML (NATO), "Communicating in Crisis" (Federal Bureau of Investigation), "Community Resilience/Shielding for the National Capital Region" (U.S. Department of Defense), and "Advice in Crisis" (Federal Emergency Management Agency). Other prior EU projects to be drawn on include: SGL for USaR, ESS, G-MOSAIC, SAFER, and CrisComScore. The SGL for USaR project addressed problems facing rescue teams following massive destruction and large structural collapses. This project focused mainly on the location of entrapped victims using chemical and physical sensors. The ESS project relied heavily on the use of multiple sources of information in order to build a picture of the situation on the ground and a real-time synchronization of information, with commanders able to communicate via group "text" (SMS) message and recorded voicemails. GMES services such as the G-MOSAIC and SAFER projects are still relevant to crisis management as they provide details of a situation on the ground within six hours of a request from crisis management agencies. CrisComScore developed guides for crisis communication strategies in order to make authorities better prepared to communicate with the public in crisis situations. ATHENA will build on these projects in a number of ways. The proposed platform could provide live updates from nearby members of the public, as well as potential victims, with regards to location of collapses, condition of structures, and possible entry/exit points to urban areas where the rescue teams may be needed. ATHENA in the hands of search and rescue teams would provide an effective and detailed platform for ensuring the flow of important information from the field to the command and control centers. The applications provided by ATHENA would provide the opportunity for pre-first and first responders to immediately send geolocated messages and mini-blogs, or provide pictures and/or live stream video to command and control centers, as well as using the information being shared indirectly between members of the

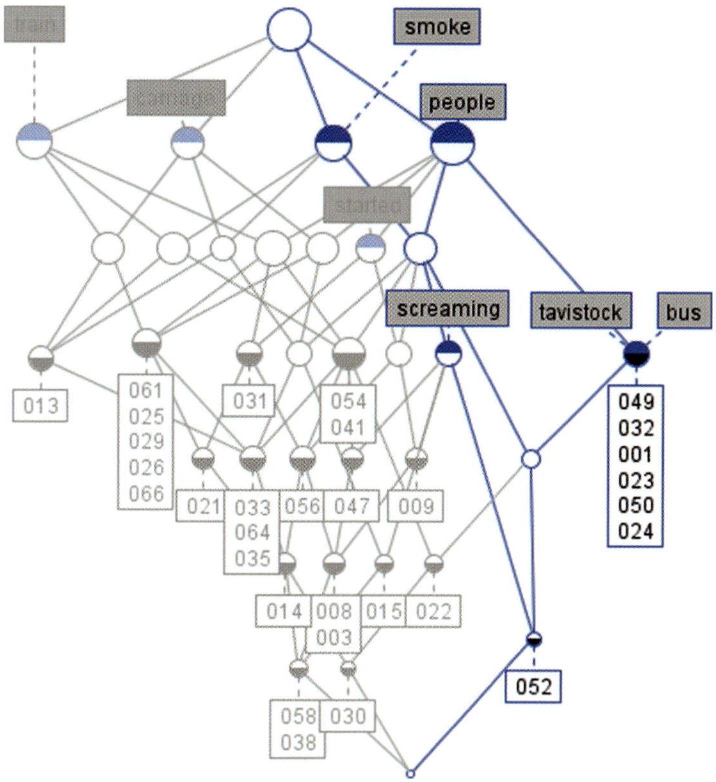

FIGURE 14.8 Account 052 links tube and bus incidents.

public via social media, providing decision makers with critical and constantly updating information regarding the layout of crisis scenes. While the public acting as pre-first responders often provide crucial help and assistance in the immediate aftermath of a crisis situation, their presence can often hinder further efforts by blue light responders. A more constant discourse coupled with the ability of mapping crowds based on geolocation information during the crisis can ensure able-bodied members of the public are provided advice on moving out of emergency areas (including safe routes, areas to avoid, and modes of transport that may still be available to them) in a timely fashion.

ATHENA Concepts in Related Domains

It is expected that the outcomes and deliverables from ATHENA will have applicability in other security and safety domains where the involvement of the public as a key resource can be enhanced and facilitated. Inherent in ATHENA is the notion of terrorist activities as crisis situations and it is envisaged that the system can be adapted to focus more on the detection and prevention of terrorism, with the public acting in a new, more coordinated and organized role, as the eyes and ears of counterterrorism LEAs. It is a small step from this to broaden the use of the system to dealing with organized crime and radical groups.

CONCLUSION

It is important to note that the FCA in the London bombing example is based on corroborative information—the key facts are revealed through the common use of words and combinations of words in information sources. Provided in real time, this analysis will give added clarity, insight, and corroboration to the information flowing into emergency services telephone and command and control operations centers. The example is limited in that it is only simulating social media scanning and does not use more sophisticated taxonomy-based text mining techniques that will be used in ATHENA. Nevertheless, the use of FCA demonstrates how key facts can be revealed and corroborated in a methodological manner. The analysis presented here was "open" in that in the criteria for key words in the text mining were limited to frequency of occurrence and the omission of stop words. In ATHENA, crisis taxonomy will allow differentiation of information into categories such as damage assessment, causality numbers and severity, and search and rescue operations. In addition, geospatial and temporal information will be acquired in real time to provide further levels of analysis and to populate the ATHENA crisis map. Automation of this analysis and tailored visualizations in ATHENA will provide crisis stakeholders, including the public as pre-first responders, with reliable, relevant, and timely information to help safeguard and rescue citizens in crisis situations.

CHAPTER

15

Exploiting Intelligence for National Security

Gregor Pavlin, Thomas Quillinan, Franck Mignet, Patrick de Oude

BIG DATA: CHALLENGES AND OPPORTUNITIES

National security can undoubtedly benefit from advanced intelligence analysis solutions capable of efficient and thorough exploitation of huge data volumes stemming from the omnipresent sensing, communication, and information processing systems. Accurate and timely intelligence is critical for the effective use of generally limited resources of law enforcement agencies combating terrorist and criminal organizations. Such organizations often operate on a global scale and, thanks to the proliferation of the sensing, communication, and computing technology, they leave useful traces in the vast digital landscape. For example, let us assume a criminal organization involved in illegal cultivation of narcotic plants (e.g., cannabis) coupled with human trafficking spanning multiple countries. As a consequence of the indoor cultivation of cannabis, the energy company will log higher than normal energy consumption in certain areas and medical services and the law enforcement staff in those areas will record higher than normal rates of certain types of crimes and violence, and so forth. It is likely that the criminal organization sets up multiple cultivation facilities throughout a larger region, further increasing the amounts of relevant data. In addition, the associated human trafficking activities will result in large quantities of heterogeneous data obtained at different locations and points in time; the vehicles used for human trafficking will be captured by video surveillance systems along highways, various types of intelligence will be gathered by national and international law enforcement and intelligence agencies on persons associated with this activity, citizens might report associated events/anomalies, communication logs will capture entries in social media, and there will be useful records of bank transactions, and so on.

While such data traces can provide invaluable indications of criminal activities and tendencies, making sense of all that data is becoming increasingly challenging. Ironically, the proliferation of the sensing, communication, and computing infrastructure has resulted in the production of huge quantities of heterogeneous data that is becoming very difficult to handle with the traditional analysis methods and computing solutions.

The information useful for the discovery and assessment of criminal activities is typically dispersed throughout many databases and hidden in large amounts of irrelevant data. The relevant data can often be considered as weak signals; that is, the same data could have been produced by many different, mostly normal activities. Therefore, we cannot make reliable conclusions by considering only one type of data. Instead, criminal activities can often be "decoded" only through patterns consisting of heterogeneous types of data generated at different locations and points in time. In other words, extraction of the relevant information requires analysis of huge amounts of often very heterogeneous data and understanding of the different types of relations between the data (i.e., domain knowledge). As one expert noted, the intelligence community in the law enforcement domain is confronted with an information tsunami. At the same time, critical information is often not communicated to the right stakeholders at the right moment in time, due to various technological, organizational, and political barriers (see the examples in Relevant State-of-the-Art Technologies). In other words, the analysts are overloaded with certain types of information while critical information is missing.

Law enforcement agencies are confronted with Big Data challenges (Eaton et al., 2012). However, in contrast to common Big Data problems, complex relations between heterogeneous data have to be considered. This is a substantial challenge that requires rich expertise about the domain. Contemporary analysis problems are usually so complex, that adequate solutions require a well-balanced combination of multiple types of complementary technologies and methods. Moreover, there is no one-size-fits-all solution. Each analysis problem requires a specific combination of data analysis methods, intelligence experts, and data sources along with the suitable access protocols and security solutions.

Therefore we introduce integration principles and a framework that supports fast creation of advanced solutions to complex analysis problems by seamlessly combining very heterogeneous analysis capabilities based on different types of algorithms and domain knowledge. This is achieved by combining the following:

1. A novel service-oriented architecture supporting integration of arbitrary analysis capabilities
2. Tools for efficient construction and maintenance of systems consisting of very heterogeneous analysis services
3. Communication mechanisms implementing multi-level security; this approach facilitates large-scale intelligence gathering and analysis in domains characterized by a great diversity of information systems, analysis methods, and collaboration between different stakeholders

Please note that this contribution is intended to provide a lightweight overview of the relevant technologies and to illustrate a novel approach to solving complex analysis problems that can be found in the national security domain. The technical details about the discussed technologies and principles can be found in the referred scientific and technical documents.

The text is organized as follows: an overview of the relevant state-of-the-art technologies and the discussion on their applicability is followed by an introduction to knowledge-based allocation of analysis services, a key principle facilitating integration of heterogeneous analysis capabilities. The integration principles are illustrated with the help of a simplified example. The remaining text introduces a combination of technologies that support advanced analysis solutions based on these integration principles and discusses their applicability in the context of the national security.

Relevant State-of-the-Art Technologies

The targeted analysis problems require inherently complex solutions and depend on many complementary methods and technologies. In this section we provide a concise overview of different types of relevant technologies supporting the following:

1. Computation, data storage, and data organization
2. Security
3. Data analysis

We emphasize the main properties of these technologies and expose how they influence each other. Each data analysis approach imposes requirements regarding the computational and data storage/organization associated with specific installation and maintenance costs. Conversely, the available infrastructure limits the choice of the analysis methods and the overall analysis capabilities.

Computational Infrastructure, Data Storage, and Data Organization. The targeted intelligence analysis problems typically require a powerful computational infrastructure that can cope with massive processing and data transfers. Given the recent trends, it is likely that the used computing infrastructure is heterogeneous, consisting of mainframe computers, large servers, desktop computers, and mobile computing devices, all connected via networks with different capacities. An important, economically efficient computational paradigm is "massively parallel computing." It supports the scalability requirements of modern large-scale processing solutions by organizing the infrastructure into "clusters" of relatively cheap, off-the shelf processing units (Beynon et al., 2002).

Another challenging aspect is physical storage of Big Data, not only because of large volumes, but also because of specific requirements introduced through distributed computation and the available infrastructure. Transferring a large dataset is time-consuming, therefore, the computation should take place as close to the data as possible (see the "data locality" concept; Chang et al., 1999). This is known as the function to data (as opposed to the data-to-function) paradigm where the location of execution of a computation is chosen as a function of where the data are physically stored. Moreover, in contemporary approaches to physical storage in Big Data systems, there is a tendency to avoid storage area networks and network attached storage. Instead, the data is stored throughout a large number of local (with respects to the processing units) disks complemented by a middleware supporting shared file system management.

Along with the physical infrastructure, data organization has a significant impact on the efficiency of the data analysis. The dominant paradigm for organizing data has been the relational model used with some variants in most currently implemented Distributed Relational Database Management Systems (RDBMS). This paradigm relies on a consistent structure of the data it holds and the implemented systems provide a number of advantages among other guarantees over the reliability (ACID capabilities) and query capabilities. However, relational models might not be well suited for some big data problems because of the following:

- The handled data cannot (always) easily be represented in the form of structured records.
- Analysis on large-scale datasets is oriented toward reading data, producing "summary" characteristics of the dataset, much less writing. Atomicity of transaction is thus less necessary and would even be a limiting factor for massive parallelization.
- Although integrity of the data is important, duplicating the data can guarantee it, without guaranteeing the consistency and enduring the corresponding penalty in computation and communication costs.
- Usage of the relation between elements in the data by an analysis algorithm is likely to constitute a serious obstacle to massive parallelization required for the analysis of large scale.
- Any query on a large quantity of data in a relational database will be time-consuming.
- Often the speed of sequential access is more important than speed of random access.

These challenges have been addressed by the Apache project Hadoop (http://hadoop.apache.org/), a middleware consisting of a programming paradigm (map/reduce) and the Hadoop distributed file system. Datasets are split in smaller blocks of key-value pairs on which the algorithms can be executed directly. To complement the purely file system-based approach of Hadoop, new types of databases have emerged lately, such as NoSql databases, lightweight database systems that abandoned the constraints of the relational system, the ACID requirements (atomicity, reliability...), and the structured query language. Many variations exist between the NoSQL and the RDBMS, depending on the way data are stored (e.g., key-value pairs, "Documents," semistructured approach similar to tables derived from the BigTable concept of Google such as HTable) and other capabilities. The recent growth of social and communication networks has motivated the development of databases especially dedicated to organizing datasets well represented by graphs. Recently Google announced new developments such as "Dremel," which borrows the concepts from both the world of RDBMS and BigTable. The approaches avoiding relational representations are usually optimized for sequential access.

System Security. There are several notable threats when dealing with Big Data, and these can be divided into two categories: data breach threats and threats to the data processing architecture. Furthermore, several types of security properties must be considered, including authentication; authorization (access control); confidentiality; and the integrity of the data, the results of any computations, and the system. For example, preserving the confidentiality of the data depends on where the data processing occurs, either on trusted hardware controlled by the owner of the data or on shared hardware. Fortunately, these are not new challenges. Extensive work has been produced in areas that can be readily adapted to the Big Data sphere, including foundational work on confidentiality (Bell and Lapadula, 1976), integrity (Biba, 1976), and authorization (U.S. Department of Defense, 1983; Sandhu et al., 1996). This fundamental work has been applied to the topic of secure distributed computing (Foster and Kesselman, 1997, White et al., 2001, Kagal et al., 2003, Foley et al., 2004).

The primary concern is that data should be accessed only by authorized users. However, by using traditional approaches to security, the requirements for sufficient security often contradict the requirements for the data throughput. The challenge is flexible creation of information flows without compromising the security aspects. Furthermore, challenges exist in how to protect the resultant information once it has left the confines of the secure data processing center. It can be readily appreciated that whenever information is processed and results are generated, these results can also be confidential.

There are two main approaches to securing Big Data: centralized and decentralized. A centralized approach implies using a single trusted compute cluster (or Cloud provider) to store and process all of the data. While recent advances in homomorphic encryption (Gentry, 2009) show that it is theoretically possible to operate on encrypted data, practical implementations of such algorithms are many years away. With the centralized approach, all data owners must trust the compute provider with the entirety of their data; there are no guarantees beyond contractual penalties and potential reputational damage to the computer provider specified in terms of service-level agreements (Quillinan et al., 2010).

A decentralized approach relies upon data owners providing access to their data only on computer systems under their control. Instead of moving the data, code is migrated to these computer resources and executed. The primary threats in such systems are twofold: the danger of malicious code executing on trusted compute resources and the danger of leaking sensitive information out of the trusted system. Approaches to resolving these threats include static analysis of third-party code (Dragoni et al., 2007) or reference monitoring inlining (Erlingsson, 2000; Dam et al., 2009). Furthermore, such a decentralized approach requires that the data must contain a record of its past, the "pedigree" (Newmann, 2002), to allow the authorization mechanism to mediate access to it properly.

Data Analysis Methods. The analysis solutions can roughly be classified into three groups:

1. Information fusion
2. Data mining
3. Visual analytics

Data/information fusion supports automated classification, detection, and prediction of phenomena of interest. Information fusion can improve the speed and coverage of the analysis processes and introduces analytical rigor (Moskal et al., 2011). Many different approaches to fusion have been introduced, each suitable for a specific type of applications (Das, 2008). In tracking applications Kalman and particle filters have been extensively used. Classification/detection problems, on the other hand, have been tackled with neural networks (NN) (Duda et al., 2001), support vector machines (SVM) (Duda et al., 2001), Bayesian networks, Dempster–Shaffer/transferrable belief models (Smets and Kennes, 1994), rule-based approaches, fuzzy logic, and so forth. Each of those approaches is suitable for a particular type of fusion problems. NN and SVM are suitable in case large amounts of training data are available. However, often the quantities of data are too small to support such solutions, especially given the complexity of the patterns. In such cases methods that support incorporation of expert knowledge, such as fuzzy logic, Bayesian networks, and rule-based approaches can provide tractable solutions. For temporally and spatially distributed observations (a tracking problem), hidden Markov models are likely to be a good solution. A recent trend in the development of advanced fusion solutions is "soft fusion," which is characterized through incorporation of human cognitive capabilities at different levels of fusion processes (Moskal et al., 2011). Contrary to the traditional approaches to automated fusion, where humans are merely users of fusion products and sometimes supervisors of fusion systems, the new paradigm emphasizes incorporation of analysts and human observers directly into the fusion processes. In other words, human experts become part of the overall fusion processes. In this way, the repertoire of real-world solutions can be significantly improved. While automated interpretation of some signals might be very difficult or even intractable, human experts can easily recognize complex patterns. However, soft fusion requires (1) algorithms and representations that support robust processing of information produced by humans and (2) efficient exchange of data and fusion products between the experts and automated fusion processes and architectures and human-machine interfaces that support seamless incorporation of human cognitive capabilities into fusion processes. Moreover, recent developments in information fusion have introduced solutions to modular, decentralized fusion (Xiang, 2002; Pavlin et al., 2010); namely, contemporary applications often require genuinely distributed processing and the heterogeneity of the used information is steadily increasing. With modular solutions, complex fusion systems can be efficiently built and distributed over multiple devices.

Any approach to fusion critically depends on suitable inference algorithms and implicit or explicit models that describe (co)relations between the data. Different types of domain models can be used, ranging from simple physical process models up to sophisticated artificial intelligence (AI) models describing dynamic domains involving very heterogeneous observations. The domain models capture the general knowledge about the processes in the domain that produce the data. They can be created by experts or through various machine learning methods. Recent developments in AI contributed powerful approaches to modeling and inference supporting advanced fusion of very heterogeneous and noisy signals in robust and efficient ways such as the modular Bayesian networks (Xiang, 2002; de Oude et al., 2010).

Often all of the relations between the data are not known; namely, the criminals are likely to change their modes of operation resulting in new types of data traces/patterns, which are not captured by the existing mental models used by the intelligence experts or automated fusion systems. In such cases data-mining techniques play a critical role. With such techniques we can identify the similarity of datasets/patterns data produced by apparently different and disparate sources or discover relations between known criminals and unknown persons by using information on the presence of people at certain places at specific times, communication between them, and so on. Such analysis is usually based on unsupervised learning techniques, such as clustering (Duda et al., 2001). When the data can be represented by graphs, proximity matrices are useful techniques to identify relationships (Song et al., 2009). In addition, data-mining techniques can be used for automatic identification of exceptional patterns and anomalies; for example, the rate of a certain type of a crime in a specific area has increased and exceeds the historical tendencies.

Contrary to data fusion methods, data-mining algorithms do not use domain models that associate patterns with a known phenomenon, such as reports on certain types of injuries and law violations with a certain type of crime. Nevertheless, data mining provides a clue for further investigation by intelligence experts. In principle, results of data-mining processes draw attention of analysts to subsets of relevant data.

Usually expert involvement is indispensable in decoding complex data. Understanding the data requires substantial domain knowledge that cannot be introduced to automated inference solutions. In this context, the visualization of data is critical. Traditionally, this task is fulfilled by the so-called online analytical processing (ONLAP) tools running above an RDBMS, allowing the visualization of multidimensional data and taking advantage of the development in representations of data, regarding representation of graphs, network, and natural language documents (e.g., word clouds). However, these tools rely on the almost instantaneous response to executing queries. The intelligibility of results will be different for each field, but examples of recent work include visualization of massive trajectory data (Andrienko and Andrienko, 2011) and financial wire transactions to detect fraud (Chang et al., 2008).

Moreover, important properties of the analysis techniques are parallelization and scalability. This depends primarily on the nature of the underlying algorithms and the dependencies between the data. In general, the more complex the dependencies between the processed data, the more difficult it is to parallelize and find scalable solutions. In case of parallelizable processes the dominant paradigm for implementation for large-scale dataset is MapReduce (Dean and Ghemawat, 2008). In this approach the dataset is first broken into smaller blocks of key-value pairs on which the processing can be applied (map phase). The results of each individual process are thus combined (reduce phase) to form the final results. However, intelligence analysis in the context of national security is likely to require processing of data characterized through complex dependencies.

Partitioning

This is especially the case with hypothesis-based reasoning and data fusion, which rely on domain models. It has been shown (Xiang, 2002, de Oude and Pavlin, 2009, Pavlin et al., 2010) that, for a significant class of fusion problems, models implemented by Bayesian networks allow efficient distribution of models and inference processes throughout a system of loosely coupled processing modules, without jeopardizing the correctness of the reasoning processes. The keys to correct distribution of fusion processes using Bayesian networks are I-MAPS, which explicitly describe the dependencies between the data.

It should be emphasized that contemporary problems are usually so complex, that adequate solutions require a well-balanced combination of different types of analysis techniques, which are seamlessly blended into a complex analysis process involving intelligence experts and very heterogeneous data sources.

Moreover, there is no one-size-fits-all solution. Each Big Data analysis problem requires a specific combination of tools, intel experts. and data sources along with the suitable access protocols and security solutions.

In the following text we introduce principles and a framework that support fast creation of advanced solutions to complex analysis, which combine very heterogeneous computing solutions on the fly.

Solving Complex Problems through Knowledge-based Process Allocation

The key to efficient solutions is decomposition of the overall analysis problem into smaller problems, each solvable through a specific analysis service and subsequent combination of the partial results. However, many analysis challenges found in the national security domain are characterized through complex dependencies between the data. In such settings, common solutions to Big Data based on distributing the computation to many identical processes (e.g., Dean and Ghemawat, 2008) are not sufficient. The data traces are typically produced by many different types of related activities. In our example of illegal cultivation of cannabis and human trafficking, each activity produces different types of data at different locations and different times. From the "birds-eye" perspective, the dependencies between the data stemming from a criminal organization involved in the cultivation and human trafficking are extremely complex. The fact that new activities are likely to be introduced in relatively short intervals makes the analysis problem even more complex. It is not practical to collect such data in a common infrastructure and then try to analyze it as a whole. Along with various technical challenges associated with moving huge amounts of data, organizing it, and configuring secure access, the analysis challenges are typically enormous.

In such cases, the detection of organized crime purely based on data fusion would not be feasible; namely, this would require complex models of relations between large amounts of different data types, which are difficult or even impossible to construct, as many of the relations between the data are likely to be unknown. On the other hand, brute force data mining on the entire dataset is likely to be intractable; in such settings it is impossible to extract the relevant relations in the data by a distribution of data to identical processes and a simple combination of the processing results. In addition, it is not clear which types of data are relevant and should be considered in the data-mining process.

However, by taking into account the underlying physical processes, we can often understand the structure of subsets of the data, that is, (co)relations between specific types of data and different types of activities; namely, the dependencies between the data result from the dependencies between the states of the physical processes producing such data (Badica et al., 2012). It turns out that the data can often be viewed as an outcome of a causal process in which the direct dependencies between the data can easily be identified by experts. For example, illegal cultivation of narcotic plants is typically associated with abnormal energy consumption, increased rate of certain types of physical abuse, hints about illegal immigrants, and so forth. Similarly, human trafficking is likely to follow certain patterns, such as the use of certain types of transportation means, typical corridors, timing, observations of physical abuse, involvement of people with contacts to known criminals, and so on.

Furthermore, while the dependencies between the relevant data are complex, it turns out that not all data directly depend on each other. This means that we can achieve correct solutions by distributing the overall analysis throughout a system of loosely coupled heterogeneous processes. The main challenge then is to assign the right analysis services to the right data. This can be achieved, if the allocation of specific services to data partitions is based on the knowledge of the relations between the phenomena in the domain. For example, an analyst specialized in the identification of illegal cultivation sites uses his knowledge of typical relations between this activity and other accompanying phenomena to determine the types of data required for his analysis. In other words, he has the knowledge to determine which data should be used by his analysis service and typically also knows where and how the access can be arranged. Similarly, the designer of an automated fusion service has the knowledge to specify the types of data that can be processed, based on the used domain models.

Such a knowledge-based approach to service allocation can support efficient solutions to complex analysis problems. We illustrate this with the help of the example mentioned in the introduction. We assume a few types of analysis capabilities provided by different experts or automated services that might be supported by various agencies involved in organized crime investigation. Please note, that the used example does not reflect the actual capabilities of the law enforcement agencies and their organization. However, it is plausible to assume that a similar organization of the analysis capabilities either exists or can easily be introduced.

Local intelligence analysts (LA), responsible for certain areas, have the expertise to investigate the existence of illegal cultivation of narcotic plants. LAs have the knowledge to collect the relevant data indicating typical symptoms of that activity, such as the data on energy consumption trends, rates of certain types of violent crimes reported by medical services and police, and so forth. By using a domain model capturing the correlations between the data and the activity, an LA can make conclusions about the likelihood of the existence of the activity. Note that this process could be fully or partially automated with the help of an appropriate information fusion algorithm using an appropriate domain model describing the relations between the observations (i.e., data) and the activity types. In Figure 15.1, analyst LA-1 makes conclusion by using outcomes of a simple fusion service and a Web crawler service. Analyst LA-3, on the other hand, applies a fully manual approach to the detection of illegal cultivation plants by using a

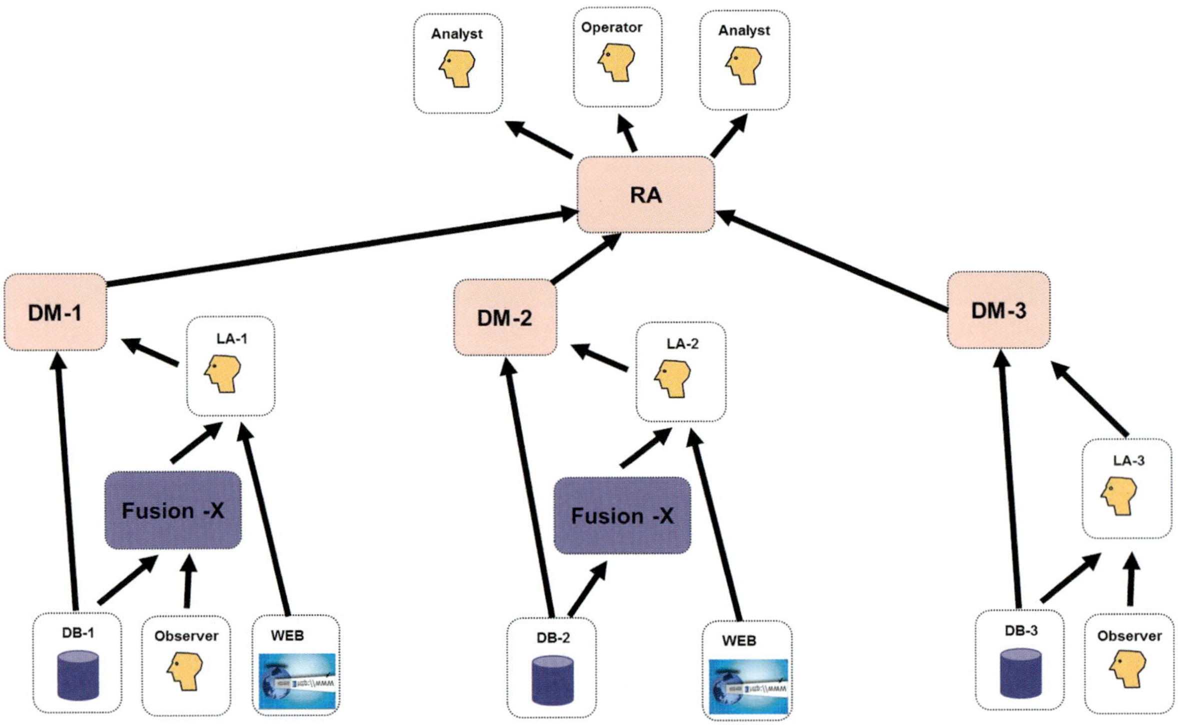

FIGURE 15.1 Collaborative detection of organized crime, a simplified example.

service providing raw data and a Web crawler service, which can be used to crawl the Web for local news reports on business and accidents.

Moreover, if the likelihood of a criminal activity is high or its existence is confirmed, a local data-mining service (DM) searching for the related data is started. The DM service takes the data used for the criminal activity detection and finds the data associated with people that are close to the people involved in the criminal activity. Such closeness could be easily obtained through the analysis of neighborhood relations based on phone records, e-mail contacts, or social media. The DM service creates an augmented dataset consisting of data associated with the involved people and those who are close to them. This dataset is delivered to a higher level relation analysis service (RA) that uses data-mining techniques to investigate the relations with other, seemingly disparate criminal activities. The RA service receives the augmented datasets from different local DM services (triggered by local LA services), each focusing on a different area. With the help of appropriate data-clustering techniques, the similarity between the discovered criminal activities at different locations is determined. This clustering is applied to the union of all augmented datasets supplied from different DM services. When distinctive data clusters are discovered, this is an indication of relations between activities from different areas. The RA service automatically sends a report on the related activities and involved persons to higher levels of the intelligence analysis system. Figure 15.1 illustrates such a collaborative approach to processing. The arrows represent the flow of intelligence products/analysis results. Note that in this example the analysts LA-1, LA-2, and LA-3, invoke different services (query them), such as specific fusion processes, Web crawlers, and human observers. This is a "top-down" approach to the creation of information flows as the LAs invoke the services and receive from them the requested information. Upon detecting the criminal activities, the LAs send a trigger to local data-mining processes (DM) along with a list of person names involved in the activity. This is a "bottom-up" creation of a connection, as the results of DMs are not used by the producer of the trigger, that is, the LAs. The local data-mining results are used as a trigger for the global RA data-mining process; the receipt of an augmented dataset from a DM triggers automated analysis of all of the data received from the different DMs within a certain time frame. This is again a bottom-up creation of a collaboration link. Similarly, the results of RA are passed on to high-level analysts. Note that this is only a simple example of how the services could establish information flows. In an extreme case, the entire activation of services could be carried out in a top-down manner. This would make sense if a hypothesis about a specific criminal network existed. The arrows represent the flow of information, that is, the analysis products.

This example exposes several advantages of the presented knowledge-based decomposition:

- A difficult analysis problem is made tractable by allocating adequate services to certain parts of data. The main criterion for the service allocation is the domain knowledge or analysis capabilities; for example, the local crime activity detection service is based on the knowledge of typical relations between the crime types and the symptoms, that is, typical observations captured by the data. In addition, such domain knowledge is required to connect the service to the relevant data sources (e.g., arrange access to specific databases, acquire reports of certain types, etc.); the domain knowledge/model used by a service determines which types of data can be processed and are, therefore, relevant.
- The detection service provides a means to reduce the data space on which higher level data-mining processes work, thus reducing the computational complexity significantly. Only the datasets that provided affirmative feedback for the crime activity hypothesis are passed on to the data-mining processes, not the actual detection result, which serves merely as a trigger for passing on selected data. While the resulting datasets processed

by data-mining services are still large, they represent only a small fraction of the entire data. In addition, the detection services have the knowledge to determine which data is relevant and should be passed on to the data mining. Note that the data-mining processes do not use any domain knowledge. They merely use domain-independent algorithms.

- The example illustrates knowledge-based data partitioning and service allocation. The domain knowledge about the relations between the types of criminal activities and the typical observable symptoms is a key to determining which information sources should be used for the detection of criminal activities. The detections draw attention of the data mining and analysis processes to specific subsets of data.
- Note that in the example the processing is based on a combination of automated processes and analysts. It is very likely that some types of analysis cannot be automated reliably.

The overall analysis structure, that is, a specific constellation of communicating analysis services, is not known prior to the operation. By targeted activation of the right services accessing and interpreting the right data partition, the hidden data structure is gradually extracted from an apparently chaotic data.

While the advantages of such knowledge-based analysis systems are obvious, their implementation is not trivial. This requires technology that satisfies challenging requirements:

- Rigorous representation of the available analysis services allows easy discovery of the right services and identification of compatible information types. Note the domain knowledge used by a service determines which types of information can be processed and produced by that service.
- Tools that support fast introduction of new services throughout the entire life cycle of the system, without any technical knowledge. The intelligence analysts should be able to declare new types of human-based or automated analysis capabilities. Due to the nature of the domain, the system will consist of a very heterogeneous, ever-growing set of analysis services.
- Automated discovery of the right services for a given problem and the available data.
- Automated creation and maintenance of secure information flows between compatible analysis services; connect only the services that can exchange information that they can process. Deliver the right information to the right analyst/automated process at the right moment in time, without breaching rigorous security mechanisms.
- Easy incorporation of arbitrary automated analysis tools.
- Easy interfacing with intel analysts with different analysis capabilities and affiliations.
- Easy interfacing with data sources, such as databases, social media, sensors, and so forth.

The following section introduces a framework that addresses the abovementioned requirements.

A Modular Integration Approach: Toward Powerful and Flexible Big Data Solutions

In this section we introduce the basic principles of dynamic process integration framework (DPIF), a novel SOA-based integration solution recently developed by Thales Research and Technology Netherlands (Pavlin et al., 2010). This system supports fast creation of advanced hybrid analysis systems by combining the following:

- A wrapper technology supporting efficient and rigorous interfacing between heterogeneous analysis services and data access services
- A configuration tool allowing easy incorporation of new analysis services; PLUG&PLAY approach to the extension of intelligence analysis capabilities
- Mechanisms for runtime formation of information flows between compatible services

The Basic DPIF Principles. In DPIF each local process (human or machine based) is encapsulated by a module that is implemented through a software agent (a DPIF agent). The agents provide a uniform interface between different local processes involved in collaborative intelligence analysis workflows. A key feature of the DPIF agents is asynchronous, data-driven processing in workflows. This is achieved through a combination of weakly coupled processes. Each module consists of at least two basic processes implemented through asynchronous threads communicating via a local blackboard (Figure 15.2). The communication engine is a thread that provides interagent communication, collaboration, and negotiation capabilities. Communication engines in different agents establish workflows between local processes in different agents by executing service discovery and negotiation (Badica et al., 2012). The processing engine, on the other hand, is a thread that encapsulates arbitrary automated or human based inference (see Figure 15.2).

A human expert is integrated into a DPIF-based analysis system with the help of a dedicated software agent, an assistant that

1. Collects all information relevant for the expert
2. Disseminates the expert opinion/estimates
3. Triggers the expert's attention in case relevant information becomes available

Such an agent continuously runs on an arbitrary server. Each expert communicates with the personal DPIF assistant via a graphical user interface, which can run on arbitrary networked computers and PDAs. Thus, DPIF services are globally accessible.

In principle, arbitrary automated reasoning techniques can be integrated into DPIF. The DPIF agents can autonomously form workflows in which heterogeneous processes support collaborative analysis. For the scenario illustrated in Figure 15.1, a dedicated DPIF agent is introduced for each analyst, fusion process, and data-mining service, as well as the sources, such as databases, Web crawlers, and the observers. Compatible agents automatically establish information flows in which the relevant information is exchanged between the data producers and the consumers that are able to process the information and meet other operational conditions. The creation of such information flows is based on a combination of service discovery and negotiation mechanisms (Badica et al., 2010) that take into account multiple conditions, such as the speed, authentication credentials, and so forth.

DPIF supports efficient and secure sharing of relevant information between the experts and automated processes; the right information is delivered to the right experts/processes at the right moment in time. One important principle is

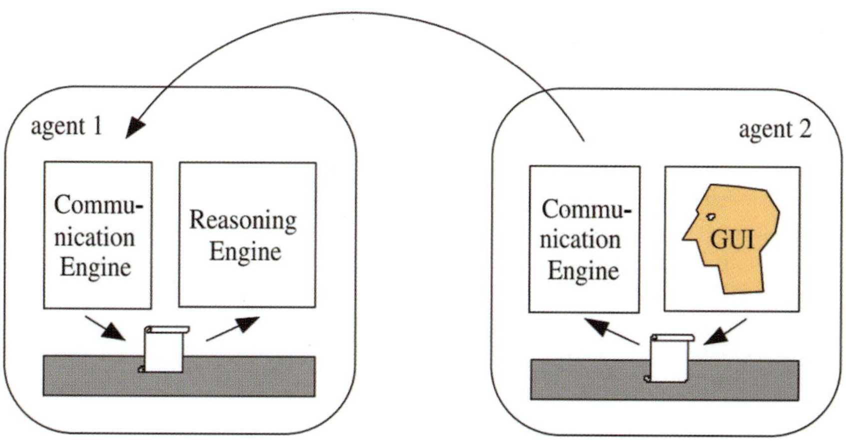

FIGURE 15.2 Interaction between agents providing heterogeneous processing services. Both agents use identical communication engines. Agent 1 encapsulates automated processing while agent 2 integrates human-based processing.

that information sharing is based on the declared capabilities and needs of the experts/processes.

From the analyst's perspective, DPIF supports information sharing and control of analysis processes. Each analyst has a software "assistant" that (1) continuously listens to the information relevant for the analyst, (2) disseminates intelligence products to the right recipients, and (3) enables simple invocation of different types of fusion and analysis services, provided by automated processes or other analysts. Note that the sharing framework is not centralized.

The DPIF uses customizable forms to facilitate (1) collaboration between experts at runtime (exchange of rich information and service invocation) and (2) description of services. Each service type in DPIF is specified through a combination of three elements:

1. A verbal description of the service capabilities (i.e., free text) and keywords
2. A form specifying the information required for the service invocation consisting of elements such as text fields, pictures, maps, audio files, and attachments
3. A form specifying the service outputs consisting of elements such as text fields, pictures, maps, audio files, and attachments.

Points 2 and 3 are based on a set of objects that represent atomic information types with clearly defined semantics and format (for more information see Pavlin et al., 2010; Penders et al., 2010).

Fast Definition of Compatible Analysis Services: A Collaborative Approach. An integral element of DPIF is OntoWizard (Penders et al., 2010), a tool that supports easy and rigorous definition of services in the DPIF-based systems. OntoWizard allows decentralized management of large-scale processing capabilities (i.e., collaborative construction and maintenance of definitions of easily composable services). The process of describing the services is delegated to the analysts and providers of automated services. They can configure the information sharing in DPIF-based collaboration systems themselves, without any knowledge of programming or formal ontologies.

Note that the OntoWizard approach is targeting organizations that can be characterized as a professional bureaucracies (van Aart et al., 2004). In such organizations an expert/service designer knows what types of information are needed in order to provide a particular service; each expert can (1) describe the service he provides and (2) specify the information required for the execution of his own service.

By using the OntoWizard approach, users can specify the supplied services and the required information, such that the processing capabilities and the needs are "understood" by the system. The system automatically generates rigorous service descriptions encoded in OWL that are used by the DPIF to (1) discover relevant analysis services and (2) compose the services into meaningful intelligence analysis systems. DPIF is using relatively simple service ontologies whose main purpose is to make analysis services easily composable and to facilitate runtime collaboration between experts and automated analysis processes. Contrary to typical approaches to runtime service composition (e.g., Sheng et al., 2002; Deelman et al., 2003; Bowers and Ludascher, 2004; Papazoglou et al., 2007; Chiu and Agrawal, 2009), the presented approach does not require any centralized ontology describing relations between services or configuration control. Instead, two types of "lightweight" ontologies are used:

- **Global service:** This type of ontology merely captures service descriptions, the semantics and syntax of messages used for (1) service invocation and (2) dissemination of service results. It is used for the alignment of the semantics and syntax of service descriptions at design time.
- **Local task:** This type of ontology coarsely describes relations between different types of services supplying different types of information. In principle, they describe which types of services provide inputs to the function used by a specific service. These relations reflect the local knowledge of each processing module/expert. Local ontologies are the key to runtime creation of workflows based on service discovery.

Note that the DPIF ontologies are not directly used for the inference about the domain. Therefore, service ontologies used by the DPIF do not capture relations between the phenomena in the domain explicitly; this is not necessary for the service discovery and composition in systems of experts. Instead, the relations between services captured in local ontologies correspond to the local domain knowledge of the experts or domain models of specialized automated fusion/analysis processes. Each expert knows which services are needed in order to provide a certain service. This assumption is realistic in a significant class of applications and allows tractable solutions with minimal ontological commitments.

Thus, by using the OntoWizard tool, each expert can describe relations between the provided and the needed services. Such a description is translated into the local ontology that is used by the corresponding DPIF agent to integrate the expert/process into a workflow. Because of this, systems exploiting complex relations between services can be built in a collaborative way, without any centralized configuration/administration authority. Note that OntoWizard enables the participants to align the semantics and the syntax of the service descriptions and the information types exchanged between these services. Each service from the example depicted in Figure 15.1 is described by the service provider with the help of OntoWizard. For each service, the provider supplies a verbal description of the service and a definition of the format and syntax of the service output. Such descriptions are stored in the repository collecting descriptions of all types of services available in the system. With this step, the service becomes visible in the system. If a service requires inputs, for example, services Fusion-X, LA-1, LA-2, LA-3, DM-1, DM-2, DM-3, and RA, the provider uses OntoWizard to (1) browse the descriptions of the service types available in the system and (2) select the most suitable existing service types. After this step, the relations between the compatible service types are captured and each DPIF agent with inputs subscribes in the overall system to the outputs of the suitable service types.

Dynamic Service Allocation. After the services and relations between the compatible services are specified with the help of OntoWizard, the DPIF framework can automatically discover (1) the services that can supply certain types of intelligence products and (2) the services that can process certain types of information. This is the basic mechanism of the service allocation to the relevant data partitions. A DPIF agent representing an analyst or an automated analysis process uses service discovery to find the relevant data sources, represented as other analysis or data acquisition services. The creation of information flows between the data and analysis services is based on a negotiation process, which takes into account multiple criteria, including authorization credentials required for the implementation of the multiple independent levels of security (MILS) mechanisms described in section Multilevel Systems Security. In other words, this process implements the data service allocation based on the domain knowledge of the analyst or the domain models of an automated fusion service.

With the help of this dynamic service allocation, complex analysis systems can be formed on the fly. The overall analysis capabilities emerge as compatible services providing data relevant for the task at hand are automatically incorporated into complex analysis workflows spanning different analysis and data acquisition services. The DPIF supports both top-down and bottom-up approaches to the service composition (see the illustration in System Security). For example, as LA-1 detects an illegal cultivation site, the DPIF system automatically delivers the results to the data-mining service DM-1, by using the service discovery and creation of a link in a bottom-up manner. On the other hand, LA-1 uses his agent to discover and connect to services Fusion-X and Web supplying the required inputs. In this case the two services are invoked in a top-down fashion.

The constellation of the analysis and data services in the overall analysis process is not known a priori, as it depends on the availability of the relevant services. Through the service discovery, all available relevant analysis and data services can be exploited. Note that the composition of services can also be implemented in a fully centralized manner, if this is required by the operators.

Multilevel System Security. In the section System Security, a number of approaches toward managing confidentiality and integrity and controlling the accessibility were introduced. The distributed and decentralized nature of the solution described previously suggests that a centralized solution, where all data is stored and processed in a single physical location, would not be practicable. Instead a decentralized solution seems appropriate. On such approach is called information-bound security where the security attributes of the data object are intrinsically linked to the data object. In such an instance, data objects are labeled with the security tokens that represent the authorization that is required to access them. Such a solution is characterized as providing MILS. Such MILS solutions allow the ordering of security domains to be specified by the owners of the information. This is in contrast to the traditional multilevel security model where an absolute order (typically Unclassified < Confidential < Secret < Top Secret) is specified for all data.

Information-bound security allows data owners to specify the security classifications of their own data and, critically, manage access to that data themselves. In the DPIF solution, a secure shared distributed data space (S2D2S) service (see Figure 15.3) supports this approach.

S2D2S provides a collaboration space where information that is relevant to the entire workflow can be shared. Note, that while the data space is not specifically prevented from storing Big Data, it is not intended to do so. However, the data space uses similar techniques as Hadoop

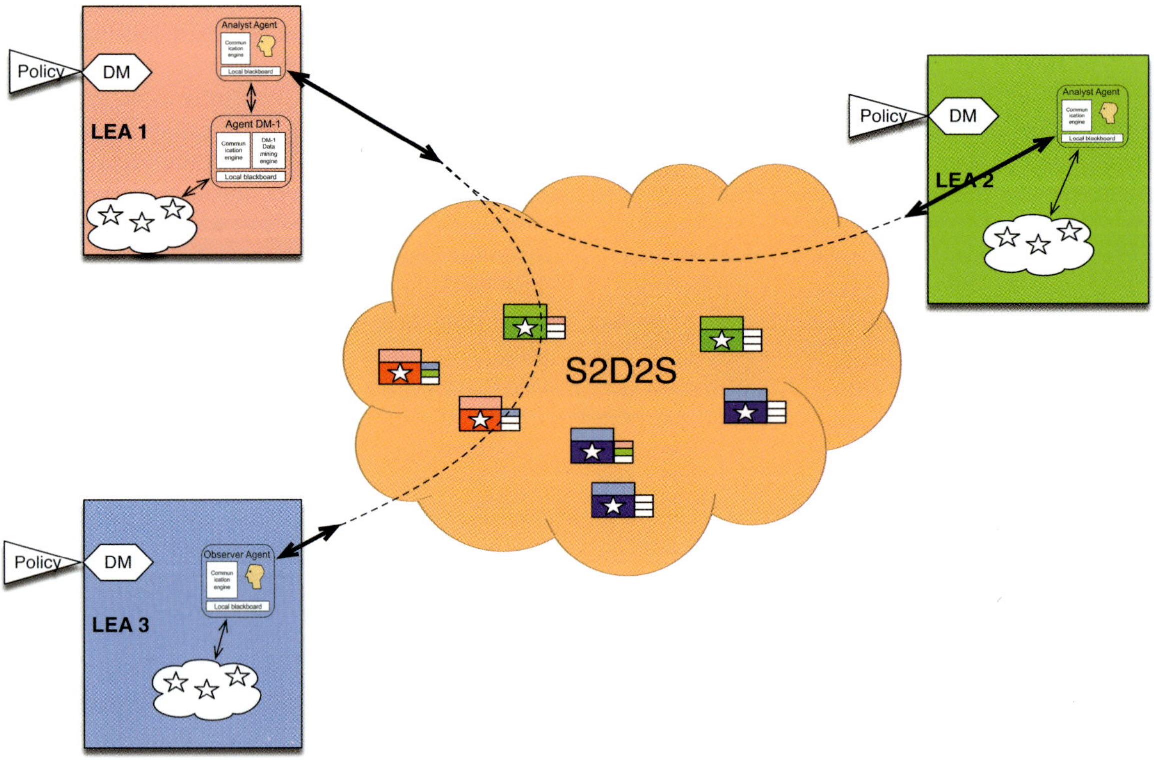

FIGURE 15.3 S2D2S service example showing three data owners (LEAs) with their requisite domain managers (DM)—agents running within the domain publish information (in a controlled manner) into S2D2S, where other authorized parties can use it.

toward supporting distributed datasets. It is intended as a collaboration space for agents to share information between security domains. Owners of the data encrypt the data with a (symmetric) key. If a third party wishes to access this information, they have to prove to the data owner that such access is allowed according to the policy of that owner. All accesses are logged and keys are rotated on a systematic basis to support revocation. Data encrypting using symmetric keys, for example using AES, does not cause significant computational overhead on modern computational systems when compared to the network operations (Chong and Deng 2006). In 2002, symmetric cryptography accounted for at most a 20% performance hit (Miller et al., 2002) and 10 years later performance gains have reduced it much further.

S2D2S uses the concept of domain managers controlling access to the information owned by their organizations. Each domain manager maintains a list of users. Users in terms of S2D2S are the producers and consumers of data and are represented by agents. When a producer publishes information in the shared space, it is encrypted and registered subscribers, who also have the key, can decrypt it. There are several distinct processes supported by S2D2S: registration; publication; subscription; revocation; rekeying, and deregistration. While there are processes in place for all of these steps, registration, subscription and publication are the most important in this discussion. Information controlled by the domain manager is split into topics, which is simply a named area within the domain.

Registration, Publication, and Subscription. S2D2S users (both producers and consumers) register with a domain manager by first looking the address up in a system-wide directory service (white pages). They then contact the domain manager using a standard protocol (SSL/TLS) and are first authenticated (by the protocol) and then provide evidence of their authorizations. The domain manager verifies that they are authorized according to the domain manager's local policy and if so adds them to the list of local users. Once this step is complete, the user can then request the ability to publish or subscribe to a specified topic, or topics, in the domain.

When a user wishes to publish in a particular topic, they send a publication request to the domain manager. This specifies the topic as well as providing any additional information that proves to the domain manager that the request should be allowed. Public keys are used to identify publishers and each publisher will use a derived key to publish. This derived key is made up of both the public key and a topic key and is provided to the publisher by the domain manager when the publication step is completed. In this way a publisher cannot read information made available by any other user unless they have a separate subscription.

Subscriptions are handled in the same basic manner as publications. However, in contrast to the publishers, when a successful subscription request is received, the domain manager sends the "raw" topic key to the subscriber. When a subscriber wishes to read information provided by a specific publisher, they must first compute the derived key for that publisher. Once this is computed the information can be decrypted.

Managing Users. If for any reason a topic key needs to be updated, the domain manager uses a specific process to ensure the keys are distributed. In particular, if a user leaves the domain (either forcefully or voluntarily), this user is first removed from the list of users. The current topic key is stored in a list of "old" topic keys in a special message by the domain manager and encrypted with the new topic key. This allows older messages to be decrypted without requiring mass re-encryption of all data. Regardless of any performance reason, it can be assumed that any information that is encrypted with an old key is already potentially compromised so no security benefit would come from re-encryption. Domain managers can initiate a key update at any time. It is used for revocation, to ensure freshness of users, in the case where they are no longer active, and periodically to ensure key freshness.

S2D2S is intended to be as invisible to the user as possible. Users can read and write to the data space without any knowledge of the cryptographic keys or the processes that take place on their behalf. Furthermore, information on the data space also contains some "meta-data" in the clear. This stores the producer ID, the topic name (so that the correct key can be used), and a time stamp indicating when it was published. Topic names are chosen by DPIF (OntoWizard described in the section Fast Definition of Compatible Analysis Services: A Collaborative Approach) to configure services during the negotiation phase. This means that it takes place transparently to the users and ensures that the communication between services and agents is seamless.

S2D2S also provides the ability for domains to join and leave whenever they wish. As domains control the information produced by users in their domain, they simply destroy the topic key and any users who do not already have access to that information.

Strong auditing controls are also in place that can prove any misuse by authorized users of the system. While information cannot be retrieved when leaked by users, contracts can be enforced

DISCUSSION

Organized crime and terrorist networks are global phenomena leaving data traces in a vast, fragmented data landscape residing in huge numbers of heterogeneous information processing systems dispersed around the globe and owned by different stakeholders. The data exposing criminal and terrorist activities must be considered as weak signals. Consequently, reliable analysis can often be achieved only by processing large amounts of heterogeneous correlated data. This, however, is not trivial as it requires (1) discovery and accessing of the relevant data and (2) the capabilities to interpret very complex patterns consisting of the relevant data.

In general, solutions to advanced intelligence analysis must consider a complex interplay between computational infrastructure, data storage and organization, system security, and analysis methods (i.e., algorithms). As it was illustrated earlier with the help of examples, contemporary analysis problems are often so complex that adequate solutions require a well-balanced combination of different types of analysis techniques, which are seamlessly blended into a complex analysis process involving intelligence experts and very heterogeneous data sources. Moreover, there is no one-size-fits-all solution. Each Big Data analysis problem in the targeted domain requires a specific combination of tools, experts, and data sources along with the suitable access protocols and security solutions.

We believe that technically sound approaches supporting integration and reuse of the state-of-the-art analysis solutions in novel ways are the key to bringing the intelligence analysis to a higher level in an economically efficient manner.

We present integration principles and a combination of technical solutions that aim to put a systematic combination of analysis services into coherent collaborative intelligence analysis systems. This integration is driven by the knowledge-based allocation of adequate services to specific data partitions. This principle is especially useful in organizations characterized as professional bureaucracies, which is likely to be a suitable form of organizing the work in systems involving different types of expertise.

This integration principle is supported by DPIF, a framework that supports fast creation of advanced solutions to complex analysis problems by combining very heterogeneous analysis capabilities based on different types of algorithms and domain knowledge. This is achieved by combining (1) a novel service-oriented architecture in which heterogeneous services can easily be represented and composed, (2) tools facilitating efficient construction and maintenance of systems consisting of very heterogeneous analysis services, and (3) mechanisms supporting multiple, independent levels of security. The presented approach satisfies the following requirements:

- Rapid and cost-efficient incorporation of arbitrary tools supporting data analysis, such as for data-mining and visual analytics tools from different vendors and various fusion approaches supporting detection of different types of events.
- Secure exchange of relevant intelligence products supplied by different analysis services owned by multiple stakeholders (various national and international agencies) with different but complementary capabilities to gather and analyze the data. This is a critical capability for solutions to complex analysis problems involving large data stemming from geographically distributed sources and analysis services.
- Fast and cost-efficient configuration of analysis capabilities across a system of different stakeholders; that is, a composition of compatible analysis services. This requires discovery of complementary services data acquisition and analysis services and fast creation of secure information channels between them. In this way we can reuse a specific data-mining or fusion tool in different contexts and system configurations, each serving a specific large-scale analysis process. This is a critical feature as there is not a single solution to the targeted big intelligence analysis challenges. Problems are complex and each intelligence analysis solution requires a specific combination of data sources and analysis methods.
- The users, that is, intelligence analysts and providers of automated data analysis services, can configure the collaborative system without any knowledge of the underlying technology. In many cases, a centralized administration of the systems and configuration authority can be avoided and the responsibility for specifying the analysis and the data access services is delegated to the users.

The presented integration framework supports a flexible combination of data-mining and model-based analysis techniques (i.e., hypothesis-based reasoning and fusion) with humans in the loop. As it was demonstrated in the earlier example, a combination of these types of analysis services can support creation of very powerful analysis solutions. The model-based services are used to zoom in on the interesting parts of the huge data volume. The model-based techniques are implemented by automated fusion services and intelligence analysts forming hypotheses based on their domain knowledge. These services exploit domain knowledge, that is, the knowledge about the relations between the observed data and the data-producing physical processes. Therefore, such knowledge-based services support linking between the different information types and services in the overall analysis system. For example, the detection results in our example can be viewed as an indication for the relevance of certain data on which the detection was based. If a dataset has resulted in a discovery of a criminal activity using a knowledge-based service, then this dataset could be further analyzed by data-mining and visual analytics methods. The data-mining services allow discovery of new relations that are not captured by domain models. Such services do not explicitly use domain knowledge and support the detection of novelties and anomalous developments. A well-balanced combination of knowledge-based and data-mining processes supports discovery of activities associated with complex patterns of weak signals.

An important question is whether the presented collaborative approach is feasible and acceptable for the organizations involved in intelligence gathering and analysis. This is the case if the organization of the analysis processes can be characterized as a professional bureaucracy (van Aart et al., 2004). In many real-world assessment problems no centralized authority is in the position to understand why certain services are compatible, as this would require substantial knowledge of the domain the services are estimating/reasoning about; such centralized approaches become intractable in many domains, including large-scale intelligence analysis in the context of law enforcement. Professional bureaucracy is a key in systems dealing with complex domains that require many different types of estimation services. In such settings the domain knowledge is dispersed throughout a system of experts. Overall, we believe that the presented solutions are relevant for organizations supporting intelligence analysis, as they often apply principles of professional bureaucracies in the analysis processes or they can easily adopt them.

Clearly, also more centralized control of the analysis processes is possible, if this is required by the involved organizations. However, the fact that important intelligence analysis problems can be solved with the help of systems characterized as professional bureaucracies has important consequences regarding the implementation of large-scale solutions involving many different stakeholders; namely, the problems are solved through loosely coupled services, where most of the data stay local to the analysis services and only certain types of intelligence products are exchanged between the complementary services. In such settings only the exchange of the critical information has to be standardized. No complex information models describing relations between all of the data are required and the service providers and users are not forced to use common databases, which is likely to grow huge in the targeted domains, and thus become very difficult to maintain and use. In addition, the security breaches are likely to have a smaller impact compared to centralized approaches due to the inherent separation of data spaces (we avoid WikiLeaks type of threats). Moreover, loose coupling between the services facilitates solutions that can efficiently reuse the legacy information systems and analysis approaches. This is important as different stakeholders already use different legacy information systems and analysis tools. It is economically and politically very difficult or even infeasible to enforce the use of a uniform information technology infrastructure and methods throughout the system of different agencies, or at least achieve a substantial coverage. Instead, the proposed solutions enable stakeholders from different agencies to gradually add new services and establish lightweight bilateral and multilateral exchange of the relevant information in a secure manner.

The presented principles, DPIF and OntoWizard, have been developed by Thales Research and Technology in Delft, the Netherlands. In the context of the FP7 DIADEM project (http://www.ist-diadem.eu), DPIF and OntoWizard have been applied to large-scale situation assessment use cases in crisis management and knowledge management domains. The evaluation of the proposed solutions indicated that the concepts and the resulting tools can efficiently cope with complex assessment problems involving heterogeneous data. In the ATHENA project, we will further develop the DPIF, OntoWizard, and MILS principles and adapt them to the national security domain. The emphasis will be on (1) improved MILS capabilities in the context of *ad hoc* information flows; (2) integration of fusion, data-mining and human-based analysis processes; and (3) implementation of operational prototype tools under the guidance of potential users.

CHAPTER

16

Re-thinking Standardization for Interagency Information Sharing

Kellyn Rein

INTRODUCTION

Data collection for intelligence purposes is vital to national security. Even more vital is the fusion of the data collected into useful information to support police and security agencies in the protection of citizens and property. However, there are a number of hurdles to this fusion.

One significant hurdle is the volume of data available, which is increasing exponentially; sifting and sorting this information to locate relevant nuggets is a daunting task. According to an IDC White Paper, "the digital universe in 2007—at 2.25×10°21 bits (281 exabytes or 281 billion gigabytes)—was 10% bigger than we thought… By 2011, the digital universe [is expected to] be 10 times the size it was in 2006…"' (Gantz et al., 2008).

Another significant hurdle is that data are produced in a multitude of formats, among these are video, audio, text, and sensor readings making automatic fusion difficult. In particular, natural language-based information, probably the most significant source of intelligence, poses a special problem as free text is only partially processable by a computer. Furthermore, multiple national agencies must work together and share information where necessary. And, as national security is no longer just "national," governments and their agencies need to work with neighboring countries and allies to share information that may written in other languages, posing legal, ethical, and security concerns for the countries involved. Standardization of data across all different sources and players would greatly facilitate information fusion, but would be very difficult, as well as prohibitively expensive, to accomplish.

Keeping up to date with criminal and terrorist activities that pose a threat to a nation's security as well in disaster response in today's global environment is increasingly complex, while at the same time increasingly important to the safety of citizens, property, and free society. There are a number of reasons for this, including, but not limited to (in no particular order), the volume of data and information available and the varying formats of the information gathered, as well as the multiple agencies both national and international involved in data gathering and sharing. Each of these agencies has a specific focus and different information needs, and each gathers structures and organizes its information differently.

As a result, there is much information available, which, if brought together, would be of great benefit. But how can this be accomplished?

There are a multitude of hurdles to jump and not just separate information systems for storing and processing data, but also privacy, ethics, and other issues, while at the same time serving the specific needs of the agencies involved.

There following aspects have to be considered:

- Multiple goals and time lines; national security is concerned with day-to-day operations (protecting borders) and with situation management (natural catastrophe, terrorist attacks, etc.), which have a limited time span, as well as with goals such as antiterrorism with time lines covering months or years.
- Information structures and their effects on data analysis, both positive and negative, for the detection of suspicious activities
- Interagency cooperation on the national level, not just police, fire, and emergency, but also relief agencies, and other nongovernment agencies in special circumstances
- Interagency cooperation on the international level, including high-level information exchange, as well as dealing with language and cultural differences

An additional overriding problem is that a huge amount of the information gathered is expressed in natural languages such as English. Unstructured natural language information (i.e., information written in free text) does not lend itself well to automatic (computer-based) information processing; in order to be optimally utilized it must be converted to a processable format by assessment and conversion by human analysts or using automated tools for analyzing and structuring such as information extraction. While a number of such tools exist and are useful, there are still some limitations in the results that they can deliver.

We look at these aspects and examine the usefulness for national security purposes of a model being developed within NATO for communication and information exchange, which is already being adapted for use in the civilian area for crisis management and forensics (organized crime) to support national security.

SITUATION AWARENESS AND INTELLIGENCE GATHERING

National security has, in fact, two goals, each has different time lines and have to be handled differently. Situation awareness and management has a more or less real-time focus; trying to deal with the immediate emergency situations of the events of the attacks the World Trade Center and Pentagon in the United States in 2001 required on-the-ground information gathering and dissemination to try to minimize fatalities. At that time, information exchange needed to be rapid and depend less on historical information and more on immediate reporting of who is where, where immediate dangers lie, where help is needed, and so on. Only later would the hunt for clues leading up to the disaster begin. Such catastrophes also require information sharing between various agencies such as police, fire, emergency, and the general populace (via television, news). Today social media such as Twitter would play an enormous role in the communication of information. In such a situation, one has no time to look for patterns or reasons, one looks for efficient exchange of information to resolve the crisis as rapidly as possible.

Intelligence gathering for such goals as the fight against terrorism and organized crime has a longer time line. Enormous amounts of information concerning individuals and groups may be gathered over periods of months or even years before concrete measures are taken. The collected data are analyzed, merged, and mined for patterns of behavior, for connections between individuals and events, and for subtle hints of collusion. In the case of disaster response, debriefing and evaluation of the crises and agency responses to them provide a wealth of information regarding possible patterns and also possible improvements to procedures for dealing with future situations.

Thus, there is an important interplay between the information needed for immediate situation management and that needed for long-term preventive actions. However, there are fundamentally different information requirements for intelligence (long-term) purposes and for situation management (short-term) for day-to-day operations as well as in the case of crisis. Information gathering and processing for both types of requirements should be compatible but often are not, often simply because individual agencies gather and store information in different ways based upon their needs. For example, harbor security authorities need to make sure that marine vessels move safely in and out of the harbor; once a ship has left the harbor, their job is done (short-term information need). However, for the purposes of combating smuggling, marine intelligence may examine the arrivals and departures of that particular vessel over a period of months or years looking for anomalous behavior (long-term information need).

SOURCES OF INTELLIGENCE DATA

Intelligence data come from myriad of sources. Much information continues to be from human sources such as informants, intelligence reports, government reports, television, newspapers (both print and digital), blogs, Web sites, and text messages. However, many nonhuman sources such as surveillance cameras, sensors, drones, and satellites provide useful information.

In order to be truly useful, data must be converted to information, in that each piece of data is part of a larger picture, and it is this picture that is of greatest use. All of the data need some sort of processing to convert it to a format that can be used for further analysis and for the fusion to give us that big picture.

CREATING VALUE-ADDED INFORMATION

Human beings are incredibly skilled data and information processors. We can take information in a multitude of forms from a huge variety of sources and synthesize that information into something much more complex. Without any real effort we recognize a face in a fraction of a second, whether the person is standing before us, is a photo in the newspaper, or is a moving figure in a video. We can decipher the meaning of words whether spoken directly, spoken on media, written, or delivered via hand signals or reading lips, with the limitations being only by the languages and writing systems known. We can take information received by one source (text) and merge it with information from another source (picture) and come up with a more complex piece of information. We can resolve ambiguity and recognize things based upon partial information. For example, Johansson (1973) showed that a human is capable of recognizing the movement of various sparse dots across a screen as the movement of another human being and even to reliably identify gender based upon that movement, which would be at this point a tremendous, if not insurmountable, task for a computer. We are capable of classifying our world based upon rules that are adaptable (a dog that has been in a car accident and has lost its leg is still recognized as a dog—although dogs are defined as having four legs—because we are able to identify "dog-ness." We are able to recognize friends by their gait, by their hair style, or a profile of their faces. Likewise we are able to recognize our pets by their size, their coloring and their markings, or the direction we are facing by our relation to the shadows cast by the sun. We instinctively can assess the speed and direction of a ball and run to catch it.

The human being is, on many levels, the ultimate information processing unit with respect to understanding, sifting and combining information from multiple sources with varying formats. With minimal effort a human can decipher and interpret visual, aural, and written information, determine relevant inputs and synthesize these into useful "value-added" knowledge about the environment of focus.

Unfortunately, human analysts are expensive and they also have a limited number of hours in the day to process new information. Furthermore, the sheer volume of new data flowing in from all of these sources, preprocessed or not, is so incredibly huge that analysts often are weeks or months behind the current influx of new data. Complicating this is that information is increasingly coming from sources that were seldom used or nonexistent a few decades ago, such as traffic cameras, surveillance videos, and use of drones or robots. This means that new ways of dealing with this inflowing information are needed. An overworked police department cannot afford to assign personnel to watch over each traffic camera 24 hours a day; instead it must rely on technology to help sort out the interesting from the uninteresting data.

Automated processing of raw data via computer to assist the analyst is necessary to reduce the volume of data into a volume that can be handled efficiently and in a timely fashion.

A great deal of information that is gathered for police and anti-terrorist units comes from human sources. Much of this is unstructured, meaning that it is written or spoken in prose. At the same time, there is a huge amount of information already stored in structured format. Text-based information, whether written or spoken, is in a free text (unstructured) format and is therefore not easy to use for automatic processing. There are two ways to convert free text to structured information. First, text may be read by analysts, who identify important elements for permanent retention in databases or summaries. Second, text-based digital information may undergo computer linguistic analysis such as information extraction in order to populate databases and ontologies of context and background information. In the case of social media, patterns showing the usage and relative occurrence of words or phrases as a means to analyze public opinion (sentiment analysis). Unfortunately, there are as yet no truly reliable and effective tools for understanding spoken language, so here an analyst, or at least a transcriptionist, is a necessity.

The analysis of device-based information likewise can be partially automated. Pattern recognition algorithms may be used on photos from traffic surveillance cameras to recognize a car plate number or model. Analysis of sound patterns from acoustic sensors may separate a human voice from background noise, although here again, ultimately, the human analyst plays a significant role. Algorithms for analyzing signals from a matrix of sensors can provide valuable information; for example, fusing the data transmitted by a series of ground sensors will warn of impending earthquakes.

The information derived through the analysis of data received from various devices and sources must be kept for further usage. Much of this information is stored in structured form in databases or ontologies. The benefits and limitations of such structured storage will be discussed later in this chapter.

RECOGNIZING DEVELOPING THREATS

Once data has been collected and analyzed, the next step is developing models of behavior observed in certain threatening situations. By matching data concerning observed behavior to the models we have developed, we hope to predict potential threats ahead of time and effectively head them off before being carried out.

Threat models, for obvious reasons, are often based upon historical evidence: once we have observed a pattern of letter bombs sent to public figures, we could look for warning signals such as an unanticipated package or an unfamiliar return address to a thick envelope, and, based upon a pattern of previous recipients, we can identify and warn potential new targets. This clearly saves lives and property.

Similar to the previous discussions regarding data structuring, a model is generally designed on the knowledge of past events, with, perhaps, some anticipation of new events. Any change in modus operandi on the part of those posing security threats requires a change to existing threat models.

Data mining and other tools that scour available (structured) information often uncover subtle but invisible patterns and connections within the data on hand, in the hopes that these patterns may reveal useful information for analysts. New models may be created from the results of data mining. However, the results of data mining are only as good as the underlying databases. Where there are gaps in the databases, there are missed opportunities for new models.

THE BENEFITS OF STRUCTURED DATA

Data collection is an extremely important foundation for security intelligence. Much of the data gathered is stored in databases and ontologies, which making finding specific information from the masses stored easier for analysts. Databases and ontologies contain structured information: a database record consists of named fields for information and a database file is a collection of records with the same collection of named fields. Where databases contain structured data, ontologies define relationships between the various objects and concepts in the working environment, including context information important for analysis.

It is easy to search structured information, to sort it, to sift it, and to merge it with other pieces of structured data in order to examine relationships between various pieces of data that are available. The ability to structure data enables rapid, relatively easy analysis of large volumes of information, and therefore plays an important and irreplaceable role.

THE LIMITATIONS OF DATA STRUCTURING

Both database and ontology structures have been determined by the builders of these information collections usually based upon their own knowledge and experience and upon input from other subject matter experts in the same and related fields. These structures are then representative of a certain snapshot of knowledge at a given point in time. Both databases and ontologies are continually updated, not just with new content, but also with new structures: files may be modified, fields added, deleted, or renamed; and new concepts or modifications to existing concepts or concept hierarchies are added to ontologies as the knowledge of the environment increases or changes. But these changes always follow the discovery that there is a gap in information needs, in other words, the changes are in most cases reactive rather than proactive.

However, one thing is common to both database structures and ontologies: by the very nature of the structures, a prefiltering of gathered information has taken place. In other words, a decision is made at some point as to which information "belongs" together; which information is important (because it is being explicitly "named" in the structure); and, implicitly, which information is unimportant (because of its absence).

While structuring information helps in the evaluation of that information, nearly every human who has had to complete a standardized form or questionnaire understands the limitations of trying to compress real-world information into a structured format that has been created by someone else. Some fields are elementary; filling in one's name and date of birth may be quite easy, but after that, there may be problems. For example, if the field for "home telephone number" has been limited to 10 digits because the designer assumed the North American (United States and Canada) 3-3-4 format for telephone numbers was used everywhere, one would encounter problems if they reside elsewhere in the word and need more digits. How often would the more appropriate answer to a yes-or-no question be "possibly" or "sometimes"? Likewise, ontologies are designed to represent and define the relationships between concepts, generally representing a sort of categorization of a particular "worldview" as determined by the designers in cooperation with subject-matter experts. The structure within ontology is based upon an interpretation of the "real world" as viewed by a group of individuals whose categorizations may vary from the perception of real-world relationships by another group.

There is a good reason why many standardized forms leave room for free text answers: not all information fits easily into a predetermined structure. Therefore, one can only assume that the information contained in any given database or ontology must contain a certain bias, simply by its design. Additionally, one must also assume there will be a certain amount of data collected that may be important but which remains unstructured because it does not fit into currently existing structures. Therefore it may remain "unprocessable" by normal data analysis methods because it is in the wrong format (i.e., "unnamed" free text).

Thus, potentially vital information may be left out if it is not in a form available for easy analysis simply because we do not know that it may be significant someday. Terrorists and criminals are always developing new schemes, so models for detection cannot remain static. Thus, we may not know today what information we are going to be using tomorrow. Therefore saving information in an easily used format but preserving the context information in which it is embedded may be important to identify previous unsuspected connections.

THE PARTICULAR PROBLEMS OF NATURAL LANGUAGE INFORMATION

Much of the information flowing into law enforcement and intelligence agencies is derived from human sources: tips from informants, interviews, reports, newspapers, government publications, blogs, e-mails, text messages, and increasingly social media such as Facebook and Twitter.

Human sources are problematic for several reasons such as the following:

- **Reliability of the source.** Devices such as sensors relay information impartially; sensors neither lie nor wish to please the recipient of their data. Sensors may be unreliable, but generally this is something that can be calibrated by extended testing of similar devices, or it can be derived from environmental clues such as temperature, humidity, and so forth. Humans cannot be calibrated. Some, however, are trained observers whose skills are focused and who may well understand how to interpret what they are seeing. Untrained observers are often swayed by emotion, lack of understanding of the situation, or simply overwhelmed by the intensity of their experiences. Law enforcement officers are well aware of the problems with eyewitnesses, even those who believe they are conveying accurate information.
- **Intent of the source.** Sources may modify their information based upon wishing to please or placate their interrogators. They may unintentionally intentionally convey incorrect or malicious information in order to shift blame away from themselves onto others or because of an emotional reaction (e.g., because they like or dislike a person involved). They may tell partial truths to win trust without giving away the game. Others just like to be the center of attention. Whatever the reason, the information must be viewed with skepticism.
- **Expressions of uncertainty in the formulation of the information conveyed.** A sensor delivers a reading, a surveillance camera captures motion in a pattern of bits; neither of these devices offers an opinion. A human, on the other hand, may state unembellished facts, but will also convey opinions, make assumptions, draw conclusions, and even pass on information from other sources. These are recognizable by the use of various linguistics clues such as "I think," "maybe," "possibility," "it appeared to be," "Mary said," and so on. It turns out that humans have very different assessments of the truthfulness implied in statements of uncertainty, even when they share a similar background. Richards Heuer of the Central Intelligence Agency discusses the problems of discrepancy in interpretation at some length in his book *Psychology of Intelligence Analysis* (Heuer, 1999).

Additional problems in formulations include such things as ambiguity or unclear references. In the fragment, "he said he would be there in a few days," it is unclear without a full context to know if the second "he" refers to the first "he" or if there are two persons implied. Furthermore, we have only a fuzzy idea of what "in a few days" means: more than one day, assuredly, and probably less than a week.

Clearly this is not a complete list, but it does convey some of the problems of using analytical tools without a surrounding context or method for acknowledging the uncertainties that exist. For example, an information extraction algorithm would select the relationship "John is a terrorist" from all of the following sentences, but the truth of this assertion is reflected quite differently in each:

"John is a terrorist"
"It is quite likely that John is a terrorist"

"It is completely untrue that John is a terrorist"
"I was told by Mary that John is a terrorist"
"For the sake of argument, let's say that John is a terrorist"

Further, even the first statement can be accepted as fact only when the source is completely reliable. This statement has completely different "truth" values when it appears in an incendiary blog posting than when it appears in the report written by, for example, the Department of Homeland Security in the United States.

AN AGENCY IS NOT AN ISLAND

The preservation of national security requires the collective efforts of multiple law enforcement, national security, and anti-terrorism agencies. Each of these agencies has its own focus and special skills. This also means that the information gathering of each of these agencies may lead to overlaps in information but, optimally, the information held by one agency should fill in the gaps for another agency in a combined effort toward a single goal.

A significant hurdle to a seamless exchange of information is the fact that there is no common representation for information collected, used, and consolidated by different agencies. Some information may be presented visually, for example, information for harbor security needs is generally overlaid on a map of the harbor, showing the current real-time location of each vessel entering, docked in, or leaving the harbor. Other information, for example, the comings and goings of ships at that same harbor over a period of months or years may be stored as records in a database or specialized logging software at the port authority. The information that an individual has been identified as a possible member of a suspect weapon smuggling group is stored in an ontology located at a national security agency. Information concerning the group's meetings near the harbor may be stored as free text in informant reports. Each instant of the results of the automated analysis of routine traffic surveillance tapes registering the abovementioned individual's car in the vicinity of a given wharf at the harbor over a period of time is stored as a record in yet another database at the department of motor vehicles. All the puzzle pieces are on record somewhere, but often stored in formats incompatible for fusion, or the fusion model does not exist because a particular scenario has not been anticipated.

Clearly, identical structuring of data would support the interchange of information. However, finding a universal model to suit all needs would be difficult, because of the varying information needs for each, and as well as exorbitantly expensive.

A NATION IS NOT AN ISLAND...

...at least not in terms of national security. Terrorist and organized criminal activities cross international boundaries. The preservation of national security today requires information collection across borders and across languages, the latter of which poses significant problems of its own. Some of this information is publicly available: newspapers, Web sites, government reports, and so forth. Other information is formally requested from corresponding international agencies. Even when all political, legal, and other considerations have been resolved, the same problems that exist with communications problems among national security and law enforcement agencies also exist between international agencies. And, of course, finally, regardless of the source, the final hurdle often remains that translation from the language of originator into the language of the receiver is needed.

SPECIFIC PROBLEMS IN WITH NATURAL LANGUAGE INFORMATION

Particularly in international endeavors, there is a further complication: human information is multiple natural languages. In general, devices deliver information neutrally; a video is not tied to a specific language (although it may be tied to a format), and a thermometer reading may be delivered in Fahrenheit, Celsius, or Kelvin.

But there are clear conversions between video formats and temperature scales. There is, unfortunately, no such clear mappings between utterances of human languages, which complicates the sharing of information.

1. **Provenance or origin of the information:** Essentially, it is the original source of the information and, if not direct information, the various persons or processes that the information passed through before it is used for crime detection. This also may include mechanisms to protect sensitive information from being publicized or shared with the wrong individuals or organizations.
2. **The reliability of the information source:** Much information comes from sources of less than sterling character or conditional reliability. For example, street sources may pass on information they believe is desired in order to better their own situation. Electronic devices may be less reliable under certain atmospheric conditions such as low light, high temperatures, or precipitation. The algorithm used for evaluation of sensor-based data delivers results that are less than 100% accurate.
3. **The reliability of the information itself.** While this seems to be a repetition of the previous point, it is not. An individual informant may be completely reliable, in that previously this person has never been known to deliver incorrect information. However, most people tend to also speculate upon things they have observed or heard, which is indicated through formulations by "maybe," "perhaps," or "I believe." Rumor and hearsay need to be identified as such.
4. **Correlation of information is occasionally relatively concrete:** The same person or organization is involved; the identical vehicle has been spotted. However, much of the correlation is indirect. It may rely on chains of information: someone knows someone who is working with a specific gang, or a series of events that happen in a particular area or within a specific time frame. Linking two events may rely on context information that is not maintained in a database.
5. **Natural language contains multiple expressions with identical meaning.** For example, "a bomb went off at...," "they blew up the...," "someone bombed...," and so forth are all ways of expressing the event "explosion." Reducing synonymic expressions to a single format makes identification of an event easier.

FINDING A COMMON REPRESENTATION

When dealing with a series of readings (data) from a single device or from multiple homogeneous devices (e.g., an array of ground sensors), the input for the fusion algorithm is in a single format, namely the format of the device's output. This format is anticipated by the algorithm, which knows exactly how to interpret the input it receives from the device(s). Furthermore, there is information concerning the device's reliability, the reliability of the output of the algorithm considering the input received, and so on. Often the algorithm has been tested on existing real or demonstration data to determine the accuracy of the results it produces. This accuracy may be represented as one or more probabilities; for example, the results of merging radar signals to determine whether an incoming air craft is friendly or hostile may be assigned probabilities for each of the possible outcomes.

When heterogeneous signals, that is, data from dissimilar devices, are fused, the algorithm will have previous knowledge of the form of the data received, as well as the uncertainties involved. Again, as in the homogeneous data format scenario, the results will generally be based upon prior information and testing, and delivered with some sort of reliability measure of the product of the algorithm.

In other words, there has been an anticipation of the types of data input as well as the results to be delivered by the algorithm. How the results will be presented to the user, in a text format, or, say, graphically on a map on a computer screen, will also be predetermined.

The problems arise when there is a need to put information together that is not in the same anticipated format. For example, in a crisis situation we may need to try to put information together from previously unanticipated sources, for example, text messages with sensor information. If we are only able fuse information using predetermined specific algorithms, we are unable to react flexibly to new situations or needs.

A standard representation for information to which the products of fusion algorithms could be mapped would be useful. In a sense, this representation would be analogous to the idea of a virtual machine in the area of computer science. A virtual machine is generally used in the design and implementation of compilers. In order to reduce the amount of work to modify a given programming language (or operating system) onto diverse hardware platform, a standard representation of a nonexistent machine is designed. All programs written in the programming language compile to this virtual machine. The intermediate code from the virtual machine can then be implemented for a specific hardware or operating system. This reduces the work of conversion from programming commands to running software on various platforms as the first conversion (commands to virtual machine) is identical from platform to platform—only the mapping from the intermediate machine to the actual machine is required.

Therefore, a mechanism by which the information derived from any source can be easily converted to a standard format for further processing would provide a clear information advantage. In this paper we will describe the use of Battle Management Language (BML) to provide a way to standardize incoming information.

A SOLUTION FOR SECURITY AND LAW ENFORCEMENT OUT OF NATO

When one views the military world, in particular in a strategic alliance such as NATO, there are a number of similar concerns. There are various branches of the military, such as air, land, and naval, gathering different types of information for their own needs. There are multiple countries involved in this coalition, with information collected in a multitude of languages and stored in databases, the contents of which, for obvious reasons, are neither identically configured nor openly stored.

However, sharing both operational and intelligence information is essential to ensure effective operations and appropriate responses to threatening situations. As a result, a model for information exchange called the Joint Command, Control and Consultation Information Exchange Data Model (JC3IEDM) has been developed with the input of NATO nations (STANAG 5525). This remains an ongoing project, undergoing continuous expansion and revision to keep up with new requirements by the partner nations. This model determines the form, structure, and names of relevant pieces of command and control (C2) information in the area of operations, so that there is a consistency in nomenclature and data types. The JC3IEDM defines terms for elements of military operations, whether wartime or nonwar, and is sufficiently expressive to formulate both military and non-military communications for a variety of different deployment types. Thus, information concerning a multitude of environmental information such the status of objects (i.e., in hostile hands, or not operational) and locations and movements of units within the area of operation, as well as events, may be seamlessly passed back and forth.

Using the data model as the basis for a lexicon, a NATO modeling and simulation group (MSG-048) developed a standardized language ("Coalition Battle Management Language") for expressing and communicating unambiguous orders, reports, and requests among C2 systems, simulation systems, and real units including robotic formations. BML has been designed as a controlled language (Huijsen, 1998) based on formal grammar (Schade and Hieb, 2006). This grammar, "Command and Control Lexical Grammar" (C2LG), is modeled after one of the most prominent grammars from the field of

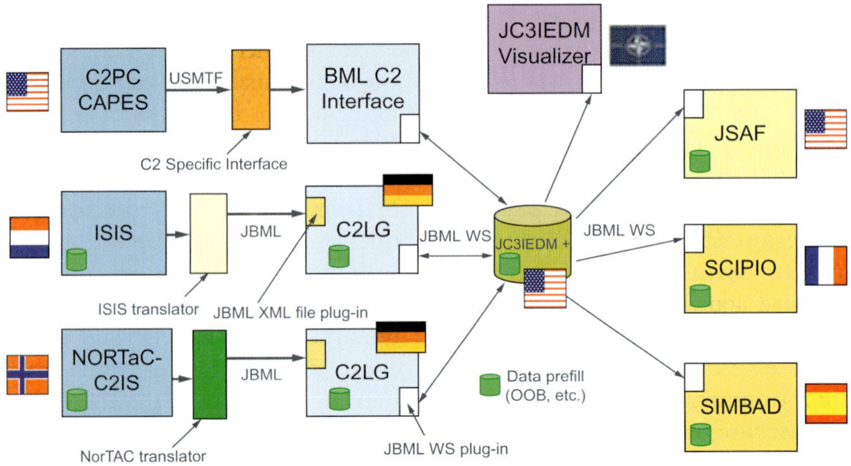

FIGURE 16.1 Architecture for the I/ITSEC demonstration shown at I/ITSEC 07 in Orlando, Florida, November 2007 *(Rein et al. 2009)*.

computational linguistics, Lexical Functional Grammar (Bresnan, 2001), resulting in BML becoming an unambiguous language which can easily be processed automatically. Various experiments with international partners have shown that this is an effective way to coordinate operations and share information (see Figure 16.1).

The communication between command and control systems of the various nations involved in the exercise via BML (J-BML or "Joint-BML," i.e., for joint forces) and the JC3IEDM allowed each nation to maintain its own C2 system as it had been developed without requiring complete restructuring of existing system components and underlying data structures. Instead, there is a mapping (middleware) mechanism that supported the interchange of orders, requests, and reports.

While many reports are generated by humans, much information is delivered by devices. In some cases, the reports are of a single value such as the temperature reading measured by a thermostat on a robot at a given time, date, and location. In other cases, an individual reading must be combined with others to provide an aggregated result. Thus, the results of fusion algorithms on device-based information may also be expressed as BML statements as shown in Figure 16.2.

Additionally, natural language expressions can be analyzed and converted into BML. Figure 16.3 shows the final steps of a conversion of the German report:"Am Dienstag wurde bei Sherabad ein Waffenlager entdeckt" to the English-like BML: "found weapon cache Tuesday at Sherabad."

For further information on the processes involved in conversion from German into English, please refer to the Additional Reading section included at the end of this chapter.

C2LG: A SOLUTION FOR SECURITY AND LAW ENFORCEMENT

Using the concepts of C2LG variants such as Crisis Management Language are being developed for special uses. However, because the underlying principles remain the same, the communication between C2LG-based languages is easily achieved. A domain-specific lexicon must be developed in cooperation with subject matter experts in security, law enforcement, and public safety, as well as emergency services, fire departments, and others who may need information in times of crisis.

The advantages of using C2LG for national security are quite clear. Regardless of the originating source, whether human or device, and regardless of the underlying natural language, information that has been garnered can be presented in an

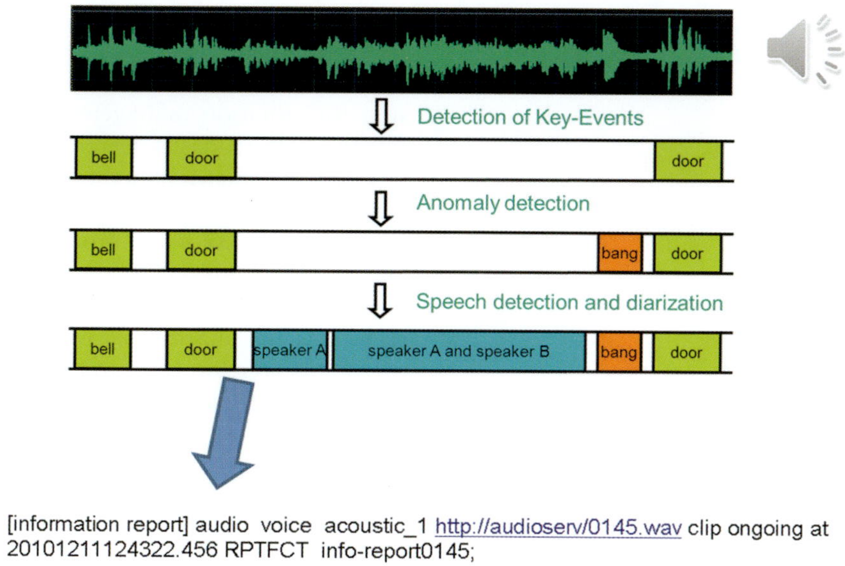

FIGURE 16.2 BML information report produced by middleware of an acoustic sensor providing date, time, and location information.

unambiguous yet flexible format, which allows for automated processing of the information contained in the statement. Integrating new information into an existing database or model simply requires a mapping mechanism, rather than the expense of restructuring.

C2LG VARIANT: A LANGUAGE FOR INTELLIGENCE AND LAW ENFORCEMENT

Clearly the advantages of a C2LG-based language for the use of security and law enforcement agencies for the sharing of intelligence information

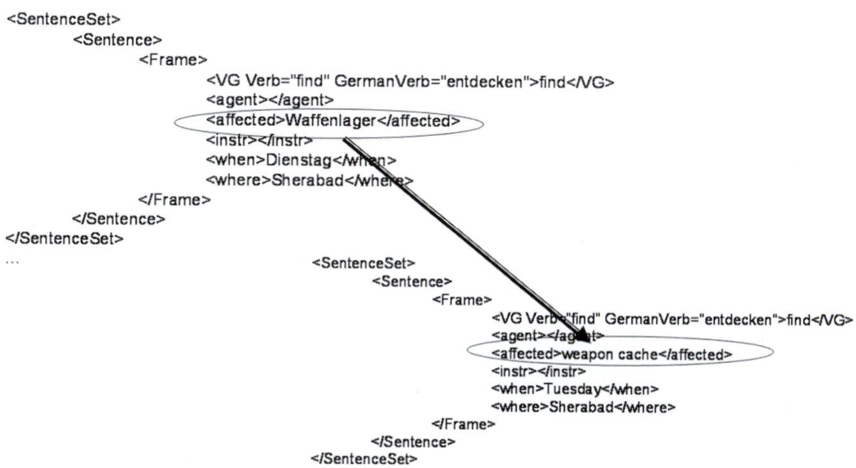

FIGURE 16.3 After information extraction, constituent identification, and mapping to BML structure, the German text is converted to a structured English-like BML.

have been outlined previously. Additionally, existing C2LG structures may be modified to speak to interagency specific needs such as the following:

1. **Provenance or origin of the information:** Labeling conventions within C2LG provide for links to the origin of the source of information.
2. **Reliability of the information source:** Much information comes from sources of less than sterling character or conditional reliability. C2LG provides for assessment of type and reliability of the source and credibility of the information contained within a statement.
3. **Each C2LG "report" is an atomic statement:** It is a single piece of information encapsulated with a certain amount of context information (time, date, location, etc.). This reduces the complexity for automatic fusion purposes.
4. **C2LG reports contain an initial label regarding the information contained therein:** This "hostility" label can be used to identify a security level identifying the sensitivity of the information in the statement to prevent accidental propagation of classified information to the wrong agencies or persons.
5. **Information concerning persons may be "anonymized":** This can be done by inserting aliases or labels into sensitive intelligence information to protect privacy.

While only in its infancy at the time of this writing, an organized crime-specific variant of C2LG will be developed within the framework of the European Commission FP-7 (Security) for the project ePOOLICE, which includes among its consortium partners EUROPOL in Holland.

C2LG VARIANT: CRISIS MANAGEMENT LANGUAGE

Crisis Management Language (CML) is currently being developed by Fraunhofer FKIE in Germany to deal with interagency disaster and crisis management. CML is being used for communication and situation awareness within the frameworks of two European Commission FP/7projects, "CATO" and "ATHENA," both of which are concerned with dealing with crises and the populace. In CATO, the main focus is on data fusion for situation management, whereas within ATHENA the focus is deeper.

CML is being designed to deal with situation management (i.e., short-term information needs). It not only allows for information gathering, but affords rapid and unambiguous communications between various agencies involved in disaster situations. A smartphone touchscreen interface to rapidly convey CML messages reduces confusion, as well as providing documentation for later debriefing and process improvement.

CML also has the advantage that its statements can be easily read by humans. For example, the following communication exchange between police, emergency, and National Guard in CML can be understood with little explanation (other than that "commissive" means "accepting task assignment"):

report event PoliceUnit13 gas-explosion at building-1312 ongoing at 201111051413 explosion_report_01;
request rescue PoliceUnit13 C2RFS one injured civilian at building-1312 start asap rescue_request_02;
commissive rescue Ambulance2 at building-1312 start at now regarding rescue_request_02 commissive_03;
rescue C2RFS Ambulance2 one injured civilian at building-1312 start at now rescue_task_04;
report position Ambulance2 at building-1312 at now reportposition_05;
request support Ambulance2 C2RFS rescue_task04 by crane request_support_06;
request support C2RFS C2_MIL rescue_task04 by crane request_support_07;
commissive support MIL_CRANCE_02 rescue_task04 regarding request_support_07 commissive_08;
support C2_MIL MIL_CRANE_02 rescue_task04 start at now support_rescue_task_09;
report transport Ambulance2 C2RFS 1 male person with fracture left leg, fracture left foot to Main Hospital start at now report_transport_10;

CONCLUSIONS

Intelligence gathering is an essential task for a nation to preserve life and property. However, because of the variety of players, the volume of information flowing in, and the diversity of sources and formats including various human languages, it is a monumental task to try to fuse details into a larger picture. Rather than requiring a standardization of underlying data structures, which may be not only prohibitively expensive, but also, particularly in the case of international cooperation, impossible, information interchange may be facilitated by mapping all information into a flexible, yet unambiguous representation of that information by utilizing the concepts of C2LG.

SECTION 4

Future Threats and Cyber Security

CHAPTER

17

Securing Cyberspace: Strategic Responses for a Digital Age

Andrew Staniforth

CYBER TERROR

When Metropolitan Police officers raided a flat in West London in October 2005, they arrested a young man, Younes Tsouli. The significance of this arrest was not immediately clear, but an investigation soon revealed that the Moroccan born Tsouli was the world's most wanted "cyber terrorist." In his activities Tsouli adopted the user name "Irhabi 007," (Irhabi means terrorist in Arabic), and his activities grew from posting advice on the Internet on how to hack into mainframe computer systems to assisting those planning terrorist attacks. Tsouli trawled the Internet searching for home movies made by U.S. soldiers in the theaters of conflict in Iraq and Afghanistan that would reveal the inside layout of U.S. military bases. Over time these small pieces of information were collated and passed to those planning attacks against armed forces bases. This virtual hostile reconnaissance provided insider data illustrating how it was no longer necessary for terrorists to conduct physical reconnaissance if relevant information could be captured and meticulously pieced together from the Internet.

Police investigations subsequently revealed that Tsouli had €2.5million worth of fraudulent transactions passing through his accounts, which he used to support and finance terrorist activity. Pleading guilty to charges of incitement to commit acts of terrorism, Tsouli received a 16 year custodial sentence to be served at Belmarsh High Security Prison in London where, perhaps unsurprisingly, he has been denied access to the Internet. The then National Coordinator of Terrorist Investigations, Deputy Assistant Commissioner Peter Clarke, said that Tsouli "provided a link to core al-Qaeda, to the heart of al-Qaeda and the wider network that he was linking into through the Internet," going on to say: "what it did show us was the extent to which they could conduct operational planning on the Internet. It was the first virtual conspiracy to murder that we had seen."

The case against Tsouli was the first seen in the UK, and it quickly brought about the realization that cyber terrorism presented a real and present danger to its national security. Law enforcement practitioners understood that the Internet clearly provided positive opportunities for global information exchange, communication, networking, education, and as a major tool in the fight against crime, but a new and emerging contemporary threat had appeared within the communities they sought to protect. The Internet had been hijacked and exploited by terrorists not only to progress attack planning but to radicalize and recruit new operatives to their cause. It was also the core and affiliated networks of al-Qaeda that were quick to realize the full potential of the global platform provided by the Internet.

It is now clear to Western national security intelligence practitioners that al-Qaeda and its global network of affiliated groups is resilient, becoming increasingly independent, mobile, and unpredictable. Of critical concern for the security of the Western world remains the way in which individuals from of our own communities are being influenced by the single narrative and extreme version of religious ideology promoted by al-Qaeda. This narrative, when combined with a complex array of social, political, and economic factors set within the specific environment of each nation, has served to manipulate individuals toward extremist perspectives cultivating a home-grown terrorist threat, which presents security concerns to the free and democratic way of life enjoyed in the West. The way in which the recruiters and radicalizers of al-Qaeda have influenced and indoctrinated the young and vulnerable across the world has alarmed national security professionals. The direct impact they have made upon impressionable members of our communities who require safeguarding from this terrorist tactic continues to damage the confidence of our communities in the ability of the state's security apparatus to police the Internet and protect their online experience. Al-Qaeda has also preyed upon school-age individuals, which reveals the extent they are prepared to go to progress their cause. During June of 2006, Hammad Munshi, a 16-year-old school boy from Dewsbury in Leeds of West Yorkshire, was arrested and charged on suspicion of committing terrorism-related offenses. Following his arrest, searches were conducted at his family home where his wallet was recovered from his bedroom. It was found to contain handwritten dimensions of a sub machine gun, taken from a book titled *Expedient Homemade Firearm*. At the time Munshi had excellent information technology skills and had registered and run his own Web site on which he sold knives and other extremist material passing on information on how to make napalm as well as how to make detonators for improvised explosive devices. While the online rhetoric of al-Qaeda cyber recruiters reached the computer in the bedroom of Hammad Munshi, authorities on this occasion were able to intervene before any critical security risks to citizens were realized. But not all individuals being recruited online would be prevented from carrying out attack planning by UK security forces.

On May 22, 2008, Nicky Reilly, aged 22, left his home in Plymouth with a rucksack containing six bottles full of nails and home-made explosives. His target was the Giraffe restaurant in Exeter, a popular place for shoppers to lunch. Reilly, who has Asperger's syndrome and a mental age of 10, was a suicide bomber, recruited online in local Internet cafes by extremists in chat rooms who had fueled a hatred of the West. Extremists had molded a home-grown terrorist and had directed him to bomb-making Web sites discussing what his target should be. As Reilly was seated in the restaurant, 44 customers had also sat down to dine. One of the eleven members of the staff working that day brought Reilly a drink, and he sat for 10 minutes before making his way to the lavatory taking his rucksack with him. Once inside a cubicle the device detonated prematurely causing injury to Reilly and damage to the restaurant. No other person was injured in the blast. A note left at his home revealed the motivation for his actions in which he paid tribute to Osama bin Laden and called on the British and U.S. governments to leave Muslim countries. The note declared that Western states must withdraw their support of Israel and that violence would continue until "the wrongs have been righted." Reilly, appearing in court as Mohammed Abdulaziz Rashid Saeed, pleaded guilty to offenses of attempted murder and preparing for acts of terrorism. At the Old Bailey on January 30, he was sentenced to life imprisonment. Mr. Justice Calvert-Smith said that "I am quite satisfied that these offences are so serious that only a life sentence is appropriate. This defendant currently represents a significant risk of serious harm to the public." He went on to say that "The offence of attempted murder is aggravated by the fact that it was long planned, that it had multiple intended victims and was intended to terrorize the population of this country. It was sheer luck or chance that it did not succeed." Defense counsel, Kerim Fraud, representing Reilly stated that "He may comfortably be deemed to be the least cunning person ever to have come before this court for this type of offence." The threat from cyber terrorism in all of its forms continues to represent a serious risk to the national security of many nations, but other criminals, extremists, agitators, and states themselves also have come to understand the unique potential of the Internet, presenting a complex malaise of new cyber-based threats to Western democracies and their citizens.

CYBER THREATS

The Internet and digital technologies continue to transform our societies by driving economic growth, connecting people, and providing new ways to communicate and cooperate with one another. The World Wide Web only began in 1991, but today 2 billion people are online—almost one-third of the world's population. Billions more are set to join them over the next decade, and there are over 5 billion Internet-connected devices with $8 trillion changing hands each year in online commerce. As a direct result the Internet is already having a profound impact on the way citizens across the globe live their lives. This social change is only set to grow and gather pace as the number of users increases. Already it appears the phenomenal growth of Internet use will be on the scale of the very largest shifts in human history, such as the coming of the railways, or even the learning to smelt metals. Real Gross Domestic Product per capita has risen by $500 over the last 15 years in mature countries enabled by the Internet. By comparison, it took 50 years for the industrial revolution to have the same effect. Given this context it is understandable why the growth of the Internet has had such a dramatic impact upon societies across the world. Cyberspace is transforming business, making it more efficient and effective. It is opening up markets, allowing commerce to take place at lower cost and enabling people to do business on the move. It has promoted fresh thinking, innovative business models, and new sources of growth. It enables companies to provide better, cheaper, and more convenient service to customers, and it helps individuals to shop around, compare prices, and find what they want. Cyberspace is an interactive domain made up of digital networks that is used to store, modify, and communicate information. It includes the Internet, but also the other information systems that support businesses, infrastructure, and critical services. Digital networks already underpin the supply of electricity and water to our homes, help organize the delivery of food and other goods to stores, and act as an essential tool for businesses. And their reach is increasing as we connect our TVs, game consoles, and even domestic appliances. In summary, the development and use of the Internet by advances in smart mobile technology is accelerating Western society's dependence upon cyberspace. Developing countries in particular stand to benefit as increasing interconnectivity makes commerce easier and allows access to information, knowledge, and education, enabling people to innovate, collaborate, and compete in global marketplaces.

The UK, like many nations in the Western world, has positively adopted cyberspace as a means of doing business. In 2009, 608 million card payments were made online in the UK, with a total spent of £47.2 billion, and in 2011 90% of high street purchases were being made using electronic transactions with ~52% of UK consumers having direct access to broadband using online shopping as an opportunity to save money. The Internet will become increasingly central to national economies, but the growing role of cyberspace has also opened up new threats as well as new opportunities. The national security machinery of governments has no choice but to find ways to confront and overcome these threats if they are to flourish in an increasingly competitive and globalized world. The digital architecture on which we now rely was built to be efficient and interoperable. When the Internet first started to grow, security was less of a consideration. However, as we put more of our lives online, security matters more and more. People want to be confident that the networks that support our national security, our economic prosperity, and our own private lives as individuals are safe and resilient. Unfortunately a growing number of adversaries are looking to use cyberspace to steal, compromise, or destroy critical data. The scale of our dependence means that our prosperity, our key infrastructure, and our places of work and our homes can all be affected. For this reason the British government's 2010 national security strategy—A Strong Britain in an Age of Uncertainty: The National Security Strategy—identified cyber attacks on the UK as a "Tier 1" threat; one of its highest priorities for action stating that "hostile attacks upon UK cyber space by other states and large scale cyber crime" presented a primary risk to national security and economic well-being.

Criminals from all corners of the globe are already exploiting the Internet to target Western democracies in a variety of ways. There are crimes that only exist in the digital world, in particular those that target the integrity of computer networks and online services. But cyberspace is also being used as a platform for committing crimes such as fraud, and on an industrial scale. Identity theft and fraud online now dwarf their offline equivalents. The Internet has provided new opportunities for those who seek to exploit children and those who are vulnerable. Cyberspace allows criminals to target countries from other jurisdictions across the world, making it harder to enforce the law. As businesses and government services move more of their operations online, the scope of potential targets will continue to grow. Some of the most sophisticated threats to cyberspace come from other states that seek to conduct espionage with the aim of spying on or compromising government, military, industrial, and economic assets, as well as monitoring opponents of their own regimes. "Patriotic" hackers can act upon a state's behalf to spread disinformation, disrupt critical services, or seek advantage during times of increased tension. In times of conflict, vulnerabilities in cyberspace could be exploited by an enemy to reduce the technological advantage of a nation's military or to reach past it to attack domestic critical infrastructures.

The threat to Western democracies from politically motivated activist groups operating in cyberspace is also very real. Attacks on public and private sector Web sites and online services in the UK in particular orchestrated by "hacktivists" are becoming more common, aimed at causing disruption, reputational and financial damage, and gaining publicity. All of these different groups—criminals, terrorists, foreign intelligence services, and militaries—are active today against countries' interests in cyberspace. With the borderless and anonymous nature of the Internet, precise attribution is often difficult and the distinction between adversaries is increasingly blurred. Assessing the actual level of threat from cybercrime is a major challenge. Law enforcement statistics are generally considered to be a poor indication of the actual level and trends. Many cyber deceptions go unreported to the authorities so the scale of cybercrime is very difficult to measure, and there are challenges in gathering accurate data because the victims have not discovered that they are the subject of a criminal act or the victims report the offense to a providing company rather than the police, while the companies tend not to disclose their levels of loss or attack to avoid loss of reputation and custom.

As the actual threat from cybercrime may be proving difficult to measure accurately, organizations also are not always aware of the new vulnerabilities that dependence on cyberspace can bring. Intellectual property and other commercially sensitive information such as business strategies and research and development data can be attractive targets. This risks undermining the strengths of countries' innovative research base, investment, and intellectual property as important drivers of economic growth leading to improved prosperity. Services relying on, or delivered via, cyberspace can be taken offline by criminals or others, damaging revenue and reputations. Cyberspace has now grown to become a domain where strategic advantage, industrial or military, can be won or lost. The Internet underpins the complex systems used by commerce and the military. The growing use of cyberspace means that its disruption can affect a nation's ability to function effectively in a crisis. Nearly two-thirds of critical infrastructure companies report regularly finding software designed to sabotage their systems. Some states also regard cyberspace as providing a way to commit hostile acts "deniably." Alongside existing defense and security capabilities, nations must be capable of protecting their national interests in cyberspace. Iain Lobban, Director of the Government Communications Headquarters (GCHQ), stated that "There are over 20,000 malicious emails on UK government networks each month, 1,000 of which are deliberately targeting them." These kinds of attack are increasing; the number of e-mails with malicious content detected by government networks in the whole of 2010 was double the number seen in 2009 and law enforcement agencies are even being targeted.

During 2012 a UK-based hacker posted online what appeared to be authentic login information for police officers in the Hertfordshire and Nottinghamshire constabularies.

The usernames, passwords, and personal identification numbers were posted to a "Pastebin" Web site along with the banner "OpFreeAssange" and a quote from WikiLeaks founder Julian Assange. In a statement, Hertfordshire Constabulary said it was investigating the breach of security and confirmed that the information had been stored on a database linked to its public facing Safer Neighborhoods pages of an external Constabulary Web site. The police force said the database was externally hosted and they were forced to disable part of its Web site as a precaution reassuring their personnel and the wider communities they served that there was "absolutely no suggestion that any personal data relating to officers or members of the public has been, or could have been compromised." With the increase in unlawful and illegitimate activity in cyberspace over the past decade, a dedicated response was required by governments to counter all cyber-related threats; a response that continues to be the greatest contemporary challenge to national security policy and practice.

STRATEGIC RESPONSES

The threat from all manner of cyber hazards was recognized by the British government as presenting a primary risk to its security and economic prosperity. As a direct result it published its first ever cyber security strategy on June 25, 2009. The strategy, Cyber Security Strategy of the United Kingdom: Safety, Security and Resilience in Cyberspace, acknowledged the UK's growing dependence on cyberspace revealing that modern globalized lifestyles become increasingly dependent upon information communications technology, and that cyberspace provided a new arena in which hostile states, terrorists, and conventional criminals can threaten UK security interests. Upon announcing the launch of the strategy, then Prime Minister Gordon Brown stated that

> *Just as in the 19th century we had to secure the seas for our national safety and prosperity, and in the 20th century we had to secure the air, in the 21st century we also have to secure our position in cyberspace in order to give people and businesses the confidence they need to operate safely there.*

To deliver the strategic aim of the new strategy designed to "reduce risk from the UK's use of cyberspace and exploiting opportunities in cyberspace through improving knowledge, capabilities and decision-making," a new cyber security architecture had to be constructed. The Office of Cyber Security and Information Assurance (OCSIA) was established, tasked with driving forward a cross-government program of work supporting the Minister for the Cabinet Office, the Rt. Hon. Francis Maude MP and the National Security Council in determining priorities in relation to securing cyberspace. The unit provides strategic direction and coordinates action relating to enhancing cyber security and information assurance in the UK, while a new

Cyber Security Operations Center (CSOC) based at GCHQ in Cheltenham, Gloucestershire, provides the coordinated protection of the UK's critical information technology systems. The OCSIA, alongside the CSOC, works with lead government departments and agencies such as the Home Office, Ministry of Defence, GCHQ, Communications-Electronics Security Group, the Center for the Protection of National Infrastructure, and the Department for Business, Innovation and Skills in driving forward the cyber security program for the UK government, which seeks to provide the UK with the balance of advantage in cyberspace.

Law enforcement agencies across the UK government also have their part to play in providing safety and security online to UK citizens and business users. The police service, under the auspices of the Association of Chief Police Officers (ACPO), developed a new E-Crime initiative to provide the strategic foundation and direction to ensure UK police services were alert to cybercrime and cyber terrorism activities and that any suspected activity, no matter how small, was encouraged to be reported by the public and commerce and recorded. This policing initiative was supported by others under the leadership of the Office of Security and Counter Terrorism at the Home Office. With primary responsibility for the cross-government strategy on tackling terrorist use of the Internet, the Home Office published guidance for those citizens who are responsible for vulnerable individuals and work within communities to help ensure that the Internet is an environment where terrorist and violent extremist messages are challenged. As part of their wider efforts to counter the threat of radicalization on the Internet, a public facing Web page was launched in February 2010 to encourage the public to take action against unacceptable violent extremist and hate Web sites and other online content. Where individuals believed that material they have located is potentially unlawful they are provided with the opportunity to complete a form on the Web page and refer it to the Counter Terrorism Internet Referral Unit (CTIRU), established by ACPO during 2010. The CTIRU provides a national coordinated response to referrals from the public as well as from government and industry. It also acts as a central, dedicated source of advice for the police service. The CTIRU initiative supports two key strands of the UK's Counter Terrorism Strategy "CONTEST" as follows:

- **PREVENT**: This seeks to stop people from becoming terrorists or supporting violent extremism by taking action against material that encourages radicalization through the glorification of violent extremism.
- **PURSUE**: This aims to stop terrorist attacks by taking action against material useful for acts of terrorism.

The CTIRU provides the UK police service with a unit of experts who can carry out an initial assessment of material located on the Internet. It is also responsible for alerting forces and the units of the UK Police Counter Terrorism Network to online terrorist offenses that may fall within their jurisdiction. Powers under UK terrorism legislation provide for the CTIRU to take a national lead in serving notices on Web site administrators, Web hosting companies, Internet Service providers, and other relevant parties within the UK, to modify or remove any unlawful content.

The CTIRU also focuses on developing and maintaining relationships with the Internet industry, an important part of ensuring the delivery of a safer and more secure online experience for its citizens. A further challenge given the global scope of cyberspace for UK law enforcement was the majority of terrorist content online being hosted in other countries outside UK jurisdiction. To counter this challenge the CTIRU continues to forge links with law enforcement counterparts abroad to help target those Web sites hosted overseas. UK Counter Terrorism and Extremism Liaison Officers based in countries around the world have a key role in supporting this work. The ACPO national coordinator for PREVENT, Assistant Chief Constable John Wright, described the role of CTIRU as

> *providing the opportunity to effectively enforce, and control, access to material believed to be extreme. In addition, the CTIRU will help to develop a culture of*

collaboration between police, partners and service providers dedicated to making the internet a safer place, particularly for young people.

The tools to tackle cyber terrorism and the criminal use of the Internet to recruit and radicalize potentially vulnerable individuals online are an important aspect of combating contemporary terrorism. Professor Sir David Omand GCB, former Director of GCHQ and UK Security and Intelligence Coordinator in the Cabinet Office under Prime Minister Tony Blair, believes that "the cyber dimension is likely to become a means for terrorists to try to cause disruption and to score propaganda points." He went on to reveal that the UK has "many cyber vulnerabilities." Such vulnerabilities include the growing threat of fraud. The British government has now come together with leading industry players to help people better protect themselves. In the first campaign of its kind involving both the private and public sectors, "The Devil's in Your Details" campaign brings together Action Fraud, Telecommunications UK Fraud Forum and the Financial Fraud Action UK—the name under which the financial services industry coordinates its fraud prevention activity—in a powerful demonstration of what can be achieved when industry and government work together. The National Fraud Authority-backed campaign is raising awareness of the importance of protecting personal information and aims to remind the public to check and make sure that who they share their details with is genuine, whether on the phone, in person, or online. The Devil's In Your Details campaign encourages consumers to suspect anyone or anything they are uncertain about, to keep asking questions, and to challenge or end an engagement if it feels uncomfortable. As an introduction to a wider campaign against fraud, this awareness activity aims to increase reporting of fraud, making it harder for fraudsters to target consumers and in the future providing evidence of the new and innovative ways in which governments must now inform and support their citizens to protect them from specific cyber-related threats.

A NEW APPROACH

In order to maximize the potential of the Internet, it is important that people feel confident that it can be used safely. As all of us make more use of the Internet in our work and private lives, it makes for a more attractive target for criminals or others. Any reduction in trust toward online communications can now cause serious economic and social harm to Western democracies. Beyond the impact upon individual citizens, the scale of the use of cyberspace means that it can now also affect society more broadly. Western governments have a strong tradition of protecting its citizens in ways that are guided by the core values of liberty, fairness, transparency, and the rule of law. These values help define who we are and what we do. The interconnected nature of cyberspace and its expansion mean that it has developed to promote many of these values, but the conventions and norms covering conduct within the cyber domain are still developing. While this helps make it the vibrant domain that it is today, it can also cause instability and uncertainty about accountability. The blurring of boundaries in cyberspace increases the risk of actions affecting larger numbers of people and organizations unintentionally. At its most serious, this leads to the potential for unpredictable and large-scale shocks.

Actions to strengthen national security must therefore also be consistent with states' obligations, such as those concerning freedom of expression; the right to seek, receive, and impart ideas; and the right to privacy. Defending security should be consistent with the commitment to uphold civil liberties. Of course, these are well-established and ongoing debates, but cyberspace can bring them into focus in new ways, and more quickly than in other areas. These changes do not affect single nations alone. The global reach of the Internet and digital technologies can provide an important means for the spread of ideas, with profound implications for societies across the globe. But like any communications medium, cyberspace can also potentially be used to restrict liberty and undermine freedoms. Some states and organizations are already seeking to control and restrict the future development of the cyber domain. These attempts are ultimately doomed to fail. But for as long as they last they are holding back progress and reducing social benefit. It is important that Western democracies continue to work with like-minded states around the world to maximize the extent to which the world can fully realize and enjoy the benefits that cyberspace will offer.

To secure the vast economic and social opportunities that cyberspace has to offer, the newly elected British coalition government under Prime Minister David Cameron during 2010 transformed its approach to cyber security, setting out a new vision toward 2015 in its cyber strategy, The UK Cyber Security Strategy: Protecting and Promoting the UK in a Digital World. The new strategy revealed that the British government believed that there was no such thing as "absolute security," indicating that its strategy to counter cyber threats was to apply a risk-based approach to prioritizing its response. The new strategic vision for the strategy was for the UK in 2015 to derive huge economic and social value from a vibrant, resilient, and secure cyberspace, where its actions, guided by its core values of liberty, fairness, transparency, and the rule of law, enhance prosperity, national security, and a strong society. To counter all cyber challenges the new strategy was divided into four strategic objectives as shown in Figure 17.1.

To tackle cyber crime the British government will work to reduce online vulnerability and restrict criminal activity online while promoting more effective partnerships. To make it safer to do business in cyberspace the strategy reveals that efforts will be made to increase awareness and visibility of threats, improving incident response and further protect information and services by fostering a culture that manages cyber risks serving to promote confidence in cyberspace. This strategy also seeks way in which to defend UK national infrastructures by strengthening defenses in cyberspace, improving

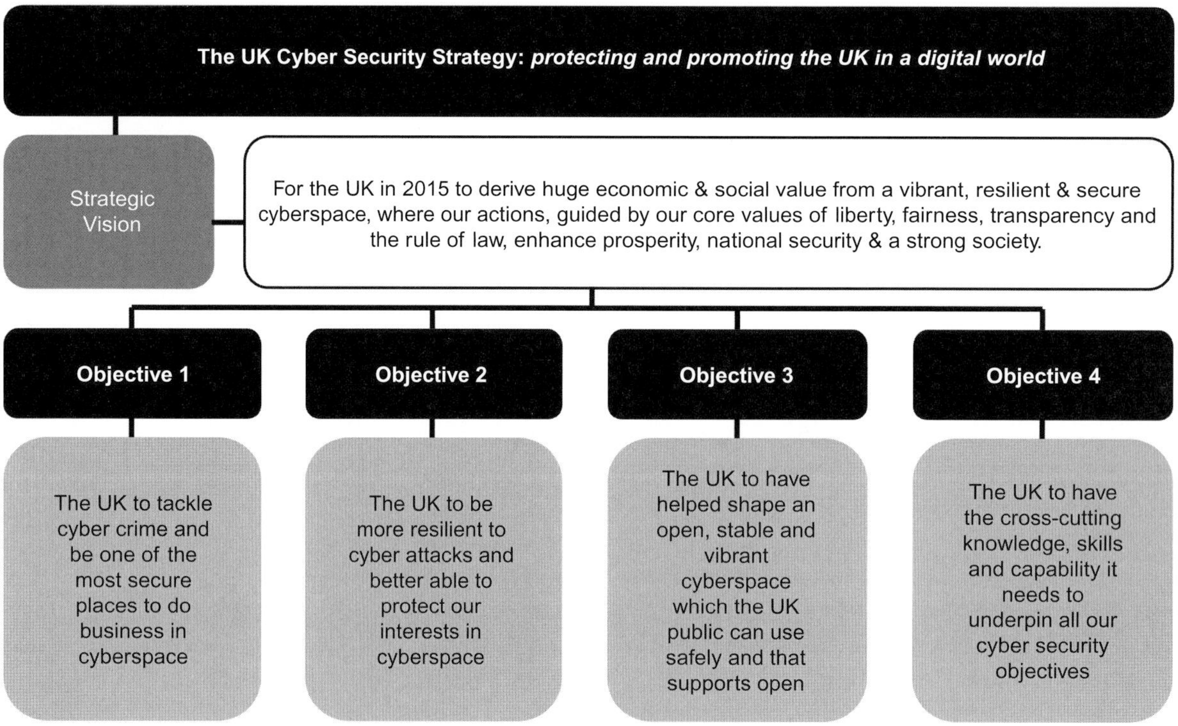

FIGURE 17.1 The UK Cyber Security Strategy: Protecting and Promoting the UK in a Digital World.

resilience and diminishing the impact of cyber attacks and countering the terrorist use of the Internet. Ensuring that the UK has the capability to protect its interests in cyberspace, national security policy makers in government ensured that the new strategic approach is set to improve the UK's ability to detect threats in cyberspace by expanding its capabilities to deter and disrupt attacks. The UK cyber strategy provides an "all hazards" approach to cyber security ensuring that all of its efforts are coordinated in a detailed program of work across the full operating landscape of government. It specifically highlights the important need to help to shape the development of cyberspace by promoting an open and interoperable environment while promoting the fundamental freedoms and rights that British citizens enjoy. This particular element ensures that the UK will pursue cyber security policies that enhance individual and collective security while preserving UK citizens' right to privacy and other fundamental values and freedoms. The strategy also reveals that internationally the UK will continue to pursue the development of norms of acceptable behavior in cyberspace, starting from the belief that behavior that is unacceptable offline should also be unacceptable online. The UK governments' position on cyber security is guided by seven principles proposed by the Foreign Secretary, Rt. Hon. William Hague MP, during February 2011.

UK CYBER SECURITY GUIDING PRINCIPLES

1. The need for governments to act proportionately in cyberspace and in accordance with national and international law
2. The need for everyone to have the ability—in terms of skills, technology, confidence, and opportunity—to access cyberspace
3. The need for users of cyberspace to show tolerance and respect for diversity of language, culture, and ideas
4. The need to ensure that cyberspace remains open to innovation and the free flow of ideas, information, and expression
5. The need to respect individual rights of privacy and to provide proper protection to intellectual property
6. The need for us all to work collectively to tackle the threat from criminals acting online.
7. The promotion of a competitive environment that ensures a fair return on investment in network, services, and content.

The British government understands that achieving its vision for cyber security for 2015 within the framework of its guiding principles will require everybody—the private sector, individuals, and government—to work together. Just as all citizens from all countries benefit from the use of cyberspace, all have a responsibility to help protect it. Therefore ordinary citizens have an important role to play in keeping cyberspace as a safe place to do business and live our lives. By 2015 the new approach by the British government will seek to ensure that its citizens know how to ensure a basic level of protection against threats online and that they have ready access to accurate and up-to-date information on the online threats that they face together with the techniques and practices they can employ to guard against them. If citizens are careful about putting personal or sensitive information on the Internet, are wary of e-mail attachments or links from unrecognized senders, and are cautious about downloading files from Web sites they know little about then they can significantly assist in countering the cyber security challenge making cyberspace increasingly resilient to all types of cyber threats posed to individual citizens. It is essential therefore that everyone, in their homes, at their place of work, and on the move can help identify threats in cyberspace and report possible offenses that make cyberspace a hostile environment for those seeking to unlawfully exploit its potential.

The private sector also has a crucial role to play in strengthening cyber security. Much of cyberspace is owned and used by private companies and it is therefore businesses that will drive the innovation required to keep pace with security challenges. By 2015 the British government will seek to ensure that companies are aware of the threat and use cyberspace in a way that protects commercially sensitive information, intellectual property, and customer data. It will also work to protect private organizations by working in closer and increasingly integrated partnerships to share best practice supported by government and law enforcement agencies. These partnerships will share information and resources to transform the response to a common challenge and actively deter the threats faced in cyberspace. The role of government will underpin the efforts of citizens and the private sector to help reduce the risk from cyber threats, and the British government has now committed to playing its full part in achieving these aims by seeking to amplify its capacity to detect and defeat high-end threats by 2015. In addition, the British government is committed to investing in the growth of a cadre of cyber security professionals and strengthening law enforcement to tackle cybercrime. One such development is already in action—the introduction of a new National Crime Agency (NCA). The NCA will be a powerful body of operational crime fighters with a clear focus on public protection with a federal approach. The NCA receives its legal footings from the Crime and Courts Bill, which was introduced into the House of Lords on May 10, 2012. Subject to Parliamentary processes, the government's ambition is that the new NCA will be fully operational by December 2013. Its mission will include tackling organized crime, strengthening borders, fighting fraud and cybercrime, and protecting children and young people.

The creation of the NCA marks a significant shift in the UK's approach to tackling serious, organized, and complex crime, with an emphasis on greater collaboration across the whole law enforcement landscape. The NCA will build effective two-way relationships with police forces, law enforcement agencies, and other partners, and will be made up of four commands including the Economic Crime Command, providing an innovative and improved capability to deal with fraud and economic crimes, including those carried out by organized criminals, and the Child Exploitation and Online Protection Centre, which will work with industry, government, children's charities, and law enforcement to protect children from sexual abuse and to bring offenders to account. Both of these primary arms of the NCA will work to significantly reduce the cyber-based risks to British citizens and protect the broader security of the nation. The NCA will also benefit from an intelligence hub, which will build and maintain a comprehensive picture of the threats to the UK from organized criminality and a national cybercrime center, providing expertise, support, intelligence, and guidance to police forces and the commands of the NCA. The new organization will confront some of the most risky and dangerous people that affect UK communities and will be one that is unequivocally focused on keeping the public safe. The British government has gone even further to ensure that all police forces efforts are marshaled to augment the NCA lead role in tackling cyber-related crime. During November 2011 Home Secretary, Theresa May, presented the shadow Strategic Policing Requirement (SPR) to Parliament. The publication of the shadow SPR set out what the Home Secretary views as the national threats the police services must address providing the appropriate national capabilities to do so. These threats included terrorism, civil emergencies, organized crime, public order, and a large-scale cyber incident. The shadow SPR received statutory effect during November 2012 and now empowers the newly democratically elected police and crime commissioners to deliver their important role of holding their Chief Constable to account for the totality of policing, both locally and nationally. The police and crime commissioners are expected to drive collaboration between police forces and to ensure that forces can work effectively together and with their partners. The SPR now provides a statutory obligation on police forces to provide the necessary resources and commitment to effectively tackle a large-scale cyber incident. It provides evidence of the commitment of the British government to focus its assets on countering cyber threats and the seriousness in which they assess the phenomenon of cyber-related threats to its national security.

CYBER COLLABORATION

Although the scale of the cyber security challenge requires strong national leadership, governments cannot act alone as they must recognize the limits of their competence in cyberspace. Much of the infrastructure that countries need to protect is owned and operated by the private sector. The expertise and innovation required to keep pace with the threat will be business driven. Similarly, although individual nations can improve their own defenses domestically, the Internet is fundamentally transnational and threats are cross-border. A global threat requires a global response. Not all the infrastructure on which countries rely is based within their own boundaries. So many nations of the Western world, like the UK, cannot make all the progress it needs to on its own. Therefore collaboration is the key to success in cyberspace, and the British government has committed to building strong partnerships with other countries that share its views and reach out to other nations where they can to those who do not.

On May 25, 2011, President Obama and Prime Minister David Cameron reaffirmed their close bilateral cooperation, and charted important new steps forward for cyber security. The United States and the UK share unparalleled bilateral cooperation and both leaders agreed on a shared vision for cyberspace, which places at its heart fundamental freedoms privacy and the free flow of information in a secure and reliable manner. Prime Minister Cameron stated:

> *Our goal is to nurture and accelerate the progress that these technologies have enabled in our economies and societies; we will continually work, individually and as partners to ensure they create jobs, enrich lives, and provide return on the sound investments we make in them today.*

Both leaders acknowledged that building consensus on responsible online behavior was an important role for their governments to tackle, recognizing that the same kinds of "rules of the road" that help maintain peace, security, and respect for individual rights internationally must equally apply in cyberspace. Through its deposit of instruments of accession in Strasbourg, during May 2011, the UK has now joined the United States and 30 other states as parties to the Budapest Convention on Cybercrime, the world's foremost treaty to combat cybercrime internationally. The Convention sets standards for national laws in dealing with online fraud and abuse, but even more importantly it permits effective cooperation between nations—a crucial tool since so many cybercrimes cross national boundaries. Noting this landmark achievement, President Obama and Prime Minister Cameron agreed to continue work to expand the reach of this important treaty.

The collaboration between the United States and the UK provides both governments with an opportunity to share experiences and develop effective countermeasures together, ensuring they are doing all they can to stay one step ahead of those who wish to unlawfully exploit and misuse the freedom offered in cyberspace. While governments across the world seek to define and then address their cyber security concerns, the technological advances of the Internet and the phenomenon of social media networks, smart mobile communications devices present both future challenges to, and opportunities for, the national security apparatus of states. By simply increasing collaborative approaches to all types of cyber threats will ensure that individually and collectively, governments are better prepared today to meet the cyber security challenges of tomorrow.

CHAPTER

18

National Cyber Defense Strategy

Paul de Souza

In trying to defend everything he defended nothing.
Frederick the Great, (Frederick II)

INTRODUCTION

A solid national cyber defense strategy must be based on the understanding that although risk can be minimized, the threat can never be completely eliminated. The attack surface will always be present no matter how many layers of defense one implements. Defense in depth in conjunction with situational awareness (SA) and active defense when properly implemented can take any nation from being reactive to proactive.

Defense in depth is nothing but the active deployment of Computer Network Defense (CND). According to the U.S. Joint Chiefs publication "Joint Pub 3-13, Information Operations" CND involves actions taken via computer networks to protect, monitor, analyze, detect, and respond to network attacks, intrusions, disruptions, or other unauthorized actions that would compromise or cripple defense information systems and networks. According to the U.S. Department of Defense (DoD), defense in depth is also "the sitting of mutually supporting defense positions designed to absorb and progressively weaken attack, prevent initial observations of the whole position by the enemy, and to allow the commander to maneuver the reserve."

In today's technology field and cyber domain, Chief Technology Officers (CTOs), cyber commanders, strategists, or even network engineers commonly add more and more of the same when it comes to implementing security. Many believe that complexity is a negative trait of systems when, in reality, complexity and variety can add real defense in depth and "progressively weakens the attack" of the adversary. The fact most users employ the same operating systems (Windows 7 and XP are the most popular operating systems in the United States) naturally gives the adversary the advantage when attacking networks running those systems. Adding less popular systems to a network will lessen the attack surface and increase the complexity of one's network.

Any nation must ensure it has the necessary capabilities to operate effectively in all domains, such as air, land, sea, space, and cyberspace. A well-developed cyber-defense strategy must cover the necessary areas that enable a nation to operate in a degraded environment. Unplugging systems from cyberspace is no longer acceptable, but fighting through and in cyberspace under attack is the key to success.

A natural overlap occurs within the realms of Information Assurance (IA) and CND. The National Security Agency defines IA as "measures that protect and defend information and information systems by ensuring their availability, integrity, authentication, confidentiality, and non-repudiation." While CND naturally focuses on systems, IA focuses on data. However, one must protect both systems and data. A healthy cyber-defense strategy will heavily target both of these.

Cyberspace now spans all other war-fighting domains, such as land, sea, air, and space. A stable national cyber defense strategy must take into consideration the enormity of cyberspace and the need to operate through and in cyberspace as a way to keep order and peace. The current national threat environment is a convergence of traditional sources of power with new social sources of power to include non-nation state actors armed with highly technological capabilities. These technological threats can and do affect physical domains. We can see the natural course being the intensification of cyber conflict from data exploitation to data disruption and to data destruction.

When designing a cyber defense strategy, one must take into consideration the operational construct of cyberspace: man-made, global, complex, contested, and mostly privately owned. The U.S. DoD released its cyberspace strategy named "Department of Defense for Operating in Cyberspace" in July 2011. Strategic Initiative Two from this strategy reads: "Employ new defense operating concepts to protect DoD networks and systems." Staying current on new concepts for operational defense is no easy task, but it is essential in order to have an effective approach in cyberspace. The DoD strategy itself states that "the implementation of constantly evolving defense operating concepts is required to achieve DoD's cyberspace mission today and in the future."

Solid defense operating concepts in the cyber domain will take into consideration the fluid nature of cyberspace and keep up with its rapid pace of change. When speaking about defense, a phrase often heard from cyber security professionals is "in near-real time" or even defense "in real time." What does it mean to be able to defend one's systems and data in real time? Is it even possible? Most cyber security vendors want potential clients to believe that such capabilities rest upon their technological solutions, when in reality the solution is a combination of technologies and human capabilities. The gap found in most cyber defense strategies is the lack of human capabilities, the lack of proper trained cyber security professionals, and the lack of SA.

TRAINING CYBER DEFENSE PROFESSIONALS

A current global trend is for cyber security professionals, so-called cyber warriors, to have their training focus on specific technologies, tools, and segmented methodologies that apply only to defensive matters. In the physical domain, soldiers are trained to both defend and attack; the same reality must be applied in cyberspace by training our cyber warriors to be battle focused.

To be battle focused in cyberspace means to be able to be cross-trained to operate in both offensive and defensive environments. In order to deploy "active defense" strategies in the cyber domain, one must fully comprehend what it means to deploy full-spectrum cyber operations. The objective of this chapter is not to delve into authorities and how the law would treat "active defense," but to give the reader the understanding of the need for a shift in thinking when approaching defensive strategies. A well-rounded cyber security professional aspiring to become well versed in Defensive Cyberspace Operations (DCO) should also be trained to create exploits and payloads, work on cyber weapons development, find new attack vectors and techniques, and finally to plan and execute Offensive Cyberspace Operations. The DCO professional of today must be able to defend his domain by understanding the mind of the adversary.

What would a national cyber defense strategy look like? The DoD's definition of strategy from the DoD dictionary of military and associated terms (US—JP 1-02) can be applied to cyberspace: "A prudent idea or set of ideas for employing the instruments of national power in a synchronized and integrated fashion to achieve theater, national, and/or multinational objectives." The following are core elements that might constitute a framework that could be used to create a national cyber defense strategy:

1. Set of ideas
2. Instruments of power (including well-trained people, the right technologies and processes)
3. Synchronized efforts
4. Objectives and direction

With that framework in mind, how is a solid cyber defense strategy created and maintained? The following are steps to a practical and objective approach to creating a cyber defense strategy that could be applied at both organizational and national levels.

1. Understand what cyber strategy is not.
2. Understand and accept the unique threats that apply to you—Know Your Enemy! (CyberINT, Attack Analysis and Strategy Analysis) CyberINT means collecting data on the Internet, before the attacks take place. It is a form of trying to predict what form of cyber attack will take place by studying online communications and connecting the dots.
3. Know yourself and how vulnerable you are. Understand your capabilities.
   - Applying actionable intelligence to the process of developing a cyber defense strategy framework can give one a better understanding of the current threat landscape. Sun Tzu, the great Chinese military strategist stated, "It is said that if you know your enemies and know yourself, you will not be imperiled in a hundred battles." The following are two main approaches to understanding your enemy.
   - Cyber attack analysis: Reactive Approach
   - Cyber intel collection: Proactive Approach
   - Define what is critical.
   - Understand what is wanted by your adversary.
   - Understand your vulnerabilities by running vulnerability assessments.
   - Engage trusted but external partners to test your systems.
4. Create a set of ideas (the WHAT of things and not necessary the HOW of things).
   - Now that I know the unique threats that apply to my organization and I know my capabilities, WHAT do I do?
   - Write down your set of ideas, much like a "risk Assessment procedure."
   - What security controls are needed to appropriately protect the information systems that support the operations and assets of the organization so that organization can accomplish its assigned mission, protect its assets, fulfill its legal responsibilities, maintain its day-to-day functions, and protect individuals?
5. Develop your instruments of power (set of skills, technology, knowledge, etc.) to counter the threat and minimize the risk (the HOW of things).
   - HOW do I develop and deploy people, technology, and processes to create "risk mitigation"?
6. Synchronize, collaborate, integrate, and coordinate.
   - A cyber strategy must keep security events in sync with time.
7. A cyber strategy must cover collaboration, integration, and coordination efforts with other cyber entities of interest. List objectives and expand on direction.
   - Strategy goals must be well defined.
   - Define success and appropriate metrics.
   - Create a strategy forecast and future direction.
8. Write down the strategy.
   - A cyber strategy is a sensitive living document that is dynamic and ever changing.
9. Repeat the cycle.

Do not rely only on third-party companies to protect your cyber assets; you create your strategy based on the level of threat you face and based on your cyber combat zone. You can use their know-how, but this is not a cookie-cutter process. It is about connections and who you know and how fast you can share knowledge and collaborate. Connect the dots in the present to understand the future and learn from the past.

The Center for Strategic and International Studies, using open-source literature, has reviewed policies and organizations in 133 states to determine how these organizations deal with cyber security and whether they have a military command or doctrine for cyber activities. The Center for Strategic and International Studies identified 33 nations with cyber warfare capabilities and strategies that directly relate to full-spectrum cyber capabilities. It is worth noting the fast advancement in acquiring such capabilities and the need to develop national cyber strategies to address such needs.

Another point to consider is the classified side of each strategy in use today by state actors. Almost every strategy has a nonpublic side, which reinforces secrecy. Most of these national cyber strategies identify the threats that directly affect their respective national security and way of life, along with initiatives to help minimize the threat surface in cyberspace.

Every nation and organization should develop a cyber defense strategy in order to help guarantee the freedom to maneuver in cyberspace, to move data through the network without being compromised, and the ability of business to thrive in modern economies. A safer cyberspace is one of the key elements for the healthy economic growth of a nation and a good indicator of an unyielding cyber defense strategy.

TYPES OF CYBER WARRIOR

At the core, it is essential that a cyber warrior be driven by a sense of patriotism or zealotry that forges unyielding vigilance. Moreover, linking this conviction to a "higher calling" or sense of purpose ties people to a set of universal ideals that focuses their energies on protecting the home front, and on constantly moving forward. A driving force must exist to motivate an individual to become a lifelong student of cyber warfare and to enter into a process of continuous self-improvement. Consistently, if one were to ask a deployed Marine if he/she were happy out in the operating environment, the response would be a resounding "Yes!" Quite simply, this is because persons with a warrior mindset yearn to answer that call to employ their special set of skills in support of the mission. The various skill sets and expertise at all levels of command (tactical, operational, strategic, etc.) each have their own unique requirements.

The first level of cyber warrior is the generalist, which is similar to the infantryman who knows how to use a rifle, basic demo, basic navigation, and so forth. At this level (Cyber Warrior—Generalist), the basic cyber warrior should have at least a simple understanding of the following concepts: hardware, networking (wired and wireless), penetration, defense, exploitation, and cryptography. Basic certifications for this level might include those such as A+, Security+, Microsoft Certified IT Professional (MCIPT), Juniper or Cisco, Certified Ethical Hacker (CEH), and Computer Hacking Forensic Investigator (CHFI).

After the Generalist comes the Cyber Warrior—Specialist. These cyber warriors have a mastery of the basics and then continue to develop their skill sets, proving mastery to a specialty. Three main specialties are computer network exploitation (CNE), computer network attack (CNA), and CND.

Cyber warriors who specialize in CNE are comparable to the Armor Scouts and Scout CAV (Cavalry Scout). These are people who know how to infiltrate networks to collect information through hardware, network, malware, and so forth. In addition to the basic education, they would have specialized training and have a deeper capacity in understanding the "big picture" of combine arms combat, such as how to envision the way little pieces fit into the whole, how to inject misinformation, how to alter in very subtle ways targeting information, and so forth. They do not disrupt networks; instead, they remain hidden like traditional Scouts to "spy" and report.

Next are the cyber warriors who specialize in attack (CNA). These are like Special Forces; they exist for the sole purpose of destroying networks, denying networks, and causing total collapse of communications. Unlike the exploitation specialist, they will further their training in penetration and disruption tools.

Then there are the cyber warriors who specialize in defense (CND). This has been argued that a CND cyber warrior knows all CNA methods and works to defeat CNA attacks. This indicates an understanding of attacks and how to perform countermeasures. This is a major difference from what is traditionally considered to be CND today. Many people, even those in the field, think that CND is little more than implementing a Security Technical Implementation Guide (STIG), maybe retina scans, but the bar must be raised when discussing what CND means as a cyber warrior.

ACKNOWLEDGMENTS

I would like to dedicate this chapter to The Cyber Security Forum Initiative members and to my close friends who support my mission "to provide Cyber Warfare awareness, guidance, and security solutions through collaboration, education, volunteer work, and training to assist the U.S. Government, U.S. Military, Commercial Interests, and International Partners."

I would also like to recognize the insights of USMC Sgt. Nicholas Andersen, U.S. Army Major Alexander Tambascia, U.S. Army Lt. Col. Mark Coffin, USMC Major Anthony Guess Johnson, USAF Col. (Ret.) Jeffrey Caton, and USAF Major General (Ret.) Harold W. "Punch" Moulton II.

CHAPTER 19

From Cyber Terrorism to State Actors' Covert Cyber Operations

Jan Kallberg, Bhavani Thuraisingham

INTRODUCTION

During the first 20 years of the Internet era there was a widespread fear of threats from the Internet, but in reality it was fairly secure. The limited abilities and resources of the early attackers contained the threat to criminal activity and marginal damage. Recent advancements in client computer security, in conjunction with the impact of time and Internet maturity, have created a population at ease and with and trusting of the Internet. In reality, the Internet has a reverse trajectory for its security, where the Internet has become more unsafe over time. The threat no longer engulfs just individuals and businesses, but also the nation state. In almost 20 years concerns have been raised about what single hackers and cyber terrorists can do to a targeted society or individual. In the mainstream media, and our collective weltanschauung, hackers and cyber terrorists have been credited with the ability to create a digital Armageddon or, in American terms, a digital Pearl Harbor. Naturally, the loudest and most graphic contributions to the public sphere have been either news media trying to get our attention or computer software companies in pursuit of marketing their security software. Fear has been the main driving source.

In reality, hackers have not achieved any significant national disturbance or damage to the nation state in the last 20 years. Successful hacker attacks mainly stole information that affected a number of individuals or companies. The few events that targeted the government, such as the highly publicized Wikileaks incident, a massive theft of federal information and communications, did not have any significant long-term impact on the targeted society. The nation state stood unaffected.

Traditionally hackers had little or no interest in destabilizing or challenging the state. The reasoning behind this could be as simple as there is no monetary gain for such activity. Cyber criminality is an enterprise that seeks to earn money through illegal activities and defraud others. That is one reason why fighting cybercrime has had such a low priority as measured by the number of prosecutions. The traditional cybercrime does not threaten the state, the government, or the societal order, and there is no sizeable harm to the general population.

Low Incentive to Attack the State

Cyber criminals are instead avoiding a state confrontation for a simple set of reasons: prosecution, forensics, and ability to extradite. For example, a criminal activity that steals $5 from 100,000 individuals worldwide by using their credit card numbers benefits from the fact that the $5 is still $5 for each victim. Only a few of these 100,000 victims will take their time to fill in a police report or report the crime to federal authorities, because they realize that they probably will not get their $5 back. Unless the federal authorities or credit card companies organize a legal counter activity, the theft of $500,000 goes unpunished. The perpetrators can increase the likelihood that they are never prosecuted or extradited overseas, because the victims are not organized and have no resourceful institutional body to take counteraction. If the cyber criminals instead attack a state, for example, the UK, United States, or France, and create significant damage, the cyber criminals face a forceful counteraction and law enforcement. Until now the aggressors on the Internet have been of minor size and limited resources, but this is changing as states become involved in a militarized cyberspace.

The Militarized Cyberspace

The militarized cyberspace becomes a contested domain when state actors enter in pursuit of an intelligence objective, power maximization, and national security concerns. The main difference now is that there are massive resources available for state actors compared to the earlier generations of independent hackers. States can engage knowledge and ability generation through the defense industry, academic research centers, and covert operations, and outsource the cyber warfare to industrial contractors.

This represents a major shift in the threats. The hackers are no longer a few people operating with a marginal budget in their spare time. Cyber attacks are becoming a well-funded operation, sanctioned from within the defense and intelligence establishment of the attacking country, using allocated resources equal to any military and intelligence operation. This serves as the argument for the comparison, and contrary to the common belief, that the first 20 years of the Internet were more secure than the cyber environment of the future.

The entrance of state actors and the creation of a militarized Internet used as a contested space for intelligence, economic espionage, information operations, and to destabilize adversarial states has radically changed the fundamentals for security in cyberspace. The state actor seeks to exploit weaknesses in the critical national infrastructure and information systems, and take advantage of the fact that our populations rely heavily on the Internet.

One major weakness in the advanced societies is the overemphasis in cyber security training and research on information assurance and the hardening of systems when the paradigm has changed toward full-spectrum cyber operations (Kallberg and Thuraisingham, 2012). By continuously hardening systems a false sense of control and security is maintained, mainly based on the earlier attacker profile with single individual or small criminal efforts to penetrate the system. Other security concerns related to cyberspace such as influencing population sentiment, information operations, and destabilizing governments by systematic attack are unaddressed. Cyber security now consists of tools and the implementation of those tools and lacks abstract theory, therefore, becoming incoherent and lacking a strategic societal system approach. This gap of consistency is an inlet for attacks.

The Growing Cyber Opportunity

The state acts in the state's best interest, unless it is confused by media. In the last decade the national security debate has oftentimes missed the distinction between national security and individual security. The attack on the World Trade Center in New York added fuel to an already established popular notion that attacks on a number of individuals are an attack on the state. Terrorism is a menace and it is the state's responsibility, as the state claims the right to maintain the monopoly on violence, to protect the citizens of the state. The blurred demarcation between national security and individual security becomes apparent for a cyber-defending nation. Cyber attacks, seen from a state perspective, are annoying and a threat to the economy until it reaches a point when it becomes a national security concern. The United States considers an attack on military networks, critical infrastructure, and main industries as an attack on the state itself.

As cyberspace matures, states are able to define their reasoning and level of thresholds for national security response. Over the last few years there has been a shift in cyber strategy focusing on the national security. During the next decade the national security concerns in cyberspace are likely to override the earlier paradigm of focusing on securing individuals and single corporations. The attacks on individuals and corporations have become solely a criminalized act; meanwhile, the state considers attacks on the national critical infrastructure, the state's core function, the state's legitimacy and authority, and its military complex as attacks on the state itself.

The increased reliance on computer networks, changes in societal sentiment influenced by the Internet, and the increased complexity create opportunities for an aggressor and terrorists. It is unlikely that terrorists will be able to represent a permanent cyber threat to a nation due to the cost and infrastructure needed. The combination of a covert state actor and terrorists as executors of attacks creates a different more likely scenario for the future.

Covert Operations by Proxies

The scenario becomes more complex if a state actor gathers information about cyber vulnerabilities in the networks of a targeted organization or other nation and then outsources the attack to a criminal or terrorist network. This innovative modus operandi creates numerous obstacles and considerations for the targeted organization. First, the attribution problem is highlighted, because even if the executing criminal network is identified, it is still unclear which actor initiated the attack. Criminal networks are enterprises and the compensation could be a range of illicit goods (Kan, 2009). States can pay to get things done. If necessary, a covert operating state can pay criminal networks cash, drugs, weapons, or any currency to act as a proxy. Terrorist organizations can finance their operation through cyber terrorism "entrepreneurship" instead of engaging in other forms of financing that are far riskier for detection such as drug dealing and credit card fraud. Second, the lack of attribution evaporates the option to initiate retribution against the initial attacker. Third, it is likely that the vehicles for the attack are dismantled directly after the attack. The computers and networks that were used for the attack are no longer in use afterward. The lack of attribution removes the risk to engage and the fundamentals of state-to-state deterrence are no longer in place (Reed, 1975).

Cyber terrorists can be a national security threat, and create significant damage to critical infrastructure and national assets for the targeted state, if the terrorists are given the toolset and pre-attack intelligence from a state actor. The covert warfare in cyberspace in many cases resembles the covert operations in the Cold War. The targeted country, or organization, could assume where the attack is coming from but attribution is not strong enough for retribution. A state engaging in retribution toward another state could face other grave unanticipated political consequences, which pose uncertainty and generate a risk-averse state actor.

The aggressor's risk is lowered if the state actor collects vulnerabilities in the opposing state's networks, builds cyber weapons, and creates a strategy to create disruption and destabilization in the opponent's networks, but uses a proxy to carry out the actual attack. In this case the aggressor is unlikely to be held accountable for its actions. The opportunity not to be held accountable is extremely inviting for countries that are covert adversaries.

If the adversary is skilled, it is more likely the attribution investigation will end with a set of spoofed innocent actors whose digital identities have been exploited in the attack rather than attribution to the real perpetrator. A strong suspicion would impact interstate relations, but full attribution and traceability are needed to create a case for reprisal and retaliation. Attribution can be graduated, and the level varies as to what would be accepted as an "attributed" attack. The national leadership can accept a lower level of tangible attribution, based on earlier intelligence reports and adversarial modus operandi, than the international community might demand, but it is restrained in taking action.

La Raison d'état

Cyberspace is already by definitions and doctrine a war-fighting domain even if only a few states are able to do any offensive cyber operations, but the strategic abilities will grow in the next decade. There are several reasons why cyber weapons are inviting.

In an era of austerity countries seek alternatives to traditional military policy options that are better suited for future conflicts, but also reduce the collateral damage that a kinetic operations creates. The pursuit of cyber abilities also drills down to pure financial numbers (Kallberg and Lowther, 2012). Militaries are expensive and require a standing force to ensure ability and deterrence. If the force is a professional army it will cost to recruit, train, pay, and pension its soldiers. In modern state reasoning, cyber warfare is a cheaper option for both covert operations and to engage and destabilize an adversary (Kallberg and Lowther, 2012).

States act in their own self-interest; therefore, it is questionable if a regulated cyberspace is in the long-term interest of the major powers, as a restrictive use of cyberspace would undermine their dominant status. Earlier efforts to create a uniform approach toward information technology security on a global scale have shown marginal progress. One example is the global standard for security certification of hardware, "Common Criteria," that has been hindered by the lack of unrestricted trust between nations (Kallberg, 2012a). If any international effort fails to create a uniform approach to securing the Internet domain we can assume by logic that major actors prefer the anarchy before order because there is a perceived opportunity and potential future gain for these powers.

Expanded Reach for Cyber Conflicts

State actors will implement cyber conflict at all levels that benefit the state. As an example, targets that had limited value for cyber criminals, such as the global space-borne information grid, are a prime target for a state actor (Kallberg, 2012b). Satellites are a major concern for any state or nonstate actor who intends to conduct operations in secrecy. Satellites gather intelligence, provide surveillance, and perform reconnaissance (Moltz, 2011). This can be extremely annoying to states that seek to avoid transparency between their international commitments, their public posture, and their actions behind the scenes.

Terrestrial cyber attacks are a single exploit on thousands, if not millions, of identical systems, and the exploit will be eliminated afterward by updates or upgrades. The difference between satellites and terrestrial cyber exploits is that a satellite is often custom made, whereas the computing design is proprietary. Cyber attacks in space exploit a single system, or limited group of systems, within a larger group of satellites (*Wired*, 2011). These space-borne assets have a variety of operating systems, embedded software, and designs from disparate technological legacies. As more nations engage in launching satellites with a variety of technical sophistication, the risk for hijacking and manipulation through covert activity increases. A satellite's onboard computer can allow reconfiguration and software updates, which increase its vulnerability to cyber attacks. The attack on the satellite is tailored, one shot, and unique.

An attributed cyber attack on the global information grid would be considered an act of war, and provide the targeted state with at least a theoretical *casus belli*, a risk that the aggressor would seek to avoid. An act of war is a tangible security risk that can have catastrophic consequences for an aggressor nation. Are attacks on the global information grid ideal for being outsourced from the aggressive state actor to terrorists and criminal network to avoid attribution? The symbiosis between a state actor and cyber terrorist can provide an ability that makes cyber terrorism a tangible national security threat at the strategic level.

CONCLUSION

The threat from cyber operations will increase in the next decade, even if we have implemented extensive information security. The Internet and the application layer become a globally contested domain where the entrance of state actors as contestants and aggressors create a radical shift. The early hackers and information thieves had limited resources and mainly a financial goal. State-run operations have a complete different set of targets and goals.

If states collect vulnerabilities in targeted systems, utilize the whole covert spectrum, and instead of attacking themselves uses terrorist groups as proxies, then cyber terrorism is a tangible and relevant national security threat.

The digital environment where critical assets can be copied, sent, and forwarded within seconds, ushers in a symbiosis between aggressive adversarial state actors and terrorist networks when the state actor can produce military-grade cyber weapons for the terrorists to use. Waltz (1990) argued that the power embedded in nuclear arms is not what you do but what you can do. The outsourced proxy cyber war from state actor to cyber terrorists operates along the same lines as military-grade cyber weapons dispersed to violent groups and militant political groups create extensive uncertainty. This uncertainty is based on what an aggressor can do— not what they actually do.

This development creates an asymmetric covert conflict with an anonymous aggressor and a reactive targeted society. Terrorists can reach their objectives, create damage, influence policy, and leverage the disproportional power relation between terrorists and the defending state.

CHAPTER

20

Cyber Security Countermeasures to Combat Cyber Terrorism

Lachlan MacKinnon, Liz Bacon, Diane Gan, Georgios Loukas, David Chadwick, Dimitrios Frangiskatos

INTRODUCTION

Any piece of work that seeks to discuss cyber terrorism must necessarily start with some definitions and descriptions to aid the reader to both differentiate and contextualize cyber terrorism from other areas of cyber security, such as cybercrime, malicious hacking, cyber fraud, and the numerous different types of system breaches, failures, and human error.

Most contemporary definitions of cyber terrorism focus on the following three aspects:

1. The motivation of the perpetrator(s)
2. The targeted cyber system
3. The impact on an identified population.

For example, the Federal Bureau of Investigation (FBI) definition (Pollitt, 2003) describes cyber terrorism as:

- Politically motivated subnational groups or clandestine agents
- Breaches in information, computer systems, computer programs, and data
- Violence against noncombatant targets

The National Infrastructure Protection Center (Garrison and Grand, 2001) defines cyber terrorism the following ways:

- As a criminal act seeking to influence a government or population to conform to a particular political, social, or ideological agenda
- To be by the use of computers and telecommunications capabilities
- To be violence, destruction and/or disruption of services to create fear by causing confusion and uncertainty within a given population.

Denning (2000) defines cyber terrorism as:

- As an unlawful activity to intimidate or coerce a government or its people for a political or social objective
- As attacks and threats of attacks against computers, networks, and the information stored therein
- As an attack that results in violence against persons or property, or at least causes enough harm to generate fear

In a real sense, therefore, we can make the argument that the key issue in cyber terrorism is the motivation to carry out an activity in cyberspace that results in violence/harm or damage to individuals and/or their property. If considered in these terms, it becomes clear that a number of existing activities in cyberspace, which result in harm to individuals and/or their property, might be constituted as cyber terrorism simply on the basis of establishing the motivation for the activity. This leads us into a current debate as to whether cyber terrorism actually exists or is simply another manifestation of existing malicious and criminal activity in cyberspace. A number of commentators have sought to make the argument that there is neither evidence nor rationale to argue that cyber terrorism exists independent of existing cyber activities (Conway, 2011). However, we would support the view put forward by a number of other authors that there is sufficient evidence, highlighted in particular by events such as Stuxnet and others described later in this chapter, to justify a consideration of cyber terrorism as a separate entity within this space (Greengard, 2010). On the basis of this argument, we would also argue that existing tools, techniques, and approaches adopted by perpetrators of malicious and criminal cyberspace activities can and should relevantly be considered within cyber terrorism. Fundamentally, if the motivation behind any kind of cyber event fulfills the criteria of seeking to promote or impose political agenda or will upon a given population identified by the various authors above, then whatever techniques are used it qualifies as cyber terrorism. Clearly, the use of these techniques by technologically advanced nations in conflict with one another would constitute cyberwarfare, which would change the nature and impact of many of the events described in this chapter. However, our focus is not on explicit cyberwarfare, although a number of the events described later in this chapter are attributed to national agencies, which does represent an implicit form of cyberwarfare.

So, What Is the Difference between Cybercrime and Cyber Terrorism?

The majority of cyber attacks are launched by cybercriminal gangs determined to steal money, credit card information, bank accounts, or personal information. The intent is to make money. A general description of the dark side of the Internet can be found in the paper by Kim et al. (2009). On the other hand not all hackers are cybercriminals. Many hackers are computer enthusiasts who take pleasure in gaining access to computers and networks just to leave their "calling card." Defacing a Web site for political motives or simply to gain acclaim among their peers is their objective.

Attack patterns seen in criminal operations differ from incidents involving cyber terrorists. Cybercriminals typically use numerous targets and do not maintain prolonged control over servers, as the risk of detection increases proportionally (Krekel et al., 2012). However, the motives for a cyber attack are to some extent irrelevant. A criminal trying to steal money or a cyber terrorist trying to cause disruption, destruction, or steal secrets (cyber espionage), will both use the same methods. The main difference lies in the purpose of the covertness: the criminal stealing money or information would not want anyone to know what they were doing, to evade capture and prosecution; whereas, cyber espionage tries not to do damage to the attacked system so that information can continue to flow out (Saalbach, 2012).

As described previously, cyber terrorists would have a different agenda and their targets are likely to be a lot less secure. Currently, banks and credit card companies go to a lot of effort to secure customer information, but these are of limited interest to a cyber terrorist. In general, they are looking for softer targets with maximum public impact. The U.S. government is increasingly aware of government-run and -controlled cybergroups originating in China and Russia. It is not too far a step, and would seem to be only a matter of time, for a terrorist group to follow suit.

The main difference between cybercrime and cyber terrorism lies in the objective of the attack. Cybercriminals are predominantly out to make money, while cyber terrorists may have a range of motives and will often seek to have a destructive impact, particularly on critical infrastructure. Cyber terrorists also want to have maximum impact with the greatest stealth. Greengard (2010) identified a range of cyber attack methods that can be deployed by cyber terrorists, including "vandalism, spreading propaganda, gathering classified data, using distributed denial-of-service attacks to shut down systems, destroying equipment, attacking critical infrastructure, and planting malicious software."

Cyber weapons are software tools used by cyber terrorists. These tools can manipulate computers, intrude into systems, and perform espionage. They are essentially the same as those used by cybercriminals (Saalbach, 2012). There is currently no evidence to suggest that terrorists are using malware or hacking into systems. However, it seems unrealistic to think that they have not identified the potential for doing so. They may even be developing a Stuxnet equivalent (described later in the chapter) for military targets at this time.

Why Are the Risks Greater Today?

The cyber landscape is very different today from only a few years ago. Now most electronic devices can be connected to the Internet—phones (IP phones, smartphones, iPhones), TVs, computers, iPads, Nintendo Wii, MS Xbox, Sony Playstation, smart home equipment (sensors, cameras, and alarms), CCTV systems—the list goes on. All of these systems have IP addresses, so they are trackable and accessible through the Internet. Devices with radio frequency ID chips can communicate with other computers and devices (Saalbach, 2012). Even systems that were never supposed to be connected to the Internet sometimes are; for example, the Supervisory Control and Data Acquisition (SCADA) systems that control water treatment plants, power grids, nuclear reactors, and production lines. Many of these systems have the ability to allow engineers to remotely log in and make adjustments to the computers that control, for example, pumps and sluice gates. The complexity of the systems connected to the Internet increases each year and with this the opportunities for security breaches also increases. In October 2011 the highest number of vulnerabilities were reported and patched by all the big vendors, such as Apple, Microsoft, VMware and Oracle (VeriSign, 2012). This is an indication of the numbers of vulnerabilities that are being found each month. Each vulnerability is a potential breach in security for anyone using that particular system. These days remote access is expected by users. People log into work machines to read e-mail and to work from home. Secure links are often provided in the form of virtual private networks, but if the computer that is connecting goes through the link that is already infected with malware, then security is compromised and the bad guys have bypassed the defenses.

There have been incidents in the past where hacker groups have broken into American computer systems. The first one identified in 2003 was code-named "Titan Rain," which been associated with an Advanced Persistent Threat. Titan Rain was the code name given by the U.S. federal government to a long series of coordinated and very sophisticated cyber attacks primarily against American computer systems between 2003 and 2005. There were thousands of files downloaded from a large number of organizations, including Lockheed Martin, Redstone Arsenal, and NASA. Shawn Carpenter, a security expert, worked for the FBI to track down the origin of the attacks. Initially the files were downloaded to servers in South Korea, Hong Kong, and Taiwan before being transferred to the southern Chinese province of Guangdong. The suspicion was that this was Chinese government state-sponsored espionage, which China strongly denies (Thornburgh, 2005).

In mid-2009 there was a series of attacks over a 6 month period on Google, Adobe, and dozens of other high-profile companies. These attacks, code-named "Operation Aurora," used social engineering to encourage a victim to connect to a malicious Web site and then "combined encryption, stealth programming and an unknown hole in Internet Explorer" (Stamos, 2012) that enabled the attacker to escalate their privileges and gain access. Google claimed that the attacks originated in China and threatened to pull out of the country (Sood and Enbody, 2012).

Titan Rain and Operation Aurora are often provided as examples of state-sponsored cyber terrorism. While this is plausible, there are a number of analysts who reject the notion that a technologically advanced state, in this case the Chinese, would leave a trail of obvious footprints leading back to the country of origin. For example, Lewis (2005) claimed that it was likely that the perpetrators of the Titan Rain attacks used poorly secured Chinese networks and systems as intermediaries. At the time, China had a very insecure information technology (IT) infrastructure due to poor security practices and the widespread use of legacy and pirated operating systems.

Possibly more worrying is the threat from the "insider." This is someone who is already a user on the network under attack and is inside the security perimeter. The insider is especially dangerous because he is far more aware of the security in place on a network and the attached servers. Insiders know about the information stored on those servers and they also know about the security that surrounds it. This is described further in the section The Insider Threat.

CYBERPHYSICAL ATTACKS

Terrorist attacks have traditionally aimed to cause considerable human loss through physical means, such as armed assaults, explosives, and biochemical agents. However, as our societies are increasingly dependent on IT infrastructures and systems that are dependent on computers and networks, a new class of potential cyberphysical terrorist threats has emerged. For example, the control systems of the Thames barrier, the flight mechanism of an unmanned aerial vehicle, the operating room of a hospital, the unmanned Docklands Light Railway, and even the typical passenger elevator contain and rely heavily on computer software, hardware, and communications. As a result, these systems are vulnerable to both physical and cyber threats. A cyber attack may facilitate a physical terrorist attack by disabling monitoring and security equipment or may cause physical damage directly. Such an attack against a gas or water management facility may require considerably less planning and resources than a physical terrorist attack with the same aim. In fact, one can easily find on the Internet detailed guides, attack tools, and specialized search engines for exploiting the computer vulnerabilities of common industrial control systems used in such facilities.

Interestingly, the concept of cyberphysical crime has been utilized in popular culture since at least the 1960s. For example, in the film *The Italian Job*, a team of robbers employs a scientist to compromise the computers of Turin's traffic control systems and help the robbers escape thanks to the resulting traffic jam. Reliable reports on real cyberphysical security incidents are rare and, to the best of our knowledge, none has been openly linked to terrorism. Nevertheless, a brief history of representative incidents can illustrate the breadth of targets and the evolution of the attack mechanisms and their complexity. It is worth noting that several were unintended accidents or the result of a hacker's curiosity without malicious intent. Yet, they have exposed cyberphysical vulnerabilities in critical systems that do not require exceptional technical knowledge to be exploited maliciously.

Notable Cyberphysical Incidents

The earliest incident that is often linked to a cyberphysical attack is the 1982 Siberian Pipeline Explosion, which has been reported to be the result of intentionally flawed industrial control software altered by the Central Intelligence Agency (CIA) and sold indirectly to the Soviets (Reed, 2004). According to these reports, the software that controlled critical pressure valves increased the effect of a pressure test of the pipeline and caused a "monumental" explosion.

The "Farewell Dossier," which was declassified in 1996, does indeed indicate that the CIA routinely fed defective technologies to the Soviet Union, but does not confirm the specific incident (Weiss, 1996). A confirmed incident involving a gasoline pipeline explosion happened in Bellingham, Washington, in 1999. The explosion caused three deaths and considerable environmental damage and was attributed in part to the slow-down of the pipeline's control software. Although no evidence of intent was identified, the control systems were found to be connected directly to the network of the building without proper access monitoring or other security measures.

Since then, cyberphysical incidents in the energy sector have multiplied. In 2003, the Davis–Besse nuclear power plant was shut down after the SQL "Slammer" worm disabled its safety monitoring systems. In 2007, the U.S. Department of Homeland Security's "Aurora Experiment" at the Department of Energy's Idaho lab demonstrated a cyber attack that blew up a power generator typically used in the U.S. domestic electrical grid. While it is not clear what type of cyber attack was used in this case, by then it was already known that critical industrial control systems were vulnerable to the same threats as Web sites and personal computers, including port scanning, SQL injection, anonymous FTP, and simple password guessing. Two years later, senior U.S. officials reported that cyber spies from foreign states had been probing the U.S. electric grid's infrastructure and had planted suspicious software for possible future use (Gorman, 2009). With the cyberphysical security weaknesses of this sector already obvious by then, it is not surprising that the first major attack, often considered the beginning of cyberwarfare, was against a nuclear facility. On November 29, 2010, Iran's president confirmed that the controller handling the centrifuges at the Natanz Nuclear facilities had been damaged by Stuxnet, an exceptionally complex worm that was designed specifically to attack this target (Falliere et al., 2011). Its complexity, the presumably high cost of development and, of course, the target, have led most analysts to suggest the United States and Israel as the originators of this new cyber weapon. Since then, at least two other worms have appeared that are closely related to Stuxnet, although with clearly different targets, and may have been designed by the same team.

The water sector has also seen a number of cyberphysical attacks over the last two decades. In 1994, a hacker used a common dial-up modem to connect to the Salt River Project's network in Arizona, and gain access to water and power monitoring information. An investigation concluded that there was no major threat to Arizona's Roosevelt Dam and there was no intention to cause harm (Gleick, 2006). As usual, the hacker had done it primarily out of curiosity. Very different was the motivation and impact of an attack in Australia in 2000 (Turk, 2005). Vitek Boden was a 40-year old employee of a firm subcontracted to install wireless control equipment for the sewage systems of Queensland's Maroochy Shire Council. When he lost his job with the firm and was also denied a job with the Council, he decided to use his technical knowledge to take revenge. He used stolen radio equipment to issue rogue commands to the sewage pumping systems and released over 800,000 liters of raw sewage into parks, rivers, and property. Although the subcontracting firm had noticed the misbehavior of the pumping stations, and had concluded that only someone with detailed familiarity of the systems could be behind it, Boden managed to connect to the pumping stations at least 46 times over 3 months. He was caught only after the police pulled him over for a traffic violation and found the radio equipment in the car. He was sentenced to two years in jail and was ordered to reimburse the Council for the cleanup.

Two years later, U.S. authorities discovered instructions on poisoning water sources on a suspected terrorist. The FBI issued a bulletin indicating that al-Qaeda agents had been seeking information on the control systems of dams, water supplies, and wastewater management facilities in the United States and abroad. While awareness of these threats has been raised since then, due to the prohibitive cost of replacing industrial control equipment there are still several vulnerable pumping stations worldwide. In fact, it was demonstrated at a 2011 hacker conference that the Internet address of the IT units controlling them are easily discoverable via common search engines, such as Google. By knowing their address, a hacker can attempt a wide range of attacks to disable them or alter their behavior.

In the transport sector, cyberphysical incidents usually cause disruption in dispatching and signaling. In the 1990s they were related primarily to the lack of user authentication mechanisms. For example, a hacker would connect via a dial-up modem to an airport network pretending to be the legitimate system administrator and would alter critical information. Later, due to the increasing use of off-the-shelf computers running Microsoft Windows, a number of incidents in the transport sector were caused by common viruses and worms that spread via the Internet and infected computers indiscriminately. One such virus disabled air traffic control systems in Alaska in 2006. Yet, in most cases, there was no malicious intent and, more significantly, there was no damage beyond frustration and financial costs due to downtime. In 2008 though, a teenager managed to take control of the tram system in Lodz, Poland, and operated its track switches, eventually causing four trains to derail and 14 people to be injured.

Since then, researchers have demonstrated that even common production cars can be targets of cyberphysical attacks (Koscher et al., 2010). Today's cars depend heavily on a variety of sensing and computing equipment that are interconnected and can affect each other in unpredictable ways. One can infect a car's electronic systems through a manipulated audio file added to its MP3 playlist or can use an infected smartphone connected to the car through Bluetooth. A car interfered with in such a manner may be forced to veer toward one direction while driving at a fast speed. Another cyber weakness of vehicles is the use of satellite navigation. These devices can be fooled to display the wrong location and traffic information and direct the driver of the vehicle toward a terrorist ambush. Interference with the satellite navigation signals over an area could cause local traffic jams, for example, to delay the emergency services following an act of terrorism. Scenarios involving such interference are increasingly likely because of the recent proliferation in the black market of GPS jamming devices that are often used by thieves to prevent stolen trucks from being tracked by their owners.

Cyberphysical attacks involving satellite systems are also becoming common in the defense sector. In 2009, militants in Iraq used off-the-shelf software, costing just $29.99, to intercept live video feeds from unmanned aerial vehicles (UAV). The software, which is still sold commercially, had been developed by a Russian company to allow interception of satellite TV, but proved to work just as well for unencrypted military surveillance feeds (Gorman et al., 2009). Since then, the military affected aircraft have been retrofitted to encrypt the video they transmit. Two years later, the U.S. military found that a number of their frontline UAVs had been infected by viruses that were logging the keystrokes of the pilots who remotely controlled them during combat missions. It is most likely that the intention behind this attack was to reveal what signals transmitted by the pilot would operate what part of the vehicle. The same year, Iranian TV showed an American UAV claiming that the Iranian army's electronic warfare unit had electronically hijacked and landed it intact. If UAVs costing millions of dollars can be interfered with via cyber means, it is more than likely that smaller civilian unmanned aerial devices, such as police surveillance cameras in major events, which receive and transmit unencrypted signals can also be hijacked and flown into a crowd. In fact, researchers from the University of Texas recently used their own mini helicopter drone to demonstrate how such an attack can be performed. The cost of the equipment they used to build their proof of concept system did not exceed $1,000.

By now, it is obvious that cyberphysical attacks can affect practically every sector that relies on a computer infrastructure, from defense and food to home automation and emergency management. Of particular interest is the health sector. Terrorist attacks against the health sector have traditionally been rare, possibly due to the moral outrage that they would cause. However, the increasingly networked infrastructure of modern healthcare systems may present opportunities for terrorists to cause damage in a more covert manner. The potential of such an attack became clear in the 1980s when massive overdoses by the Therac-20 computerized radiation therapy machine caused four deaths (Leveson and Turner, 1993). The machine's designers had faith in the computer software's reliability without the necessary hardware safety mechanisms and interlocks that were found in previous versions of the machine. In 2008, scientists demonstrated that common cardiac devices could be operated remotely without authorization, allowing a malicious user to deliver remotely a life-threatening shock (Halperin et al., 2008). In 2009, 10% of Sweden's healthcare IT infrastructure, including MRI machines and heart monitors, were disabled by an Internet worm originally designed to affect normal personal computers. The same year, a medical clinic's security guard in the United States was arrested for cyber intrusions that intentionally tampered with the air conditioning systems putting patients and pharmaceuticals in danger (FBI, 2009). A terrorist organization could potentially adopt such approaches to impede the emergency response operations after a physical attack and thus cause maximum damage.

MALWARE CANDIDATES FOR CYBER TERRORISM

As hacker attacks are on the increase, it is not unreasonable to assume that terrorist groups around the world also have their eye on the "low hanging fruit" that litters the Internet and that can be accessed using current cyber attack tools. The creators of worms and viruses have not had specific targets in their sights when they released their malware into the wild. However, there have been reported incidents where malware has gained access to critical systems by accident. Such an event occurred when the MS SQL Slammer worm gained access to the Davis–Besse nuclear plant in Oak Harbor, Ohio. The worm bypassed the firewall that was in place and flooded the network with worm traffic, blocking the safety systems for nearly 5 hours and the computer that controls the processing plant for over 6 hours (Byres, 2004). The Slammer worm also got onto ATM machines and into airline reservation systems (Chen, 2010).

Critical infrastructure is defined as water treatment plants, oil refineries, power grids, gas pipelines, and so forth. These are considered by governments to be essential assets without which society cannot function. SCADA systems are used to gather data and control these systems, particularly where it is difficult or dangerous for humans. This is usually done in factories and industrial plants, where there may be production lines or for monitoring nuclear plants, gas pipelines, or water treatment facilities. SCADA systems were originally designed to be closed systems, that is, not connected to the Internet. However, it has been found that they are increasingly routinely connected to the Internet. Remote access by engineers to make minor adjustments does have some merit. However, security should be the top priority. It was found that a number of SCADA systems that could be accessed via the Internet still had the four-character default password in use. Many SCADA systems were also connected to a back office network (Ten et al., 2010). This was a recipe for disaster, as normal users on such a network are generally not security aware and may pose a particularly serious threat to this type of network. This also gives an idea of the scale of the threat and the exposure of these systems to attack from the Internet.

Currently, the main contenders for malware that could be used as a cyber weapon are Stuxnet, Duqu, Flame, and Shodan. An overview of each of these is presented below.

Stuxnet

The biggest threat to SCADA systems has been the Stuxnet worm. The earliest reported appearance of Stuxnet was in June 2009 (Falliere et al., 2011).

This version was relatively harmless, but Stuxnet rapidly evolved and the next variant reported early in 2010 was using a valid signed certificate obtained from Realtek Semiconductor Corps for a Stuxnet driver, which enabled it to trick users into downloading it as it appeared to be legitimate. Throughout 2010 Stuxnet continued to evolve until by mid-July it was able to exploit a Windows shell vulnerability (Exploit MS10-046) that permitted remote execution of code. The certificate from Realtek was quickly revoked by VeriSign, but Stuxnet replaced it with another valid one from JMicron Technology Corp. Within days reports began to come in of the first infections of WinCC and PCS 7 SCADA software running Siemens SIMATIC software that ran on a programmable logic controller (PLC). The time between each of these improvements in the malware's capability has been progressively shorter, from months between events at the start, down to days, as Stuxnet evolved.

From July to September 2010 Microsoft issued patches in an attempt to stop Stuxnet from spreading. Stuxnet exploited at least four zero day exploits (Chen, 2010), which is quite remarkable. Most malware writers would only have used one at a time, so as not to waste future opportunities. Analysis of the Stuxnet code revealed that it was attempting to inject and hide code in a PLC found in Siemens systems. These PLCs interface between the control systems and the equipment that is being controlled, such as robot arms or elevator doors. Stuxnet only infected specific systems and did not activate if the victim computer was not connected to a SCADA system. As Stuxnet is a worm, it can install itself in the operating system and travel between systems. The method of propagation used was via USB sticks, as not all these systems were connected to the Internet. To maintain stealth and avoid detection, after a number of successful infections it deletes itself. It used Siemens default passwords to gain control before injecting code into the PLC.

The aim was to find the right kind of system to infect, such as a nuclear power plant, and then to begin to slow down and speed up the centrifuges. Any engineer called out to diagnose this fault would find it very difficult to identify the problem. The aim was to cause physical damage to these systems (Chen, 2010). According to statistics collected, it was estimated that by September 2010 there were around 100,000 infected hosts around the world and the majority were in Iran. This indicated to many security experts that Iran was the primary target (Falliere et al., 2011).

The work done for Siemens by Langner (2011) to decompile the Stuxnet code was very revealing. The code was found to be well engineered and sophisticated. It was atypical in terms of malware code, as it was quite large and written in a number of different programming languages, which was unheard of in all previous worms and viruses. It also appears to have been written by a number of different individuals. The method Stuxnet uses to attack specific pieces of equipment shows that the writers of the code had detailed knowledge of these plants and the systems that control them. It is the view of Langner (2011) that Stuxnet was not the work of hackers, but of a government-funded team of programmers, and that the biggest cyber superpower was the prime candidate, that is, the United States. The prime motive appeared to have been to disrupt Iran's nuclear program.

Stuxnet continues to spread and infect computer systems via the Internet, although its power to do damage is now limited by effective antidotes, and a built-in expiration date of June 24, 2012 (Farwell and Rohozinski, 2011).

Using freely available search engines (see Shodan) it is relatively easy to find the IP addresses of the SCADA systems, which manage and control the critical infrastructure of almost every nation (Naraine, 2010). That leaves a number of critical infrastructures vulnerable to cyber attacks. The worry among the cyber security communities regarding Stuxnet was the level of sophistication and the types of systems targeted.

Duqu

Duqu is referred to as the son of Stuxnet. How does it differ from Stuxnet? It is clearly based on the Stuxnet code but Duqu does not contain any code that could affect industrial control systems. Its mission seems to be to collect information such as design documents from the same systems that Stuxnet attacked. The purpose is assumed to aid the development of the next version of the attack tool (Symantec, 2011).

Duqu used a different approach to Stuxnet. It was delivered via e-mail with a Word document, which contained a zero day exploit that enabled Duqu to install itself. The aim was to gather information on system configurations and also to collect the keystrokes of users with the use of a key logger. For SCADA systems that are connected to office systems this seems like a very efficient way for Duqu to propagate. There have been a number of variants and the code seems to still be evolving.

The Duqu code comprises a configuration file and a driver file (dll), which has a valid (although stolen) digital certificate. This is the same technique used by Stuxnet. Duqu also needs an installer to load the dll. Forensic analysis of the configuration file showed that the time and date of the infection is stored in the file. It appears that Duqu will only be active for 30 days and then it removes itself, presumably to reduce the chances of detection. Having installed and collected intelligence, Duqu then attempts to communicate with a number of command and control (C&C) centers. C&C centers have been identified in India, Belgium, and Vietnam. These centers are acting as proxies and merely forwarding the traffic on, so it is very difficult to identify the real C&C center. The files transferred look like jpg files but have the data collected appended and lightly encrypted and compressed within them. As of March 2010 there have been 15 variants of Duqu identified (Symantec, 2011).

Duqu has serious implications for any network that requires top security. It hides itself on the infected system. It has the ability to log everything that a user types. It also collects information about the network and the infrastructure. All of these data are then encrypted and sent out disguised as an image file, which is sent to a C&C center somewhere on the Internet.

Flame

The next contender in the cyber weapon arsenal is Flame. It is unclear how long Flame has been around and opinions differ. It was first identified by Kaspersky in 2010. However, there is evidence to suggest that Flame was operating as an espionage tool prior to this (Lee, 2012).

Flame used social engineering to trick people into downloading it by spoofing the Microsoft's Windows update service using fake certificates. Users would then click on the update link and become infected by Flame (Whitney, 2012).

Analysis by Kaspersky has shown that Flame is a sophisticated attack toolkit with cyber espionage capability. It is significantly larger than Stuxnet (20 times bigger) and more complex than Duqu. Flame is coded using the object-oriented language C++. This makes it difficult to analyze due to the compiler and the way the language is constructed. It also appears to have been written in such a way that it is difficult to follow the logic of the code (Matrosov and Rodionov, 2012). It is made up of a number of attack tools, which include taking screenshots at regular intervals, recording audio conversations, key logging, and packet sniffing on the network. Flame has many ways to steal data. It has no similarities with the Stuxnet/Duqu code, but it does use C&C servers to upload the stolen information. Once Flame has installed there are more modules that can be added to improve the data-stealing capability. It would appear that at this time Flame is still undergoing further development, although the authors are still to be identified. Interestingly, the files within the code have false creation dates (starting in 1992) to hide the actual "age" of Flame.

Flame was clearly designed to steal information and not money from banks, making it a prime candidate for the cyber weapon of choice (Gostev, 2012). The cyber security coordinator for the United Nation's Geneva-based International Telecommunications Union, Mr. Obiso, told Reuters in May 2012, that he considered Flame to be a "dangerous espionage tool that could potentially be used to attack critical infrastructure" (Bozorgmehr, 2012).

> *Flame can easily be described as one of the most complex threats ever discovered. It's big and incredibly sophisticated. It pretty much redefines the notion of cyberwar and cyberespionage.*
>
> ***Alexander Gostev (2012), Kaspersky Lab Expert***

Shodan

The Shodan search engine was launched in November 2009. Shodan, named after the Sentient Hyper-Optimized Data Access Network of science fiction, was developed by a teenager called John Matherly who wanted to see how much he could find out about devices connected to the Internet. He was surprised to find that a large number of industrial control computers were in fact accessible from the Internet. To make it worse, many of these systems had little or no security at all. These vulnerable systems controlled water plants and power grids around the world.

How is Shodan different from other search engines that crawl the Web looking for data in Web pages? Search engines such as Google and Bing search through the text on Web pages to find what the user is looking for. Shodan searches the World Wide Web interrogating ports and grabbing banners to identify vulnerable devices. It identifies the IP addresses of devices and then tries to connect to them, and if it succeeds it "fingerprints" that device. All of the information collected, including geographical location, software, and any banner information displayed is stored and then available for anyone to download. It also searches for default passwords or nonexistent security controls. It is estimated that information about 10 million devices was collected each month, which are then available for anyone to query in the same way that you would with Google. It is reported that a Shodan user "found and accessed the cyclotron at the Lawrence Berkeley National Laboratory" (O'Harrow, 2012). Other users have found thousands of unsecured Cisco routers. It is therefore not unexpected that hackers are using Shodan to search for SCADA systems that are connected to the Internet (Naraine, 2010).

While Shodan is not a cyber weapon on its own, it is certainly a facilitator for cyber terrorism.

THE INSIDER THREAT

A very serious threat to any network comes from the insider. Who is the insider? This is a person who is not affected by any security that keeps intruders out of a network, because they are already inside the perimeter. This could be someone who is permitted to access the network because they have a legitimate login and ID. They could be an employee or a contractor working for the company, or anyone who has a formal business relationship with the company. They could be a bank customer who can access their own account details or someone who has stolen the credentials of a user. They could be someone who is forced to aid an outsider to perform some action. They could be a former insider who has retained their login credentials (Bellovin, 2008).

Many organizations focus their security on addressing potential attacks from outside the organization and give insufficient consideration to threats from insiders. Statistics quoted publicly on insider threats vary significantly; however, there is no disagreement that the threat is very real. The 2007 E-Crime Watch SurveyTM, conducted by the United States Secret Service, the CERT Coordination Center (CERT/CC), Microsoft, and *CSO Magazine*, found that where the perpetrator could be identified, 31% of attacks were committed by insiders and 49% of their survey respondents (671 security executives and law enforcement officials) had experienced at least one deliberate insider attack in the previous year. It is, however, important to clarify what we mean by insider threats. Jones and Averbeck (2011) defined three types of insider threats:

1. **Trusted unwitting insider**: This is someone who has no malicious intent but accidentally, through an error of judgment, supports or initiates an attack. For example, by opening an inappropriate e-mail releasing malware or, more classically, opening up a USB stick, which they think has been lost. In reality it has been planted for them to find, and unwittingly open up with the best of intentions to try and find the owner, releasing malware into the system. Inadvertent threats are as real and as important to address through education and so forth, but are not the focus in this section. Attacks of this type are generally referred to as access control failure attacks.
2. **Trusted witting insider**: This is someone who has legitimate access to systems and makes a conscious decision to, for example, release unauthorized data to a third party. Attacks of this type are generally referred to as misuse of access attacks.
3. **Untrusted insider**: This is someone who has gained access illegally, for example, by fooling someone with a lost USB stick, who now has internal access and can now act as though they are a trusted employee. Attacks of this type are generally referred to as defense bypass attacks.

What motivates someone to spy and steal information that could potentially aid another country? This is a complex issue and there are numerous factors. The motivation could be money, revenge, blackmail, or even anger at not getting promoted. There could be divided loyalties or they may simply want the thrill of living a James Bond type fantasy (Moore, 2008). Insiders can be current or former employees, contractors, or other parties who have or have had access to privileged information and include business partners and employees from companies to whom work has been outsourced. Insiders have a huge advantage over outsiders in that they are aware of company policies and procedures, how they are applied, and where the vulnerabilities and weaknesses are in their setup and use. For those with more technical skills, they will know how the technology is used, the level of security, how firewalls are set up, and if they are programmers, they then may have access to directly edit code. All this makes combating attacks by insiders more challenging.

A study was performed by the U.S Secret Service and CERT in which cases of insider attacks on U.S. critical infrastructure sectors were analyzed. Of this group 54 cases were followed up by CERT. It was found that 86% of the subgroup held technical positions and 90% routinely had administrator system access as part of their job (Keeney et al., 2005). These people are in a position to compromise security either by setting up secret accounts or by abusing their login privileges to access confidential or top-secret information.

It has been found that attackers using identity theft to masquerade as valid users often exhibit abnormal behavior (Salem, 2008). This would be a possible method for use in the detection of masquerades on the network. However, the perpetrators of attacks such as Titan Rain did not make any mistakes or exhibit any unusual behavior as they covertly stole information and more important, no one even knew they were there.

Examples of Insider Attacks

There have been a number of high-profile insider attacks over the years where information had been stolen and delivered directly to foreign governments. In 2007, Chi Mak was convicted of stealing U.S. Naval secrets and delivering them to China using members of his family as couriers. He confessed that he had been sent to the United States in 1978, by the Chinese government, to work in the defense industry and to gain a position of trust (Claburn, 2008).

An engineer, named Greg Chung, who worked on the U.S. space shuttle and other sensitive projects, was found to have been spying for China from 1979 until 2006. Chung had the highest level of clearance and managed to remove more than 225,000 pages of documents relating to Boeing-developed aerospace and defense technologies. Some of these were extremely sensitive at the time. Greg Chung was arrested in February 2008 and convicted of spying (Scherer, 2009).

An American seaman called Hassan Abujihaad converted to Islam in 1995. He was serving on a missile destroyer deployed to the Gulf and was found to be sending classified documents to a London-based organization called Azzam Publications, which had links to terrorism activities via e-mail and Web sites (Former U.S. Navy Sailor, 2009). The FBI alleged that "the Azzam websites were among the first to successfully utilize the internet on a global scale to propagate the call to jihad" (Mahony, 2010). Abujihaad had leaked classified information to al-Qaeda, which included the vulnerabilities of a number of battleships and also their movements in the Gulf during that time.

The insider threat is not new as demonstrated by the case of Walter Kendall Myers and his wife Gwendolyn. Walter had worked at the Bureau of Intelligence and Research in the State Department where he had one of the highest security clearances. It came to light that he had spent 30 years spying. Both were arrested in June 2009 and subsequently convicted of supplying classified documents to the Republic of Cuba and of committing wire fraud (Wilber and Sheridan, 2009).

Elliot Doxer worked for Akamai and had been leaking the company's trade secrets for an 18 month period. Fortunately, the undercover Israeli intelligence officer that he thought he was dealing with turned out to be an undercover federal agent. He was arrested in 2010 and charged with foreign economic espionage (Bray, 2010).

In March 2011 the U.S. Department of Defense (DoD) announced that 24,000 files had been downloaded from military contractor systems. DoD Deputy Secretary William Lynn stated, "It is a significant concern that over the past decade, terabytes of data have been extracted by foreign intruders from corporate networks of defense companies. In a single intrusion this March, 24,000 files were taken." The U.S. DoD has seven million computers located in hundreds of countries and operating over 15,000 networks. They are currently taking action to try to stem the massive leakage of information that is currently taking place (Dignan, 2011).

Research on Insider Threat

The research done by Moore et al. (2008) was based on 49 insider sabotage cases. They attempted to identify common patterns within these cases. Seven general observations to help to identify insiders were proposed as a result of this work. The main conclusion was that disgruntled employees were the most likely candidates, for whatever reason. But they were also facilitated by a general lack of access controls (Moore et al., 2008).

Detecting the insider is a challenging problem as these attacks are often very sophisticated. The insider's familiarity with the networks and systems of the company that they work for makes it easy for them to cover their tracks and very difficult to catch them. It is estimated that approximately one-third of all data theft is due to insiders (Pfleeger, 2008).

One of the leading authorities on insider threats is CERT, the Software Engineering Institute of Carnegie Mellon University. They have accumulated data on hundreds of cases of insider attacks over the years for analysis. As of 2011 (Cappelli, 2011), their database contained 123 cases of sabotage, 196 cases of fraud, 86 cases of intellectual property theft, and 43 miscellaneous cases. What follows is a discussion of the key findings from some of their recent work on financial fraud (Cummings et al., 2012) and intellectual property theft (Moore et al., 2012).

Motives for an attack vary. Cappelli et al. (2009) analyzed 196 cases of insider attacks that occurred in the United States and observed their cases falling into the following categories (noting that some cases fell in to more than one category):

1. **IT sabotage**: These occur through individuals who are motivated to harm the organization, its data, or an individual. They misuse their access to systems, data, or networks and account for 45% of cases. Attacks were primarily committed by former employees and males; however, the fact that males were the majority is unsurprising as 74% of employees in this field are males. Motives identified from this group were disgruntled employees and revenge for some negative event such as termination, disputes, new supervisors, transfers or demotions, and dissatisfaction with salary. The majority who committed this type of attack did not have authorized access at the time of the attack. Thirty percent used their own username and password, others used a compromised account, an unauthorized backdoor they had created, systems or database administrator accounts, and so forth. Attacks included logic bombs and sabotaging backups. Most attacks were carried out through remote access, out of normal working hours, and in most cases system logs were used to identify insiders.
2. **Theft or modification for financial gain**: These occur where insiders intentionally exceed their authorized levels of access with the intention of stealing confidential or proprietary information for financial gain and occurred in 44% of cases. Targets focused in the banking and financial sectors followed by the government sector and then the IT and telecoms sector. The vast majority of these crimes were committed by current, not former, employees working in lower level, nontechnical positions and split evenly between males and females. Collusion with other insiders and outsiders was high, a recurring pattern was an outsider recruiting an insider. Ninety-five percent stole or modified information during normal working hours and 75% used authorized access, with 85% using their own username and password. The majority of the cases were detected through nontechnical means such as data irregularities or customer alerts and were typically caught through system, database, and file access logs. Within the financial sector (Cummings et al., 2012), it was noted that:
   - Criminals who executed a "low and slow" approach accomplished more damage and escaped detection for longer: on average fraud started over 5 years after hiring and it took an average of 32 months to be detected.
   - Insiders' means were not very sophisticated; very few held a technical role or used technical means and in more than half the cases, authorized access was used in some form.
   - Fraud by managers differed substantially from fraud by non-managers by damage and duration. Fraud by managers caused nearly twice the financial damage than non-managers and lasted almost twice as long—33 months compared to 18 months.
   - Most cases do not involve collusion: 16% involved collusion and of those 69% involved outsiders.
   - Most incidents were detected through an audit, customer complaint, or coworker suspicion; routine or impromptu auditing was the most common route for detection.
3. **Theft or modification for business advantage**: This is where insiders intentionally exceed their authorized levels of access with the Intent to steal confidential or proprietary information for business advantage and occurred in 14% of cases. The vast majority of crimes were concentrated in the IT and telecoms sector; however, the banking and financial sectors, chemical and hazardous materials and the defense industrial-based sectors were also affected. All of the attacks analyzed were carried out by males, 71% in technical positions, 29% in sales, 25% former employees, and 75% current employees. Nearly 80% had accepted positions with another company or had already set up a competing company. In 25% of cases information was passed on to a foreign company

or government and 88% had authorized access to the information. The majority of the cases occurred within a one month period and in approximately half the cases the insider colluded with at least one other insider. Cases were detected through emergence of competing products, informant, and so forth, and were typically proven through system, database, and file logs.

4. **Miscellaneous:** This is where insiders intentionally exceed their authorized levels of access with the intention of stealing confidential or proprietary information for purposes other than financial or business advantage and occurred in approximately 9% of cases.

As identified earlier, many people relate insider attacks to a disgruntled employee; however, the CERT team has noticed the following recent trends and issues related to insider threats:

1. **Collusion with outsiders:** Half of the insiders who stole or modified information for financial gain colluded with outsiders.
2. **Business partners:** The number of insider attacks from trusted business partners who have been given authorized access is increasing.
3. **Merger and acquisitions:** There is an increased risk from employees who are working in an uncertain climate from both the acquiring and acquired organizations.
4. **Cultural issues:** It is important to recognize that cultural issues can influence employee behavior.

Clearly, the range and scope of the events described in this section demands that there must be equivalent levels of countermeasure, otherwise our existing systems might fail in the face of such pressure. The next section sets out a range of countermeasures that are currently in use to address these issues.

COUNTERMEASURES TO COMBAT CYBER TERRORISM

There are a number of standard computer security measures that have a significant effect in countering cyber terrorist activity, if they are properly implemented and maintained. These include properly installed, managed, and regularly updated firewalls; packet-sniffer software; virus checkers; access control lists; and user validation systems. However, by far the greatest threats to any security system are the human users, who accidentally, forgetfully, lazily, ignorantly, or maliciously breach the security of systems on a daily basis. For the vast majority of cybercriminals, and cyber terrorists, they do not need sophisticated software or hardware tools to break into systems, as long as the user issues remain unaddressed. Therefore, the establishment of good cyber hygiene must be a priority for every organization, together with clear, well-defined, standards-based policies and protocols, and training systems, aimed at every level of user, establishing security as central to organizational culture.

Once these issues are addressed, consideration can be given to software measures to address more sophisticated threats, including diversionary tools such as honeytraps and dummy sites for hackers, sandboxing to trap malware, and bounties to trap bugs and security holes.

Policy

"How many of the recent, high-profile data breaches at blue-chip companies could have been prevented with better governance? While corporate governance is common practice, often obligatory, in many aspects of business, governance is not always present in information security. Yet it plays a vital role in reducing risk and speeding response" (ISF, 2011).

It is not sufficient to deal with cyber security by *ad hoc* application of tools and procedures as and when problems arise; indeed, it is often then too late. An organization needs to be proactive and to be ready, organized with a set of controls, trained personnel, and a written security policy, known by all staff, with defined rules and roles. Such a management policy should be based upon principles of good IT governance and be based upon recognizable standards that give assurance to all stakeholder parties.

Standards bodies such as International Standards Organization (ISO), American National Standards Institute, and British Standards Institute devise formal sets of rules by which processes and activities should be undertaken to achieve optimum performance. Relevant standards for cyber security might be ISO27032 CyberSecurity (draft standard), which is to be the defining standard for cyber security requirements, ISO27033 Network Security (draft standard), ISO27034 Application Security, and ISO27035 Information Security Incident Management (draft standard), as well as the already well-established ISO27001.

The use of recognized standards to form a cyber security policy is important as standards give trustworthiness to other parties, such as supply-chain partners, regulators, and law makers. Supply-chain partners such as suppliers, clients, and other trading partners are reassured about using electronic business transactions. In fact, a further useful standard here might be ISO27036 Information Security for Supplier Relationships. Regulators, too, may require reassurance on the security of network/Internet transactions especially in certain industries such as finance; for example, in the United States the Securities and Exchange Commission and in the UK, the UK Financial Services Authority. Lastly, compliance to standards shows due diligence and commitment when possible litigation arises in such areas as data protection, copyright, and computer misuse.

It has to be acknowledged that cyber security is a moving target; hacktivism, fraud, and denial of service attacks are constantly changing their modus operandi. Controls should therefore be monitored regularly using audit techniques. Auditing assures that the requirements of a cyber security policy are being met in practice. In practice, controls, both technical and administrative, may be ignored (deliberately or accidently), totally removed, or adapted to be less effective. Auditing identifies the effectiveness of the controls in place (the right control doing the right thing?), how efficient they are (are they used properly and quickly in practice?), and how economic they are (cost-effective?). In addition, auditing identifies whether new controls may be required and whether there exists a gap between the reality and the requirements of the adopted standard. This gap analysis shows what and where the shortfalls are and indicates how far the standard is being met. The gap may be used to measure the extent of compliance to the standard, to reassure a regulator, as a benchmark to compare the organization with other organizations in the same industry, to reassure supply-chain partners, or simply as part of a calculation of return on investment to reassure the accountants.

Cyber security auditing is as much an art as a science and needs careful planning, execution, and reporting. Auditing standards, methods, and tools may be found at the Information Systems Audit and Control Association and the Institute of Internal Auditors.

Training

Cyber terrorism is considered a top-tier national risk for many governments given the potential harm and disruption it can cause due to the world's increasing dependency on IT systems. While the obvious targets might be governments, banks, and utilities (e.g. water, oil, electricity, gas, chemical, and communication infrastructure), as attacks on these have the ability to cause the most economic, political, and physical havoc and damage to the critical national infrastructure, cyber terrorism groups are becoming more coordinated and sophisticated in their attacks and will make use of any computer connected to the Internet to support an attack. Cyber terrorism therefore affects everyone from large organizations to all citizens who own or use a computer connected to the Internet. The following list provides a brief summary of the different categories of people involved and a brief analysis of their training needs.

1. **Members of the public:** The single definitive source of advice for UK Internet users is Get Safe Online, which is a Web site sponsored by a cross section of organizations including the UK government. In November

2011, their Get Safe Online Report (Get Safe Online, 2011) stated that 87% of users surveyed had virus protection software and 41% of them updated it every time they switched their computer on. Clearly a lot more is needed to educate the public with a growing trend in cybercriminals making use of a wide variety of techniques including the use of personal information from social media cites to tailor realistic information more able to fool people into allowing a variety of forms of malware into their computers to clickjacking, and so forth. Training needs to start at an early age and more work needs to be done in educating school-age users as well as adults.

2. **IT support personnel within organizations:** These are staff who are technically trained to deliver IT services to an organization. Many have not received the level of training in security required or have misunderstood the threat to their organization. Over 80% of attacks could be dealt with through basic cyber hygiene, such as patches, passwords, anti-malware, and firewalls; however, even when used, many do not keep them up to date. Relevant training through certifications and Chartered Status should be required and monitored by senior managers.
3. **IT developers:** Many developers write poor code through laziness or a lack of understanding of how to protect their code from things such as SQL injection attacks. Education and training programs need to provide more of a focus on security issues, and organizations need to invest in regular CPD for their developers in this area.
4. **IT project managers:** It is not uncommon for large organizations to use staff with good project management skills, but limited technical capability, to manage and take oversight of IT projects; however, they frequently lack the technical knowledge to ensure the systems they manage are developed and maintained in a secure manner. These staff need to be trained to understand the risks to the organization, the questions to ask, and how to ensure that their IT projects are providing the right level of security required.
5. **IT users within an organization:** Most IT users within an organization find security an irritation as it makes systems less usable. As a result, they invariably find workarounds, not understanding the potential risks that they may be introducing into their organization's systems. This includes issues related to the use of personal devices at work (Bring Your Own Device; BYOD), which can be used by the entire family at home, introducing malware and other assorted risks.
6. **CEOs, Senior Board-level personnel:** Organizations are spending millions on security yet many still end up in the media as a result of security breaches. Most CEOs and board-level directors do not understand the security risks, how to manage them, or the behavior of their employees, which may result in security breaches (Lumension, 2011). All CEOs and senior board-level directors need to understand as much about the dangers of IT as well as how to exploit IT for business purposes in addition to who in their organization needs what type of training. They need to be able to adequately assess their vulnerability to a cyber terrorist attack, understand how to assess their risk, and drive appropriate policies. Should an attack occur, they need to consider how they will deal with data losses, downtime, the impact on infrastructure, and their customers, including the loss of their information, costs, reputational damage, how to address future issues of security versus privacy, risks of outsourcing and off-shoring, and so forth. Depending on the potential impact, senior staff may need crisis management training to help them deal with the media and management of a breach, which may take months or years to fully uncover and resolve. Use of training systems such as Pandora (Bacon et al., 2012), which can simulate realistic crisis training using an event-based time line model to allow different scenarios to be explored, could prove particularly useful.

Cyberphysical Security Challenges

The vast majority of cyberphysical systems have been designed and tested with physical safety but not cyber security in mind. More significantly, computer-controlled equipment in our critical national infrastructure, such as dams and nuclear plants, usually have an expected lifetime of 30 years and are too expensive to replace. Also, they have not been designed with modularity and upgradability in mind. A modern personal computer can be protected against most cyber threats by upgrading its software and applying security patches. This is not straightforward for 20-year-old industrial control equipment. A system upgrade may need months of planning and may cause prohibitively long downtime. In addition, modern software security packages are usually too demanding for the large number of legacy components found in such systems (Cardenas et al., 2009).

Still, the fundamental difference between cyberphysical systems and conventional IT systems is the interaction of the former with the physical environment. Unavailability of a corporate network or individual computer may cause frustration and may delay operations, but is unlikely to cause lasting damage. Real-time availability is more important in cyberphysical control systems, as was demonstrated at the 1999 gasoline pipeline explosion in Bellingham, Washington. On the other hand, this interaction between the physical and cyber world may also provide opportunities, as otherwise undetectable cyber attacks may become detectable though their physical manifestation. Yet, scientists still have not taken advantage of these interactions and all current detection mechanisms take into account only cyber traces to determine whether a system is under cyber attack or not. We expect this to change thanks to new, dedicated cyberphysical test beds that are currently being built in research centers around the world in response to increasing governmental interest in cyber security. The focus of these test beds and corresponding research varies from power networks (Edgar et al., 2011) to aviation cyber security (De Cerchio and Riley, 2011) and emergency response infrastructure.

Cyberphysical attacks may be attractive particularly to state-backed terrorism, since they can cause significant physical damage in a more covert manner with less risk to one's own troops and diplomatic status. However, development of exceptionally potent cyber weapons like Stuxnet is unlikely to be within the technical reach of terrorist organizations. To put things into perspective, the scientific team behind the cyber attacks that compromised a production car in 2010 spent two years of world-class academic research to achieve it, and the Stuxnet attack against the Iranian nuclear facility was most probably organized by a technical superpower. For this reason, we do not believe that a cyberphysical attack alone will be used soon by terrorists to cause considerable human loss. It is more likely that a common cyber attack will be used to facilitate a traditional physical attack by disabling cameras and other security systems or to disrupt emergency response by causing an artificial traffic jam and interfering with local communications. In that sense, conventional cyber security mechanisms, such as antivirus software, intrusion detection systems, and firewalls, can protect to a certain extent against cyberphysical attacks too. More important, promoting a culture of cyber hygiene and vigilance, with people and organizations following security policies, using strong passwords, regularly applying security patches, and so forth, would make a cyber terrorist's work more difficult.

Insider Threat Countermeasures

CERT has identified some practical countermeasures against the insider from their Common Sense Guide to Prevention and Detection of Insider Threats (Cappelli et al., 2009).

In addition to analyzing employee behavior in order to develop counterstrategies, there is a body of research around counterproductive work behavior (CWB), which has been recognized as a key factor in helping to identify factors influencing an insider to commit an act, along with the indicators and precursors that lead to those malicious acts (Cummings et al., 2012). CWB covers a variety of behaviors, but specifically encompasses sabotage, stealing, fraud, and vandalism. Sackett (2002) categorized the antecedents of counterproductive work behavior into the following groups: personality, job characteristics, organizational culture, work group characteristics, control systems, and perceived injustices. The primary personality model used in CWB research is the Five Factor Model (Costa and McCrae, 1992), which analyzes people's personality on five dimensions: openness to experience, extraversion, conscientiousness, agreeableness, and emotional stability. Salgado (2002) showed that levels of conscientiousness and agreeableness were significant predictors of workplace deviance.

Computer simulations have been used to simulate insider activity and test different detection mechanisms; however, these cannot be relied on as in the case of financial fraud, only 6% of fraud cases were detected by software and systems and in only 20% of cases transaction, access, and database logs were useful for incident responses. It is therefore vital that all organizations implement policies and procedures covering all aspects of the organization. Sixteen best practice recommendations from CERT (Cappelli et al., 2009) are outlined below:

1. Consider threat s from insiders and business partners in an enterprise-wide risk assessment: A balance needs to be found between trusting employee and protecting assets.
2. Clearly document and consistently enforce policies and controls: Many of the cases analyzed by CERT could have been prevented through this approach.
3. Institute periodic security awareness training for all employees: Employees must understand that policies and procedures exist for a good reason and that they must be enforced.
4. Monitor and respond to suspicious or disruptive behavior, beginning with the hiring process: This includes dealing appropriately with repeated policy violations (which could escalate) and the effect of personal and professional stress indicators.
5. Anticipate and manage negative workplace issues: This should run from pre-employment to termination, consequences of policy violations should be clearly communicated and enforced. Employees should be encouraged to discuss workplace issues without fear of reprisal and terminations should be handled with care as most insider IT attacks occur after termination.
6. Track and secure the physical environment: Access to physical and virtual areas should be restricted to those who need it and all attempted violations and so forth should be logged and monitored.
7. Implement strict password and account management policies and practices: Ensure all activity from an account is attributable and provide an anonymous reporting mechanism to report unauthorized access including social engineering attempts; perform audits regularly to ensure expired accounts are disabled.
8. Enforce separation of duties and least privilege: Train employees and ensure critical functions are divided across employees so collusion is required to carry out an attack. Authorize each individual only for the access they need and be sure to remove access when an individual's job changes.
9. Consider insider threats in the software development life cycle: Ensure an appropriate separation of duties; more insider threats occur during maintenance than system development. Be sure to protect and restrict access to backup systems and so on.
10. Use extra caution with system administrators and technical or privileged users: Technically competent individuals are more likely to use their technical knowledge to exact revenge for perceived wrongs. Employ techniques such as separation of duties, two-man rule for critical system administrator functions, and so forth, should be employed.

11. Implement system change controls: Unauthorized modifications were a key feature of insider compromises so employ stronger change control mechanisms and alerts.
12. Log, monitor, and audit employee online actions: Logging and periodic monitoring will help detect suspicious activity such as the downloading of confidential files.
13. Use layered defense against remote attacks: Insiders are more confident when not scrutinized by coworkers, so restrict access to work-based machines unless external access is required, in which case monitor logs closely.
14. Deactivate computer access following termination: Whether termination was favorable or not, have procedures and policies in place to ensure fast deactivation of accounts and access.
15. Implement secure backup and recovery processes: Ensure secure backup and recovery procedures are in place, single points of failure are eliminated, test processes regularly, and so on.
16. Develop an insider incident response plan: This is required to control the damage. Should an attack occur, it is important that robust evidence is appropriately gathered and not corrupted, and that lessons are learned.

Sandboxing

A sandbox is a security mechanism for separating running components of a system. It was described in 1996 but is now used more and more. It is worth mentioning that HTML5 has a "sandbox" attribute for use with iframes. A sandbox is often used to execute untrusted software from unverified, or even verified, sources. Sandboxing offers prevention of manipulation, reverse-engineering, and reprogramming of systems and components, and is usually a purely software-based protection. A sandbox can be a virtual machine (e.g., VMware based), which has been set to emulate a complete host computer, on which a conventional operating system may boot and run as on actual hardware or something more specialized. In a more advanced scenario multiple sandboxes can take the place of multiple parts of a system targeted by multiple threats. The large majority of Web sites today embed third-party JavaScript (in many cases obfuscated) into their pages, coming from external partners. Most of this is benign and comes from trusted sources, but it is not unlikely that these scripts could come under the control of an attacker. It is now usual practice for security researchers to run such scripts into a sandboxed environment to establish how an attacker can perform unwanted actions safely.

The easiest way to understand how sandboxing can be used is to think of an example where an e-mail sent to your inbox has an executable attached. Assuming that this is a malicious application, once run it could stealthily harm your system and potentially any other systems that you are connected with. This would happen in most cases in the background and would not be noticed until it is too late. To stop such a threat it is imperative to understand how it operates, but this is very difficult to do after it has completed its operation. If, however, we were able to run this attachment in a protected environment then we could examine how it attempts to access and harm our system and carry out a step-by-step dissection of its operation. Traditionally the tamper proofing of programs relied on tamper-resistant hardware, but this is not always easy to use due to cost limitations and complexity of the required underlying configuration (Goldberg et al., 1996). Sandboxing offers a lower cost option to tamper proofing, as long as it is applied properly.

Bug Bounties

In 2004, Mozilla started a bug bounty program. This offered money to anyone who discovered a new bug or security flaw in software. Since that time a number of companies have followed suit. In 2011, Facebook joined the bounty program and reported that the submissions they receive have enabled the social Web site to improve security (Robertson, 2012). Does this make the bad guys turn into "white hat" hackers? This is uncertain, but there is clearly money to be made by discovering bugs but not exploiting them. If money will motivate people to report bugs, and by inference security holes, then this can only help to secure the networks connected to the Internet.

THE FUTURE

In 2000, the threat to SCADA systems used by the North American electric power grid was clearly identified by the U.S. National Institute of Standards and Technology. If this was known then, one must ask why the Stuxnet attacks were able to succeed. The report cited a number of reasons for the increased vulnerability. Nine factors that influence the likelihood of a cyber attack were discussed. The first, and quite significant one mentioned, was the shift to open protocols and standards from proprietary mainframe-based computer control systems. Items 2 to 5 related to factors that increased the likelihood of insider attacks. Items 6 to 9 are of interest to this discussion and are quoted below (Oman, 2000):

1. Increasing incidents of international and domestic terrorism targeted against North America.
2. Increasing number of countries with government sponsored information warfare initiatives.
3. Rapid growth of a computer-literate population.
4. Widespread availability of hacker-tool libraries.

The conclusion was stated by Oman (2000):

> *Increasing reliance on automated control systems with remote access (via phone or internet) and the growing global economy have expanded the number of potential attackers with access to substation controllers and SCADA systems, and therefore magnified the risk electric utilities have from sabotage and espionage.*

This warning has clearly not been heeded.

The United States has tested its capability to respond to cyberwarfare. In 2002 the U.S. Navy conducted an exercise called electronic Pearl Harbor, in which a massive attack on critical infrastructure was simulated. Since then three more "Cyber Storm" exercises have been run. In 2010 a new tool that could shut down the Border Gateway Protocol was launched. This was known as the "kill switch" and was designed to be defensive by shutting down the Internet to prevent a terrorist type cyber attack. This has never been properly tested as at that time it was felt that the disruption to the Internet would be too great (Saalbach, 2012).

The evidence for government-sponsored cyber espionage points to China and Russia. "In an unusually blunt report issued last year by U.S. intelligence agencies, the Obama administration said that massive cyber espionage operations by China and Russia posed a 'significant and growing threat' to U.S. national security, yet other countries often view U.S. complaints as hypocritical given its own cyber activities" (Dyer, 2012). However, if the speculation regarding Stuxnet is true then the United States and Israel may also have a place in this line up.

Ralph Langner (2011) described Stuxnet as a military-grade cyber missile that was used to launch an "all-out cyber strike against the Iranian nuclear program."

> *What Stuxnet represents is a future in which people with the funds will be able to buy an attack like this on the black market. This is now a valid concern.*
>
> ***Lagner in Clayton, 2010***

While there is no doubt that Stuxnet did cause damage to equipment at the Iranian nuclear facilities, it is also clear that the disruption only temporarily delayed Iran's nuclear program, and was quickly repaired.

The United States considers the threat to their military operations from the Chinese very real. The Peoples Liberation Army (PLA) relies on the Chinese commercial sector research and development (R&D) that could be subverted for use in cyber terrorism. Foreign partners share the cost of the R&D of technology. Telecommunications hardware manufacturers based in China provide the PLA with access to cutting edge research. This means that microelectronics manufacture destined for the U.S. military, civilian government, defense, and telecommunications industry are potentially at risk from compromise even before they have been installed or exposed to the Internet. State-sponsored activities target data that do not translate into hard cash. The target is information that could be of use to a foreign power. This could be technical defense information or military data relating to ongoing operations. All United States businesses that have manufacturing partnerships with China are concerned about intellectual property theft, according to a survey conducted in 2011 by the United States–China Business Council. (Krekel et al., 2012)

As we move into an era of smart environments, smart homes, smart devices, and the Internet of Things, the level of interconnectedness of all our systems increases exponentially, and the threat of cyber terrorist attacks bringing these systems down increases at the same level. Perhaps the most worrying aspect of this is the number of developments that are taking place without appropriate regard for security, while critical infrastructure providers and military and financial organizations are now clearly aware of the need for better cyber hygiene and security standards; there are a large number of organizations that are softer targets. The fact that we would regard as anathema an attack on life-support services in hospital systems does not make them safe from attack, and from a cyber terrorist perspective the ensuing outrage would be a desired result.

The growth of hacktivisim, tracing its roots from groups such as the Chaos Computer Club and the Cult of the Dead Cow, and now allied to a number of widespread societal protest organizations, also presents a further problem here. Clearly, within free societies, the rights of citizens to protest and promulgate a point of view different to the government of the day, or the accepted norm, has to be protected. However, the point at which this infringes the rights of others, by damaging or bringing down systems of target organizations or bodies, means these have to be regarded as cyber terrorist activities. If not, they will rapidly become a front for more radical groups utilizing their activities to achieve their own ends, as indeed the Chaos Computer Club did in the late 19080s (Anderson, 2006). However, the growth of such movements is also evidence of a growing radicalization of youth on a worldwide basis, and there has to be concern that terrorists will seek to establish a route into hacktivist groups, not just as a front for their activities, but also as a recruiting ground for even more radical political and religious ideologies.

So, we face a very uncertain future, with our growing societal dependence on advanced, interconnected technological solutions offering potentially both our greatest advances and our greatest threats. As the famous saying goes "there is no such thing as a free lunch," and the cost for our technological advances has to be paid in ever greater vigilance in the protection and management of our systems, and ever greater awareness by organizations and individuals of the need for good cyber security. Trustwave, in their 2012 Global Security report, identified the two most important security goals for organizations for 2012 as "Education of Employees" and "Identification of Users" (Trustwave, 2012)—we now need to make it happen.

KEY ISSUES

If we are to tackle the issues of cyber terrorism identified in this chapter, then we need to address these from several perspectives as seen in the following sections.

Political/Policy Issues

The issues of cyber terrorism are not limited by national boundaries, nor do they lend themselves to purely local solutions. In considering actions that will be effective, there is a need to address local legislation, to ensure that there is an appropriate response to local events, stunts, or attacks. However, since the majority of the events that we are concerned with have international, or at least cyberspace, links there is clearly a need for concerted and consistent international legislation and action. Clearly, in such a space, there is a need for action from an international coordinating body, to date there has been no such initiative from the United Nations, but NATO has developed tools and capabilities to support international action against cyber terrorism. NATO offers powerful tools in four key areas:

- Operational ability to monitor networks, in particular international Internet traffic
- Intelligence gathering and knowledge of a large number of world arenas, particularly conflict arenas
- Partnership of 69 countries (more than one-third of the world); it tries to integrate existing analytical capabilities to build cyber defenses.
- Operational capabilities, particularly in monitoring and analyzing groups and the impact of Web site information on the radicalization of youth on a worldwide basis

In a worldwide marketplace, where technology companies sell access and expertise in the use of their systems in huge numbers (Cisco issues over a million certifications per year for courses on their technologies), security can only be enforced by similar levels of international cooperation, legislation, and action. The use of NATO systems, and national engagement with the NATO agenda offers some potential for future coordinated international response to cyber terrorism activities.

Research Issues

While any improvement of our cyber defenses would be beneficial, there are a number of technological research challenges with increased focus on cyber terrorism. We have chosen one for each of the four strands of the UK government's Pursue, Prevent,. Protect, Prepare strategy (Home Office, 2011).

Pursue. Pursue refers to activities that can stop terrorist attacks. Most cyber attacks against critical national infrastructure need substantial online research and active probing for a considerable length of time to identify vulnerable components. The technological challenge is to develop early warning mechanisms that monitor a system and its cyber surroundings and spot signs of preparations for future attacks against it. A relevant project that targets specifically botnet attacks has been piloted with the Seattle, Washington, in the United States (DHS, 2011), and the European Commission has recently published an open call for feasibility studies on technologies toward a Europe-wide early warning system.

Prevent. Prevent refers to activities that can stop people from becoming terrorists or supporting terrorism. Research has shown that radicalization is increasingly facilitated through the use of mainstream online platforms, such as social networks, forums, and media-sharing Web sites (Bermingham et al., 2009). The challenge here is to develop technologies that can identify pockets of radicalization and relevant online material without infringing the privacy of individuals.

Protect. Protect refers to activities that strengthen our protection against a terrorist attack. In the context of cyber terrorism this may refer to authentication, detection, or response mechanisms against a range of possible attacks. Of particular interest are technological mechanisms that could identify the intended aim and ultimate target of an attack. For example, denial of service attacks are often launched indiscriminately by amateur hackers without a specific goal, but such an attack may also be the first step that blocks monitoring equipment before a coordinated act of cyber terrorism (Loukas and Oke, 2010). Being able to identify the real target of an attack in real time rather than forensically postmortem would be a significant step forward for the defense against cyber terrorism. Initial work in this area has focused primarily on prediction of the next step of an attack (Zhang et al., 2009).

Prepare. Prepare refers to activities that mitigate the impact of a terrorist attack. Rapid

TABLE 20.1 Cyber Security Framework

Issue	Action	Reference
Organizational policy	Develop a clear and well-defined organizational policy on all aspects of cyber security, and based on identified international standards.	ISO and ANSI standards on data and information security (see the section Policy)
Recruitment	Develop a recruitment policy that explicitly addresses issues of cyber activity, radicalization, and extreme views. Work out how you might exclude a radicalized individual from employment.	Rather worryingly, there are currently no national guidance reports on this issue. Develop your own, based on the models provided in this report.
Training	Create an institutional training program that promotes organizational awareness and support, and explicitly addresses issues of cyber security.	Build a program based on the advice given in the section Training.
Insider threat	Develop institutional policies and practices that address the issues of insider threat and can be validated to provide support for your policies, and management buy-in.	Use the CERT Common Sense Guide to Prevention and Detection of Insider Threats (Cappelli et al., 2009), described in the section Insider Threat Countermeasures.
Software/hardware tools	Ensure that systems are up to date and secure, and develop an update and replacement strategy that fits the organization.	Current virus checkers, packet-sniffers, network pattern identifiers, hardware detection tools, and a myriad of other tools can be utilized. Ensure systems are in keeping with organizational policy.
Cyber hygiene	Training staff and developing policies is insufficient, without the development of a cultural model of cyber hygiene, led from the top. This model has to clearly identify cyber security as a fundamental priority for the organization.	U.S. DoD has identified models of organizational structure and activity that constitute good cyber hygiene. http://www.defense.gov/news/d20110714cyber.pdf
Organizational risk appetite	Organizations have significantly different risk profiles, based on their sphere of operation. Develop a risk profile model and operational plan, based on your organizational requirements, but reflecting the national and international risks that you face. Identify the level of risk that your organization can comfortably accommodate.	Base your work on Neutze (2012)). Cybersecurity in Germany——Toward a Risk-based Approach. AICGS, Johns Hopkins University.

self-healing features have been developed and tested with success against attacks that target the underlying network infrastructure, both wired (Sakellari, 2010) and wireless (Gungor and Hancke, 2009). In such systems, the network infrastructure is able to monitor itself and adapt in a manner that minimizes the impact of the attack. The challenge is to extend the self-healing concept to include all components of the critical national infrastructure, from industrial control equipment to satellite navigation systems and medical devices.

Practitioner Issues

Perhaps the key argument to emerge from this chapter should be a framework of issues and remediating actions that can be undertaken by security practitioners, in any situation or role that can be utilized to address cyber security issues, whatever their source. In keeping with our introductory arguments that addressed the problem of distinguishing the rationale for a cyber attack, at the time of the attack, so the cyber hygiene and countermeasures we introduce should not concern themselves with the rationale for the attack, but rather with preventing, resolving, or mitigating the impact of the attack on the systems involved. Table 20.1 provides a framework, based on the information provided in this chapter, to address issues of cyber security, with specific reference to cyber terrorism, in any organizational system.

Above all else, we should understand and accept that cyber security is a common responsibility that needs to be fundamental to the culture of all organizations and activities utilizing this technology to further their aims; if this is not the case then the cyber terrorists will undoubtedly win.

CHAPTER

21

Developing a Model to Reduce and/or Prevent Cybercrime Victimization among the User Individuals

Hamid Jahankhani

INTRODUCTION

It appears that the current thinking in cybercrime prevention puts the technology first and concentrates on protecting computers and devices with the hope that the users will not fall victim to cybercrime and forget about the human element in crime.

Unlike traditional crime, which is committed in one geographical location, cybercrime is committed online and it is often not clearly linked to any geographical location. Therefore, a coordinated global response to the problem of cybercrime is required. This is largely because there are a number of problems that pose a hindrance to the effective reduction in cybercrime. Some of the main problems arise as a result of the shortcomings of the technology, legislation, and cyber criminology.

This research is primarily based on the hypothesis that understanding the characteristics of the users of the computer systems will allow for the creation of more effective cybercrime prevention strategies, which will also result in reduced impact on the users as compared to a more broad-brush approach to crime prevention. A secondary hypothesis of the research is that the particular characteristics of the users have an impact on their vulnerability to cybercrime. A consequence of this hypothesis is that if this is true, then it is important to identify the relationship between the particular characteristics of the users and its linkage to the type and severity of cybercrime. This kind of linkage will allow for the development of more effective crime prevention techniques, which will be tailored to the type of user. The entire population of users of the Internet can then be classified on the basis of the particular characteristics and the appropriate techniques deployed for the particular class of users. This will, in turn, serve to increase the effectiveness of these crime prevention programs further, as they will enable the more efficient use of resources.

CRIME PREVENTION THEORIES

The theories of crime prevention are distinguished from theories of crime causation, and while the two may be linked in some cases in the literature; however, they are seen as rightfully being regarded as distinct because of the following:

1. There is no established theory of crime causation that can be accepted and used as a basis for preventative theories.
2. The issue of crime prevention is not a theoretical pursuit at all. While criminality theory focuses on causality, prevention, and transcends etiology, it proceeds into the field of strategy and tactics.

Therefore, while theories of criminality are abstract, and based on deduction and observation, prevention is practical and must seek methods of controlling human behavior in practice. The two fields, therefore, differ substantially at a theoretical level.

It is widely argued that many of the conclusions reached from a theoretical approach to criminology had failed when applied to practical scenarios, both due to the difficulties experienced when attempting to put theory into practice and also because in the sociological field society constantly changes during the period between the development of a theory and the time of its practical application. Therefore any consideration of the theory of criminology must be conducted with a view to the strategy and tactics that would be used to put that theory into practice in the field. Crime prevention theory must focus on its goals of preventing crime before it occurs, and preventing persons who have committed crimes from re-offending. Therefore, application of theory that does not focus on this goal would result in fruitless efforts.

The two factors that are generally used to explain the causes of crime are environment and personality.

Looking first at environment, Lunden (1962) argued that attributing crime to environment would deny the social reality that there is very little meaningful community existence in modern society. Therefore, any criminological theory that concludes that the community must adapt to prevent crime, in Lunden's view, is useless; because, in the field, working with criminals and working in areas of high crime, there is no distinct and specific community with which to work. It was noted that a crime prevention theory implies some kind of social reform that would be implemented in favor of the status quo. Therefore, theory comes up against resistance in practice, as most people distrust reform. The theories must, therefore, be willing and able to overcome such resistance to any type of change in the social sphere if they are to have any chance of making a difference in practice.

The second common theoretical approach views crime as the result of complex internal and emotional reactions to the environment. Another criticism Lunden made of criminological theory is that it tends to view crime as a pathological or abnormal behavioral trait, rather than as a fairly normal aspect of human behavior. Lunden went on to conclude that it is impossible to prevent all crime, although it is possible to decrease the amount of it. It was also noted that there are and have been in the past, societies with only a minimal level of crime. However, not many people would choose to live in such societies. As long as there are human beings, there will be crime. Therefore, criminological theory must deal with this fact as it applies to today's society, one that has little and weakening community.

Geason and Wilson (1988) examined a number of criminological theories including rational choice theory, which portrays criminals as economic decision makers who make conscious choices to commit crimes because the benefits outweigh the costs or are perceived as such. Situational crime prevention, which entails "the use of measures directed at highly specific forms of crime," was also examined by Geason and Wilson (1988). This involves the management, design, or manipulation of the immediate environment in as systematic and permanent a way as possible. This system, often referred to as primary prevention, seeks to reduce the chances of committing crime. Situational approaches to crime prevention are based on rational choice theory in that they view the crime as the result of a conscious decision that the perpetrator of the crime was able to make, in response to the circumstances and environment they are presented with. This assumes that the motivation to commit crime is not beyond control. Geason and Wilson (1988) suggested a link between situational factors and personal internal factors. That is, the decision to consider committing a crime in the first place would depend on the personal characteristics of the individual, while the final decision on whether or not to commit the crime in this particular context, and in this particular moment, would depend on the situation.

Geason and Wilson (1988) went on to discuss the issue of displacement, which is a major criticism of situational crime prevention. This argument holds that situational preventive measures do not prevent the crime from occurring absolutely, but merely force the criminal to reconsider the options. For example, the installation of a house alarm is a situational preventive measure but if the burglar simply decides to burgle a neighbor's house instead, then it cannot really be claimed that a crime has been prevented, it has simply been moved to another more accommodating location.

Situational Crime Prevention Theory

Beebe and Rao (2005) looked at the application of situational crime prevention theory as a means of explaining information systems security and its effectiveness. The value of information as an asset in today's economy and the role that information systems security technologies could play in protecting such assets were noted. They argued that it would be necessary to understand the factors that could contribute to the effectiveness of such security systems. What Beebe and Rao (2005) achieved is an extension of the theoretical study of security in relation to computer data, using situational crime prevention theory. The issue of extending a criminal justice theory designed to cover physical crimes to computer-based crimes was addressed, and a conclusion was made that situational crime prevention theory may be nonetheless effective, and that there are opportunities for the theory to improve the effectiveness of information systems security approaches by reducing the anticipated amount of crime.

Lifestyle Routine Activity Theory

According to Cohen and Felson (1979) the lifestyle routine activity theory (LRAT) suggests that crime is likely to occur when the following factors converge:

1. Motivated offenders
2. Suitable targets
3. The absence of capable guardians against violation

Cohen et al. (1981) tested the effect of LRAT variables (exposure, proximity, and guardianship) on criminal incidents (burglary, assault, and personal larceny). Using National Crime Survey statistics of U.S. households, they found that LRAT variables have a significant effect on predatory victimization.

LRAT would suggest that all criminal events occur in a particular place at a particular time. Cohen and Felson (1979; cited in Kyung, 2008) asserted that "the synchronization of a perpetrator's rhythms with those of a victim's" facilitate a convergence of a potential offender and a target. Cohen and Felson (1979) also believed that examining how and why criminal offenses occur in particular places may be useful and important to a study of cybercrime victimization. This argument relies on the notion that cybercriminals often search for suitable and valuable targets in specific types of social arenas (Piazza, 2006).

However, the Internet as a part of human environment and the physical measures taken to prevent the traditional crime are similar to those used to increase cyber security. The software, such as antivirus and firewalls, should not be ignored by anyone as a part of the cybercrime prevention strategy, especially if the computer is part of the network. As the physical prevention measures, the software is protecting and guarding the user devices (PCs) in the Internet.

Jahankhani and Al-Nemrat (2010, 2011) have tested the LRAT and identified the importance of three elements of LRAT, in particular the discussion of the three subelements of capable guardianships. It can be concluded that none of the strategies can play an effective role in reducing cybercrime victimization or can guard against cybercriminal activities when operating in isolation.

Furthermore Jahankhani and Al-Nemrat (2010, 2011) recognized the importance of awareness as an element in reducing cybercrime victimization in the society. It therefore seems appropriate to introduce "user awareness" as a new category within theoretical capable guardianship, accompanying formal social control, informal social control, and target hardening.

This new pillar will strengthen the guardianship discourse in its fight to reduce cybercrime victimization. In order to support the argument that "user awareness" is a crucial key element that needs to be added to the capable guardian, it must work in coalition with other elements. The model of Jahankhani and Al-Nemrat (2010, 2011) shows how user awareness may play a vital role in reducing cybercrime victimization in light of LRAT (Figure 21.1).

CRIME PREVENTION MODELS

Cyber Terrorism Prevention Model

Fiore and Francois (2010) developed a cyber terrorism prevention checklist for use by organizations. This model consists of actionable steps that management and information technology (IT) security departments could put in place to prevent an organization from becoming a victim of or its infrastructure being unwittingly used for cyber terrorism. The model comprises actions that relate to intelligence gathering, an area that was claimed to include three possible security lapses that would lead to penetration of an organization's system and loss of confidential or sensitive data. Strategies include the avoidance and actionable prevention steps of identity impersonation or identity theft and Spyware. Some of the preventive steps include access controls, document controls, information procedures that will protect data and identities, scanning programs, the installation and maintenance of firewalls, intrusion detection systems, and the use of third-party software audits.

To avoid internal threats from an organization's own employees, Fiore and Francois (2010) recommended carrying out stringent background checks. Also, policies and procedures are required to deal with disgruntled employees and to control any backdoor threats. The importance of testing all backup systems was also highlighted.

Within the area of systems damage, Fiore and Francois identified four areas of security lapse. The first is breakdown in the human firewall, which can be reduced by using inquiries, controlling points of contact, and ensuring awareness of people in the building or accessing IT equipment. Bounds checking and code reviews, system patches, and the use of alternative heterogeneous applications and platforms can reduce threats. The IT department can also be filtering any executable file attachments that are received from outside and taking steps to educate users on the methods that can be used to reduce the risks.

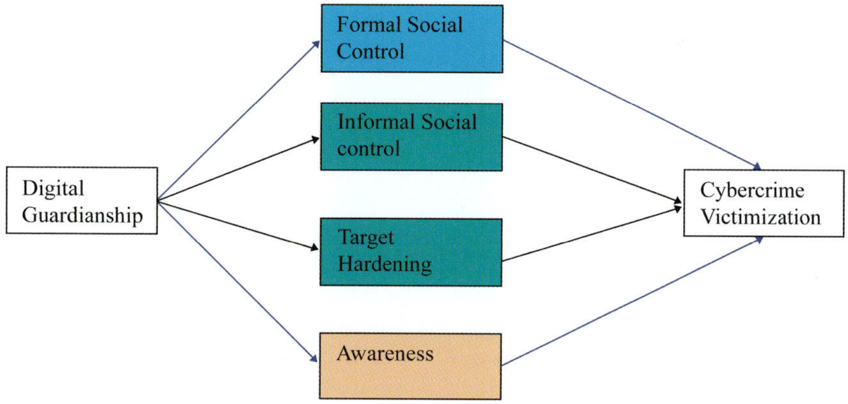

FIGURE 21.1 The structural model proposed by Jahankhani and Al-Nemrat (2010, 2011).

Wireless network strategy together with strong user authentication procedures is essential and must be in place. Organizations can also make use of Virtual Private Network (VPN) technology that makes encrypted data difficult to access. The use of Wireless LANs and wireless demilitarized zones can make wireless networks more secure. In order to reduce the risk from denial of service attacks, recommendations are to filter RFC 1918 addresses, controlling spoofed addresses, monitoring bandwidth usage, and scanning internal hosts and devices. System hijacking, which allows others to communicate securely using an organization's network, has been linked to steganography and can be controlled by checking for unauthorized software. Scanning both inbound and outbound e-mails to ensure that unusual files are not being attached is also essential.

Organizations often assume that using proxies or firewalls will prevent unauthorized Web surfing or unauthorized passing of information to external recipients. Using a SOCKS server or port mapping, HTTP tunneling can get around TCP and UDP. It is important to review logs of traffic to ensure that corporate espionage is not taking place. IT departments should monitor systems closely to ensure that corporate security policies are not being bypassed. Also, it should be ensured that unauthorized VPNs are not being used to mask unauthorized access to the system. Worms, Trojan horses, and viruses are becoming more prevalent, sophisticated, and capable of ever more intelligent attacks on systems. Scanning for unauthorized software is important in reducing the risk of such attacks, as is the use of up-to-date antivirus software, and perhaps considering the use of alternative heterogeneous applications or platforms that are less susceptible to attack. Educating users and ensuring proxies and firewall filters that are working effectively are also important steps.

Disinformation, which could include the dissemination of false information from an organization, or the insertion of false information into databases to reduce the effectiveness of such databases, is also potential threats. Such potential threats can be dealt with to reduce the risk of their occurrence.

DNS poisoning and domain hijacking involve falsifying IP addresses or stealing a domain from a registrar, and the risk of this can be reduced by using the latest security features of DNS, ensuring passwords are in use, and requiring SSL encrypted Web page or PGP signatures from e-mails.

DNS information should also be controlled and prevented from being taken from a system. In order to avoid Web site defacement, which is the unauthorized alteration of Web site content, staging servers should be read only, user authentication should be required for "sign-ons," software patches and security policies should be maintained and kept up to date, DNS should be hardened, and code should be reviewed to ensure weaknesses are weeded out of the system.

Cybercrime Execution and Analysis Model

Hunton (2009) demonstrated the opportunities for law enforcement when investigating the cybercriminal by defining an emerging cybercrime execution model.

Such models will allow for the transfer of conventional policing models into the cybercrime environment, which is often seen as being abstract and too technical for the application of such models. Before setting out the specific components and characteristics of this model, Hunton (2009) reviewed the background to the issue. The aim was to simplify cybercrime investigations so that investigators and analysts would cooperate and work together better when investigations are taking place. Such cooperation is currently hampered by the complexity of investigations. By providing specific points of reference, it is hoped that cooperation will be more practical. The components of cybercrimes that are in common with other crimes will be identified and this will provide areas of common ground upon which cooperation can be based.

Hunton's (2009) model aimed at assisting investigators in planning their investigations, regardless of the level of complexity of the investigations, with due regard to the techniques and technology that would be used, and in a manner that would allow for a consistent examination of each element of the crime. The intention of developing the model was to provide a tool that could facilitate further innovation of both practices and policies in the field of cybercrime investigation. This involves breaking down the technical tasks that are required to be performed into more manageable and use of the Internet and other networks to facilitate the process. Given the difficulties of cybercrime that include both technical and technological complexities and legal issues surrounding the fact that there is no common legal system across the Internet, there is a growing urgency surrounding development of such models. Hunton's (2009) claim that the model proposed allows for specialist skills to be identified rapidly and ensure that the knowledge necessary to secure relevant and admissible, high-quality evidence is present at the stage that they are required. This will also facilitate more simple communication of the elements of the execution of the crime and its investigation, in courts and to juries, as well as to victims of cybercrime.

Virus Prevention Model

Wang et al. (2009) looked at the lack of effectiveness of traditional antivirus methods to prevent virus infection. They proposed a behavior-based virus prevention model. The behaviors are defined by observing dynamically linked libraries and application programming interfaces. Information was then filtered to ensure that redundant behavior attributes were identified and informative features for training a virus classifier were selected. The performance of the model on a database of 1,758 benign executable files and 846 viruses was measured. It was noted that the results of the experiment were promising with 99% of known viruses and 96.66% of previously unknown viruses detected.

Cybercrime Prevention and Detection Model

Shiva Kumar (2003) examined the types of cybercrime and identified the types of harm they caused. Shiva Kumar also discussed the preventive steps that governments and organizations could take to reduce the risks from these harms. Provisions of cyber law were examined and the elementary problems associated with cybercrime were noted.

Cyber Trust and Crime Prevention Model

Collins and Mansell (2004) examined the issues of cyber trust and crime prevention. In a report commissioned by the UK government, they examined the position of crime prevention in this field by making reference to science journals. Their report stated:

> *provide a synthesis of theoretical and empirical work in the sciences and social sciences that indicates the drivers, opportunities, threats, and barriers to the future evolution of cyberspace and the feasibility of crime prevention measures. It was based on ten state-of-the-art science reviews commissioned by the Foresight Project. Each of the papers highlighted the current state of knowledge in selected areas as well as gaps in the evidence base needed to address issues of cyber trust and crime prevention in the future*

Collins and Mansell (2004) examined the issues of complexity of system behavior, managing identities in cyberspace, cyberspace usability and risk management security, cyberspace and crime prevention strategies, building forensics into data management tools, and cyberspace market evolution.

Cybercrime Reduction and/or Prevention Model

Efforts to reduce and/or prevent cybercrime victimization among user individuals have put the technology first and concentrate on protecting computers and devices with the hope that the users will not fall victim to cybercrime. These technical interventions differ significantly from mainstream crime prevention models that focus on the human element in crime. The field of criminology is yet to catch up with the explosion of the Internet and cybercrimes. Jahankhani and Askerniya recently proposed a grid model for classifying cybercrime prevention strategies along four different dimensions (Figure 21.2). These include (axis *X*) the level of "Tech-savviness," (axis *Y*) individuals at different levels of risk, (axis *Z*) the cognitive developmental stages of the individuals, and (idea/theme axis) the prevention program goal.

This model considers social aspects as an important element within crime prevention techniques and link education and awareness as a key player in crime reduction.

The Level of Tech-savviness. The first dimension on the grid model represents the level of tech-savviness targeted by the intervention.

Strategies for reducing and or preventing cybercrime have been focused at risk and protective factors required for the different types of Internet activities in which individuals engage.

Specific activities included in the grid model are based on the most common habits/activities individuals' exercise on the Internet. However, this does not mean that the individual user is engaged with all the listed activities, but definitely with at least one of them. In order to reduce risk and improve protective factors, interventions have focused on improving participant's awareness, education, and training with regard to specific skills related to the specific listed activities.

These interventions need to be implemented on different developmental stages of the individuals in an attempt to help the users not fall victim to cybercrime.

Risk Level of the Participants. The second dimension in the grid relates to the users' levels of risk and the extent to which interventions need to concentrate on participants' level of knowledge, awareness, and training.

Low risk is allocated to users who have extensive knowledge of technology and the Internet exposure risks. Medium risk is for users who do not know enough about Internet exposure risks and are above average risk of falling victim to cybercrimes, including online

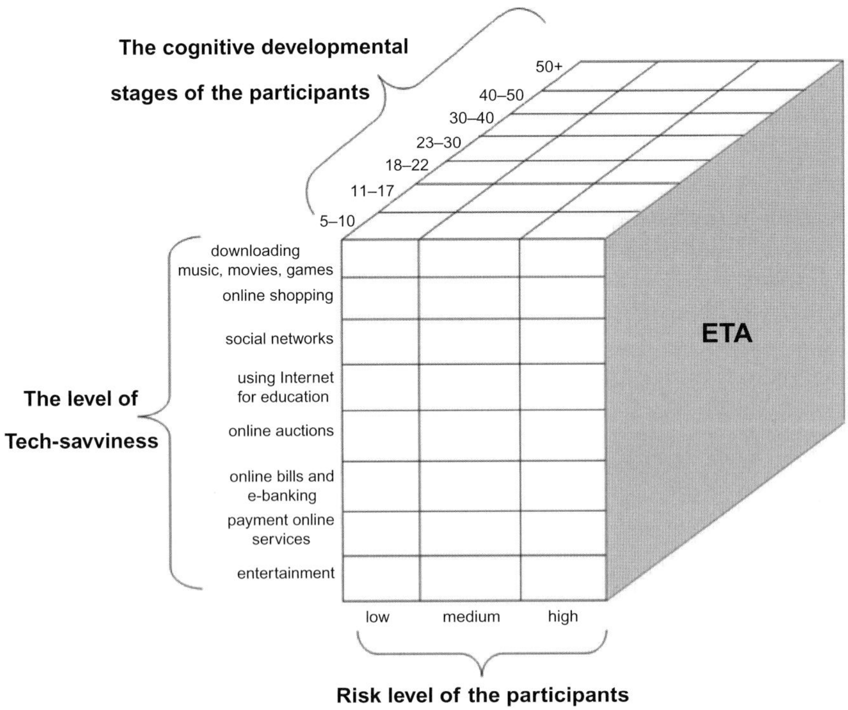

FIGURE 21.2 Cybercrime reduction and/or prevention model *(Jahankhani and Askerniya, 2012)*.

credit card fraud, identity theft, and computer viruses. This group of users has an up-to-date knowledge of securing computers and devices, but they do not have a deep understanding of how to change their behavior when they are online. The high-risk users are the people who surf the net aggressively and continuously expose themselves to the elements. This group of users puts both themselves and the people around them at high risk.

Cognitive Developmental Stages. The stages of cognitive development are represented in the third dimension in the grid. Risk and protective factors have different effects at different developmental stages. Therefore, aim of the prevention would be different depending on the age of the user.

Prevention Program Goal. The fourth dimension in the grid model represents the prevention program goal.

The most effective strategy to reduce and/or prevent cybercrime victimization among user individuals is to improve on cognitive development and behavioral skills by developing a set of education, training, and awareness programs specific to Internet exposure risks and cyber behaviors.

The term "awareness" is very popular in the academia and the industry. It is used massively in different areas of computer science and Internet security.

Barney and Anselm (1964) singled out following four different classical types of awareness;

1. An open awareness means that each member of society is aware of the others' true identity and his or her identity among the others.
2. A closed awareness refers to when someone does not know either the other's identity or the other's view of his identity.
3. Furthermore, we have suspicion awareness, which is a modification of the closed one. That is, someone suspects the true identity of the other or the other's view of his own identity, or both.
4. Pretence awareness is a modification of the open one. Here, both "interactants" are fully aware but pretend not to be (Barney and Anselm, 1964).

Liechti and Sumi (2002) in their research "Editorial: Awareness and the WWW," formulated the definition of awareness on the Internet as:

> *True, awareness is often meant as awareness of other people, and refers to the ability to maintain some knowledge about the situation and activities of others. Having a general idea of what is happening, or merely that something is happening, is often already very valuable.*

The model has a number of significant strengths. First, it is simple, and therefore easy to understand. It is important that a model is easily understood if it is to be applied widely and effectively. Second, the model is able to respond to changes in the external environment, such as changes in the nature of cybercrime. This is because the content of the education and training programs can be changed according to the new threats. Another advantage of the model arises from the reliance on situational crime theory. As highlighted earlier, situation crime theory does not have boundaries; this means that the theory develops with the manner in which the crime itself develops. If, for example, cybercrime turns out to become a major threat to users of tablets and smartphones, the training and awareness programs can be developed such that they educate users about how to guard against cybercrime on these platforms. This ability of the model to be relevant and adaptable to change in the external environment further lends to its practicability. Models that are highly specific may not be as flexible and able to respond to changes in the external environment. It would, therefore, be impractical to spend a significant amount of money in implementing models that might become obsolete and useless after a short period of time. This model, however, can be made to be relevant even when there are significant changes in the external environment. However, the unavoidable downside of this adaptability is that the implementation details of the model have to be regularly updated. That is, the contents of the training and awareness programs have to be continually updated for it to be effective and actually prevent and or reduce cybercrime.

The situational crime prevention theory would require the use of measures directed at highly specific forms of crime, which should also be as permanent as possible. The three measures suggested by the model, namely education, training, and awareness, are all able to effect changes that are permanent in nature. The training and education of users to make them aware of the current threats and how they can protect themselves creates users who are more technically savvy. Hence, even if constant reinforcement training is not provided, the users will already have a basic amount of technical knowledge and are able to continue to keep themselves updated. This is a permanent change for the better.

Furthermore, the entire model is based on the adoption of preventive measures to deter crime. Educating users' results in them taking better precautions and also serves to make their systems better protected against attacks. Better protected systems are less likely to become targets. A large number of attacks are instituted on systems that have no protection.

CHALLENGES FACING PREVENTIVE MEASURES

Many computer systems are protected by the security mechanism of user accounts that can only be accessed with a valid username and password. The username and password combination method of securing user accounts and computer systems are one of the, if not the oldest, methods of providing security to a computer or a computer network. However, the username and password method is not a very secure method. Usernames and passwords can be easily guessed. Password-cracking programs, which run all permutations and combinations of the ASCII characters in order to guess the username and password of an account, are easily available. Furthermore, users often forget their passwords, which means the computer systems have to provide mechanisms by which the passwords of the user accounts can be reset. This provides opportunities for criminals to gain control of these accounts by posing as legitimate users who have forgotten their login credentials. Clearly, more secure technology for regulating access to computer systems is required. Currently, more secure technology such as facial recognition and biometric identification systems are being developed. Many new computer systems already use these technologies to secure access. When these technologies mature and are used more widely, it can be expected that the traditional username and password login credentials will become obsolete. In addition to shortcomings with the technology for regulating access to computer systems, other shortcomings of technology exist, which allow attackers to gain control of a computer system. These are often called "zero day attacks." Therefore, attackers are continually looking for new vulnerabilities and take advantage of systems that have not been patched already.

Legislation is a major issue, but the attitude toward cybercrime and its victims needs to be addressed if we are to see an effectual clamp down on cyber villains and the like. Investment is essential, accurate recording of cybercrime does not exist, and international legislation and bureaucracy are major barriers for investigators.

There is no supreme regulatory body with adequate power to enforce the required regulations. In fact, the regulations do not really exist, other than those written on the statute books of individual countries. It seems therefore, that although certain jurisdictions are concentrating on resources to build an infrastructure around formalizing what constitutes cybercrime, little effort or funding is put into supporting law enforcement to tackle the proliferation of offenses committed.

Yar (2006) articulated the shortcomings of criminology in the arena of cybercrime as being the problem of where and who. Many criminological perspectives define crime on the social, cultural, and material characteristics, and view crimes as taking place at a specific geographic location. This definition of crime has allowed for the characterization of crime, and the subsequent tailoring of crime prevention, mapping, and measurement methods to the specific target audience.

However, this characterization cannot be carried over to cybercrime, because the environment in which cybercrime is committed cannot be pinpointed to a geographical location or distinctive social or cultural groups. For example, traditional crimes such as child abuse and rape allow for the characterization of the attacker based on the characteristics of the crime, including determination of the social status of the attacker; geographical location within country, state, district, urban, or rural residential areas; and so on. However, in the case of cybercrime, this characterization of the attacker cannot be made, because the Internet is "anti-spatial." Yar (2006) explains that identifying location with distinctive crime-inducing characteristics is almost impossible in cybercrimes. This, in turn, serves to render the criminological perspectives based on spatial distinctions useless.

Criminology allows for the understanding of the motivations of the criminals by analyzing the social characteristics of the criminals and their spatial locations. For example, poverty may be considered to be a cause of crime if poor areas exhibit high crime rates, or a high percentage of criminals are found to come from poor backgrounds. According to Yar (2006) criminology helps to understand the reasons behind the preponderance of crimes committed by people with particular characteristics, such as the overrepresentation of offenders from groups of people who are socially, economically, or educationally marginalized. It was further explained that the association between geographical location and social characteristics has led to the association between crime and social exclusion in mainstream criminology.

However, in the case of cybercrime, such a correspondence appears to break down. One of the most important points to consider is that access to the Internet is disproportionately low among the marginalized sections of society who were considered to be socially excluded and therefore more likely to commit a crime. Furthermore, the execution of a cybercrime requires that the criminal have a degree of skill and knowledge that is greater than the level of skills and knowledge possessed by the average computer user. It can, then, be said that cybercriminals are those who are relatively more privileged and who have access to the Internet, knowledge, and skills at a level above the average person. Therefore, the relationship between social exclusion and crime that had been widely accepted in traditional crime could not be true in the case of cybercrimes, and that cybercriminals are fairly "atypical" in terms of traditional criminological expectations. Hence, the current perspectives of criminology that link marginality and social exclusion to crime have no use in explaining the motivations behind cybercrimes. Without an understanding of motives, it is difficult for law enforcement agencies and government to take effective measures to tackle cybercrime.

Brenner (2010) also raised the important issue of privacy versus the need to prevent crimes and collect evidence of criminal activity in the digital world. In the real world, there are a number of safeguards against the infringement of personal privacy by law enforcement, such as laws regulating the tapping of phones and monitoring of digital communications such as e-mails and chats. Most people expect that their e-mail is private. However, the technological basis of e-mail is such that it is in fact not private. In normal letters, the information contained in the letter is transmitted in a sealed envelope. In order to access the information it is necessary for the carrier to gain access into the sealed envelope, which involves breaking the seal and perhaps resealing the envelope, and this can be classified as deception. However, in the case of e-mail, the information is not transmitted in a sealed envelope. It is transmitted in plain text, and therefore, can be intercepted.

The issue of privacy in the digital world is one that is becoming increasingly important with the rise of Web 2.0. Social networking Web sites such as Facebook and Twitter collect an enormous amount of personal information about the individual. Yet the privacy controls on these Web sites are often not easy to understand, making the less technologically savvy person a more likely candidate for attack.

CONCLUSION

The worldwide growth of personal and business use of the Internet over the past two decades has been almost exponential. Restrictions on the use of the Internet or how it is used is limited to how you access it, that is, how easily is the Internet accessible to an individual or business. The ability to be able to gain access 24/7 has presented untold opportunities for potential scammer, hacker, and identity thief or just about anybody in the World that has an unlawful intention to commit cybercrime.

The opportunity to generate income streams that were not previously possible combined with anonymity has increased the opportunity for crime.

Cybercrime offenses are crimes that have an underlying element of dishonesty that have been (or were) in existence before the Internet existed. What has changed is the following:

- Increased targeting of victims
- Cost reduction in targeting those victims
- Increased global capacity and the ability to cross international boundaries with ease
- Complete lack of legal process to deal with the issues
- Increased anonymity
- Extremely small chance of the crime being reported
- Lack of suitable international legislation to bring the suspect to justice

Total lack of understanding of the gravity of such offenses is as interpreted by the Courts of Justice.

Just as it is important to understand the characteristics of the criminals in order to understand the motivations behind the crime and subsequently develop and deploy crime prevention strategies, it is also important to understand the characteristics of the users of computer systems in order to understand the manner in which cybercrime makes these users victims, and also to develop and deploy effective crime prevention strategies at the user side.

ACKNOWLEDGMENTS

The author thanks Dr. Ameer Al-Nemrat and Dr. Imran Askerniya whose research helped to form the basis of the material here.

CHAPTER

22

Conclusion: National Security in the Networked Society

Simeon Yates

We argued in Chapter 1 that the current constellation of digital media and networks; networks of regular international trade and travel; and the complex networks of global political, economic, and social systems provide a new context for the formation of National Security Strategy. Key to this new context are the opportunities, threats, and challenges this new constellation provides. In the rest of the book we tried to explore specific aspects of these challenges.

NATIONAL SECURITY TODAY AND IN THE FUTURE

Section 1 explored current thinking on national security and contemporary threats to that security. Importantly, a key theme in this section was the interdependence of nation states regarding both the prevention of and vulnerability to security threats. As we noted in the introduction, a model of national security based on "unit level" and anthropomorphic model of nation states as actors on a global stage cannot hold in the age of the networked society. As Buzan (2009) noted, the interconnectedness of contemporary global societies requires that security be looked at via the lens of systems, networks, and institutions that function within, between, and among nation states; this point is reiterated by Stanniforth in Chapter 2. In Chapter 3 Saathoff et al. reminded us that threats to security may arise from inside the state. This does not mean they are disconnected from the networked world. Far from it, as the examples discussed by Saathoff et al. indicated, access to networks provided the information, knowledge, materials, and substances that influenced behavior, as well as being the target or vectors for auctioning threats. It is also notable that failures to communicate between law enforcement networks may play a role in such insider threats going undetected.

Chapters 4 and 5 counterpoint the argument that nation states may be of less relevance in a networked age. In Chapter 4 Lehr focused on the response to asymmetric threats in a naval context, for example, the rise in piracy off the horn of Africa. This chapter strongly argued that such challenges to security fit a networked or "post-modern" view of our global world, and that ironically, this represents a very Western cosmopolitan view of global politics. Lehr concluded with the concern that other nations may not share this view and that Western nations may have underestimated the old-fashioned national security threat posed by direct action by other nation states. Stanniforth reminded us in Chapter 5 that nations remain and that a key feature of the preservation and functioning of modern nation sates lies in the policing and security of borders. Stanniforth noted that these borders are ever more difficult to police, whether through transnational agreements such as Schengen or through their penetrations by ever-growing "network flows," to use Castells' term. The need to police, protect, and derive intelligence from the people, goods, and information crossing borders will remain as long as there is a need to secure the nation state.

THE PUBLIC, THREATS, AND NEW MEDIA

Section 2 examined the issues and challenges facing citizens in a networked society. In Chapter 6 Rogers and Pearce clearly articulated the case for robust risk communication strategies. As is noted in many chapters, the networks in which many citizens now live their daily lives can provide access to ever-greater levels of information. Ensuring that risks are proportionately communicated and that policy makers have a good understanding of the levels of panic or compliance during moments of crisis is key to the operationalization of strategies. In Chapter 7 Krieger and Rogers applied the ideas of Chapter 6 to the case of a CBRN incident. A key feature of their discussion is the importance of citizens trusting the information they are provided, the institutions interacting with them, and the nature of the communication. Networks are not just made of communications channels, incidences of economic exchange, or of social interaction. They are also strongly influenced by our perceptions of those links. Such perceptions are closely tied to how we value the link itself, the channel, the persons or organizations involved, and the form of the message. Strategies to support the resilience of citizens at such challenging times need to be based on a deep understanding of these issues. Chapter 8 builds on this issue of trust. It notes that in the UK there is a key idea of "policing by consent" that builds upon a historical approach dating back to Saxon times. In this context the current PREVENT strategy for addressing terrorist threats builds upon this community-based approach. Yet the Internet and digital media open up citizens to networked flows from beyond the confines of their local communities. A key part of ensuring trust- and consent-based policing of potential terrorist threats must entail understanding how these "dark" flows can be countered in a community context.

In Chapter 9 Manso and Manso developed the themes of this section further by exploring the possibilities provided by networked digital media to support both citizens and intuitions to respond to threats and crises. In this case it is the ability to deploy ubiquitous mobile and social media in response to crises. In this chapter it is shown that the networked society provides a resource to support the resilience sought in Chapter 7. There are often complexities and compromises in technology implementations, and in Chapter 10 Dastbaz et al. reminded us that many new media are "Janus faced." Although Manso and Manso made the merits of mobile information communication technologies (ICT) clear, these media can compromise many of the rights Western citizens take for granted. In embracing digital media as tools for communication and potentially to provide resilience in crises, citizens may also open them up to greater surveillance by the state. Networked media, especially mobile ICT and social networking, are challenging both legal and cultural assumptions of public and private. The balance point between freedom, privacy, personal and public safety, and the role of the state in a networked age is clearly something still up for debate.

DEPLOYING NETWORK TECHNOLOGIES FOR NATIONAL SECURITY

Section 3 marked a change in focus within the book. Here we explored the use of new technologies in strategic responses to national security threats. Importantly, we have explored how current and future ICT tools might be developed and deployed. In Chapter 11 Stedmon et al. provided a very strong case for the use of user-centered methods in the elicitation and implementation of security system requirements. This is an argument that has been made recently in many areas of ICT development and product design, yet is still often overlooked in major national and public sector system development. National security "users" are of course often the citizens the state wishes to protect. Koraeus and Stern reminded us in Chapter 12 that although we may exist today in a mass of information flows, it is the extraction of learning from prior experience (knowledge) and the management of it that is key to improving responses to potential and actual threats.

Although many new technologies support processes of knowledge extraction, storage, and management; for example, see Launders and Polovina's discussion of semantic tools in Chapter 13, the key task remains making this knowledge effectively available to managers, policy makers, and practitioners. As Saathoff et al. noted, actual examples of many types of security threats or crises may in fact be quite rare. Capturing the key learning from such events is therefore essential, whether it is through the use of ICT or through appropriate networks of experts. In Chapters, 14 and 15 Andrews et al. and Pavlin explored how the huge floes of information within networked societies might be deployed both operationally and in the development of strategy. Here again we find that ICT tools and solutions may both create these flows and allow us to extract key knowledge and information in ways not previously possible. Rein reminded us in Chapter 16 that actually implementing such systems is challenging. Despite the large amounts of data collected in key hubs by the institutions of nation states, generated and shared by citizens through their networks and produced in response to crises, there remains a need to find standardized ways to store and exchange this material within the networks of security agencies.

THREATS TO THE INFRASTRUCTURE OF THE NETWORKED SOCIETY

What then of the core infrastructure of the networked society itself, within national borders or external? Section 4 explored the issues of national cyber security and the threats to nations and networks. Stanniforth provided an overview of the emerging challenges of cyber security in Chapter 17 while Chapter 18 laid out the current key strategic goals of the United States cyber security strategy. In both cases the recognition of this as a transnational threat that requires a transnational response is made clear. Kallberg and Thuraisingham reminded us in Chapter 19 that as with conventional threats, asymmetric or state based, there are a variety of players within global networks and that individuals, groups, or even nations may act to threaten each other's cyber security.

MacKinnon et al. combined many of the issues discussed in this book, user needs, technological solutions, and implementation methods to build proposals for general cyber security strategies to be employed by end users, organizations, and institutions. A key feature of the argument in Chapter 20 is that in a networked society cyber security has to be a shared responsibility across a wide range of citizen, institutional, national, and international actors. Finally, but very importantly, Jahankhani asked us in Chapter 21 to reflect on both the criminal and citizen behavior in the context of cyber crime. This provides a basis for modeling behavior and looking to methods to prevent threats to citizen's networks and ICT systems.

CONCLUSION

In our introduction we set out three key issues:

1. The development of strategy in the context of national security driven by national interest
2. The importance of high-quality information and knowledge in the development of strategy—strategic intelligence
3. The nature of globalized networks based on ICT systems and people that are concurrently locations for threats to national and global security, tools to defend or to threaten national security, and sources of information for both nations and citizens about threats and events

In essence, the strategic intelligence management vision for national security is released through connection between national security and national interest, dynamic collaboration among stakeholders, and proactive communication strategies and enabling technologies. We hope that in the four sections of this book we have provided examples of how current security practitioners, academic researchers, and ICT system developers have sought to address these issues at the citizen, organizational, and national levels.

Abelson, H., Sussman, G.J., 1986. Structure & Interpretation of Computer Programs. Michigan Institute of Technology, Dpt. of Electrical Engineering & Computer Science: Video lectures. [Online]. Retrieved from http://swiss.csail.mit.edu/classes/6.001/abelson-sussman-lectures/ (Last accessed 24.04.07.).

Acton, J.M., Rogers, M.B., Zimmerman, P.D., 2007. Beyond the Dirty Bomb: Re-thinking Radiological Terror. Survival 49 (3), 151–168.

Adger, W., 2000. Social & ecological resilience: are they related? Progress in Human Geography 24 (0), 347–364.

Akhgar, B., Yates, S., (Eds.), 2011. Intelligence management: Knowledge driven frameworks for combating terrorism and organised crime. Springer Publication, 1st edition, X, 224 p. 48 illus.

Alavi, M., Tiwana, A., 2003. Knowledge Management: The Information Technology Dimension. In: Easterby-Smith, Lyles (Eds.), Handbook of Organisational Learning and Management. Malden Blackwell Publishing Ltd, pp. 104–121.

Alizadeh, S., Ghazanfari, M., 2009. Learning FCM by chaotic simulated annealing. Chaos, Solitons & Fractals 41 (3), 1182–1190.

Allegra, P.C., Cochrane, D., Dunn, E., Milano, P., Rothman, J., Allegra, J., 2005. Emergency department visits for concern regarding anthrax – New Jersey, 2001. MMWR Morb. Mortal. Wkly. Rep. 54 (Suppl.), 163–167.

Alvesson, M., Kärreman, D., 2001. Odd couple: Making sense of the curious concept of knowledge management. Journal of Management Studies 38 (7), 995–1118.

Alvesson, M., 1993. Organizations as Rhetoric: Knowledge-intensive Firms & the Struggle with Ambiguity. Journal of Management Studies 30 (6), 997–1015.

American Association of Poison Control Centers (AAPCC), 2012. Bath Salts Data. Updated February 8, 2012. [online]. Retrieved from http://www.aapcc.org/dnn/Portals/0/Bath%20Salts%20Data%20for%20Website%202.8.2012.pdf (Last accessed 21.02.12.).

American Red Cross Multi-Disciplinary Team, 2011. Report on the 2010 Chilean earthquake & tsunami response. U.S. Geological Survey Open-File Report 2011-1053, v. 1.1, 68 p., [online] retrieved from http://pubs.usgs.gov/of/2011/1053/.

Anderson, B., 2010. Pre-emption, precaution, preparedness: anticipatory action and future geographies. Progress in Human Geography 34 (6), 777–798.

Anderson, B., Adey, P., 2011. Affect and security: exercising emergency in "UK civil contingencies". Environment and Planning D: Society and Space 29 (6), 1092–1109.

Anderson, K., 2006. Hacktivism & Politically Motivated Computer Crime. [online]. Retrieved from: http://www.aracnet.com/~kea/Papers/Politically%20Motivated%20Computer%20Crime.pdf (Last accessed 01.09.12.).

Andrews, K., 1971. The concept of corporate strategy. Dow-Jones, Irwin, CA.

Andrews, S., 2011. In-Close2, a High Performance Formal Concept Miner. In: Andrews, S., Polovina, S., Hill, R., Akhgar, B. (Eds.), Conceptual Structures for Discovering Knowledge - Proceedings of the 19th International Conference on Conceptual Structures (ICCS) 2011. LNAI 6828. Springer, Berlin, pp. 50–62.

Andrews, S., Orphanides, C., 2012. Knowledge Discovery Through Creating Formal Contexts. In: Xhafa, F. (Ed.), International Journal of Space-Based and Situated Computing, vol. 2(2). Inderscience, Switzerland, pp. 123–138.

Andrienko, N.V., Andrienko, G.L., 2011. Spatial Generalization & Aggregation of Massive Movement Data. IEEE Trans. Vis. Comput. Graph. 17 (2), 205–219.

Ansoff, H.I., 1965. Corporate Strategy: An analytical approach to business policy for growth and expansion. McGraw-Hill, New York.

Aradau, C., Munster, R., 2007. Governing terrorism through risk: taking precautions, (un)knowing the future. European Journal of International Relations 13 (1), 89–115.

Argenti, A., 1968. Corporate planning: A practical guide. Allen and Unwin, London.

Argyris, C., Schön, D.A., 1978. Organizational Learning: A Theory of Action Perspective. Addison-Wesley, Reading, MA.

Argyris, C., Schön, D.A., 1996. Organizational Learning II: Theory, Method, Practice, second ed. Addison-Wesley, Reading, MA.

Arquilla, J., Ronfeldt, D., 2001. Networks & Netwars: The Future of Terror, Crime, & Militancy. Rand, Santa Monica, CA.

Ashour, O., 2010. Online De-Radicalization? Countering Violent Extremist Narratives: Message, Messenger & Media Strategy. In Perspectives on Terrorism 4 (6), 15–20 Terrorism Research Initiative.

Bacon, L., MacKinnon, L., Cesta, A., Cortellessa, G., 2012. Developing a Smart Environment for Crisis Management Training. Special edition of the Journal of Ambient Intelligence & Humanized Computing, entitled

Smart Environments & Collective Computational Intelligence for Disaster Management 3 (2).
Badica, C., et al., 2012. Multi-Agent Systems, Ontologies & Negotiation for Dynamic Service Composition in Multi-Organizational Environmental Management. Software Agents, Agent Systems & Their Applications 2012, 286–306.
Ball, K., Haggerty, K., Lyon, D., 2012. Handbook of Surveillance studies. Routledge.
Bargh, J.A., Mckenna, K., 2003. The Internet and Social Life. Annu. Rev. Psychol. 55 (0), 573–590.
Barlow, J., 2002. Filtering vs Globalization. The Journal of Education, Community and Values 2 (10).
Barney, G.G., Anselm, L.S., 1964. Awareness Contexts & Social Interaction. Am. Sociol. Rev. 29 (5) Oct.
Bastings, I., Oosterhout, A., Nieuwenhuizen, M., Leeuw, M., 2007. Assessment of the vulnerabilities of modern societies to terrorist acts employing radiological, biological or chemical agents with the view to assist in developing preventive and suppressive crisis management strategies. In: Consortium, A. (Ed.), Final report ASSRBCVUL project. Brussels.
BBC, 2012. Flame malware makers send 'suicide' code. [online]. Retrieved from http://www.bbc.co.uk/news/technology-18365844.
Beck, P.W., 1999. Nuclear energy in the twenty-first century: Examination of a contentious subject. Annual Review of Energy and the Environment 24 (1), 113–137.
Beck, U., 1992. Risk Society— Towards a New Modernity. Sage, London.
Beck, U., 1999b. World Risk Society. Polity Press, Cambridge.
Becker, S.M., 2004. Emergency Communication and Information Issues in Terrorist Events Involving Radioactive Material. Biosecur. Bioterror. 2 (3), 195–207.
Beebe, N.L., Rao, V.S., 2005. Using situational crime prevention theory to explain the effectiveness of information systems security. In: Proceedings of the 2005 SoftWars Conference Las Vegas, NV, Dec. 2005.
Bell, D.E., La Padula, L.J., 1976. Secure computer system: unified exposition and {MULTICS} Interpretation. The MITRE Corporation Technical report (ESD-TR-75-306).
Bellaby, P., 2003. Communication & miscommunication of risk: Understanding UK parents' attitudes to combined MMR vaccination. Br. Med. J. 327, 725–728.
Bellovin, S.M., 2008. The Insider Attack Problem Nature & Scope, Insider Attack & Cyber Security—Beyond the Hacker, Series: Advances in Information Security, vol. 39, 2008, XII, 223, Chapter 1, p. 1.
Benjamin, R.I., Rockart, J.F., Morton, M.S., Wyman, J., 1984. Information Technology: A strategic Opportunity. Sloan Manage. Rev. 25 (3)(Spring 1984).
Berezuk, G., McCarthy, G., 1992. Investigational Drugs & vaccines fielded in support of operation Desert Storm. Mil. Med. 157, 404–406.
Bermingham, A., Conway, M., McInerney, L., O'Hare, N., Smeaton, A.F., 2009. Combining Social Network Analysis & Sentiment Analysis to Explore the Potential for Online Radicalisation. In: Proceedings of the International Conference on Advances in Social Network Analysis & Mining. IEEE, Athens, Greece, pp. 231–236, 20–22 July.
Bermingham, A., Conway, M., McInerny, L., O'Hare, N., Smeaton, A.F., 2008. Combining Social Network Analysis & Sentiment Analysis to Explore the Potential for Online Radicalisation. CLARITY Centre for Sensor Web Technologies & School of Law & Government, Dublin City University. School of Law & Government, Dublin City University, Dublin.
Beynon, M., et al., 2002. Processing large-scale multidimensional data in parallel and distributed environments. Data Intensive Computing, Parallel Computing 28 (5).
Bhatt, G.D., 2001. Knowledge management in organizations: examining the interaction between technologies, techniques & people. Journal of Knowledge Management 5 (1), 68–75.
Biba, K.J., 1976. Integrity Considerations for Secure Computer Systems. The MITRE Corporation Technical report (ESD-TR-76-372).
Biernacki, P., Waldorf, D., 1981. Snowball sampling. Problems and techniques of chain referral sampling. Sociological Methods & Research 10 (2), 141–163.
Birchard, K., 1998. Experts still arguing about over radiation doses. Lancet 354, 400.
Blackett-Review, 2011. Blackett Review of High Impact Low Probability Risks (GSO, Ed.). Government Office for Science, London.
Blackler, F., 1993. Knowledge & the Theory of Organizations: Organizations as Activity Systems & the Reframing of Management. Journal of Management Studies 30 (6), 863–884.
Blackler, F., 1995. Knowledge, Work & Organizations. Organisational Studies 16 (6), 1021–1045.
Blackwell, C., 2009. The Insider threat: Combatting the enemy within. IT Governance Publishing.
Blanchard, J.C., Haywood, Y., Stein, B.D., Tanielian, T.L., Stoto, M., Lurie, N., 2005. "In Their Own Words: Lessons Learned From Those Exposed to Anthrax". Am J Public Health 95 (3), 489–495.
Blandford, A., Rugg, G., 2002. A case study on integrating contextual information with analytical usability evaluation. International Journal of Human-Computer Studies 57, 75–99.
Boin, A., Ekengren, M., Rhinard, M., 2006. Functional Security & Crisis Management Capacity in the

European Union, Forum for Security Studies (EUROSEC) April 2006. [online]. Retrieved from http://www.ssrnetwork.net/document_library/detail/3750/functional-security-&-crisis-management-capacity-in-the-european-union.

Boin, R.A., 't Hart, P., Stern, E.K., Sundelius, B., 2005. The Politics of Crisis Management—Public Leadership under Pressure. Cambridge Univer-sity Press.

Bord, R., Connor, R., 1990. Risk communication, knowledge, & attitudes: Explaining reactions to a technology perceived as risky. Risk Anal. 10, 499–506.

Bornemann, M., et al., 2003. An Illustrated Guide to Knowledge Management. Sammer, Martin (Eds.), Wissenmanagement Forum, Graz, Austria.

Bowers, S., Ludascher, B., 2004. An ontology-driven framework for data transformation in scientific workflows. In: DILS. pp. 1–16.

Bozorgmehr, N., 2012. Iran Attacks Israel over Cyberattack. Financial Times. 20th May 2012. [online]. www.ft.com/cms/s/0/cb1326b0-a9a-11e1-9772-00144feabdc0.html (accessed July 2012).

Bray, H., 2010. Akamai employee charged with trying to sell secrets to a foreign government. The Boston Globe, October 6, 2010, [online]. Retrieved from http://www.boston.com/business/ticker/2010/10/akamai_employee.html.

Brenner, S.W., 2010. Cybercrime: criminal threats from cyberspace. ABC-CLIO.

Bresnan, J., 2001. Lexical-Functional Syntax. Blackwell, Malden, MA.

Brint, S., 1990. Rethinking the Policy Influence of Experts—From General Characterisations to Analysis of Variation. Sociol. Forum 5 (3), 361–385.

Brown, J.S., Duguid, P., 1991. Organizational Learning & Communities-of-practice: Toward a Unified View of Working, Learning & Innovation. Organization Science 2 (1), 40–57.

Bruce, J., 1995. A New Arms Race in the Gulf. Jane's Intelligence Review 007 (01), 37 (January 1995). Passim.

Bueno, S., Salmeron, J.L., 2009. Benchmarking main activation functions in fuzzy cognitive maps 2. Expert Systems Appl. 36, 5221–5229.

Burners-Lee, T., Hendler, J., Lassila, O., 2001. The Semantic Web. Sci. Am. 284 (5).

Bush, G., 1986. Public report of the Vice President's Task Force on Combating Terrorism. Government Printing Office, Washington, D.C.

Buzan, B., 2009. What is national security in the age of globalisation? [online]. Retrieved from. http://www.regjeringen.no/nb/dep/ud/kampanjer/refleks/innspill/sikkerhet/buzan.html?id=493187.

Buzan, B., Wæver, O., Wilde, J., 1998. Security: A New Framework for Analysis. Lynne Reinner Publishers, Boulder, Colorado.

Byres, E., Eng, P., Lowe, J., 2004. The Myths & Facts behind Cyber Security Risks for Industrial Control Systems. In: Proceedings of the VDE Congress. VDE Association for Electrical Electronic & Information Technologies. (October 2004) [online]. Retrieved from http://nealsystems.com/downloads/Myths%20 &%20Facts%20for%20Control%20System%20Cyber-security.pdf (Last accessed July 2012).

Cabinet Office, 2008a. Learning Lessons from the 2007 Flood: An Independent Review by Sir Michael Pitt. [online]. Retrieved from http://webarchive.nationalarchives.gov.uk/20100807034701/http://archive.cabinetoffice.gov.uk/pittreview/_/media/assets/www.cabinetoffice.gov.uk/flooding_review/pitt_review_full%20pdf.pdf (Last accessed 30.08.12.).

Cabinet Office, 2008b. National Risk Register. [online]. Retrieved from http://webarchive.nationalarchives.gov.uk/+/http://www.cabinetoffice.gov.uk/media/cabinetoffice/corp/assets/publications/reports/national_risk_register/national_risk_register.pdf (Last accessed 29.08.12.).

Cabinet Office, 2011a. Communicating Risk Guidance. [online]. Retrieved from http://www.cabinetoffice.gov.uk/sites/default/files/resources/communicating-risk-guidance.pdf (Last accessed 29.08.12.).

Cabinet Office, 2011b. Strategic National Framework on Community Resilience. [online]. Retrieved from http://www.epcollege.com/EPC/media/MediaLibrary/Knowledge%20Hub%20Documents/A%20Strategic%20Framework/Strategic-National-Framework-on-Community-Resilience_0.pdf?ext=.pdf.

Cabinet Office, 2012. National Risk Register of Civil Emergencies 2012 Edition. [online]. Retrieved from http://www.cabinetoffice.gov.uk/sites/default/files/resources/CO_NationalRiskRegister_2012_acc.pdf (Last accessed 29.08.12.).

Cangelosi, V.E., Dill, W.R., 1965. Organizational Learning: Observations Towards a Theory. Adm. Sci. Q. 10 (2), 175–203.

Cappelli, D., 2011. Insider Threats: Actual Attacks by Current & Former Software Engineers. Presentation by CERT, Software Engineering Institute, Carnegie Mellon University.

Cappelli, D., Moore, A., Randazzo, M.R., Keeney, M., Kowalski, E., 2004. Insider threat study: Illicit cyber activity in the banking and finance sector. [online]. Retrieved from http://www.cert.org/insider_threat/study.html.

Cappelli, D., Moore, A., Trzeciak, R., Shimeall, T., 2009. Common Sense Guide to Prevention & Detection of insider threats 3rd Edition–version 3.1. CERT, Software Engineering Institute, Carnegie Mellon University.

Cappelli, D., Moore, A., Trzeciak, R., 2012. The CERT guide to insider threats: How to prevent detect and respond to Information Technology Crimes (Theft Sabotage Fraud). CERT Software Engineering Institute Carnegie Mellon University. Addison Wesley.

Cardenas, A., Amin, S., Sinopoli, B., Giani, A., Perrig, A., Sastry, S.S., 2009. Challenges for Securing Cyber Physical Systems. In: Workshop on Future Directions in Cyber-physical Systems Security, DHS, Newark, New Jersey, 22–24 July.

Carneiro, A., 2001. The role of intelligent resources in knowledge management. Journal of Knowledge Management 5 (4), 358–367.

Castells, M., 2000. The Rise of the Network Society. Oxford: Blackwell.

Chalk, P., 1997. Grey-area phenomena in Southeast Asia: Piracy, drug trafficking & political terrorism. Canberra Papers on Strategy and Defense, No. 23, Australian National University, Canberra.

Chang, R., et al., 2008. Scalable & interactive visual analysis of financial wire transactions for fraud detection. Information Visualization 7 (1), 63–76.

Chang, C., Ferreira, R., Sussman, A., Saltz, J., 1999. Infrastructure for building parallel database systems for multi-dimensional data. In: Proceedings of the Second Merged IPPS/SPDP Symposiums. IEEE Computer Society Press, April 1999.

Chauvel, D., Despres, C., 2002. A review of survey research in knowledge management: 1997–2001. Journal of Knowledge Management 6 (3), 207–223.

Chen, H., Chung, W., Qin, J., Reid, E., Sageman, M., Weimann, G., 2008. Uncovering the Dark Web: A Case Study of Jihad on the Web. Journal of the American Society for Information Science and Technology 59 (8), 1347–1359.

Chen, T.M., 2010. Stuxnet, the Real Start of Cyber Warefare? ieeexplore, Network 24 (6), 2–3 November/December 2010.

Chiu, D., Agrawal, G., 2009. Enabling ad hoc queries over low-level scientific data sets. In: SSDBM. pp. 218–236.

Chong, D.J.T., Deng, R.H., 2006. Privacy-enhanced superdistribution of layered content with trusted access control. In: Proceedings of the ACM workshop on Digital rights management. ACM, New York, NY, pp. 37–44.

Christopherson, K.M., 2007. The positive & negative implications of anonymity in Internet social interactions: "On the Internet, Nobody Knows You're a Dog". Computers in Human Behavior 23, 3038–3056.

Chua, A., Lam, W., 2005. Why KM projects fail: a multi-case analysis. Journal of Knowledge Management 9 (3), 6–17.

Claburn, T., 2008. Engineer Gets 24 Year Sentence For Trying To Steal Navy Secrets. InformationWeek, March 25, 2008, [online]. Retrieved from http://www.informationweek.com/engineer-gets-24-year-sentence-for-tryin/206905727.

Clarke, R.A., Knake, R., 2010. Cyber War: The Next Threat to National Security and What to Do About It. Harper Collins, New York.

Clayton, M., 2010. Stuxnet malware is 'weapon' out to destroy Iran's Bushehr nuclear plant?, September 21, 2010, [online]. Retrieved from http://www.csmonitor.com/USA/2010/0921/Stuxnet-malware-is-weapon-out-to-destroy-Iran-s-Bushehr-nuclear-plant (Last accessed August 2012).

CoGui, 2012. A Visual Conceptual Graph tool for building graph knowledge representation. [Online]. Retrieved from www2.lirmm.fr/cogui/ (Last accessed 11.07.12.).

Cohen, L.E., Felson, M., 1979. Social Change & Crime Rate Trends: A Routine Activity Approach. American Social Review 44 (August), 588–608.

Cohen, L.E., Felson, M., Land, K. 1981. Social inequality and predatory criminal victimisation: An exposition and a test of a formal theory. Am Sociol Rev 46, 505–524.

Cole, B.D., 2008. Sea Lanes & Pipelines. Energy Security in Asia. Praeger Security International, Westport, CT, and London.

Cole, E., Ring, S., (Eds.), 2005. Insider threat: Protecting enterprise from sabotage spying and theft (Trans.). Syngress Publishing, Rockland, MA.

Collier, P.M., 2006. Policing & the intelligent application of knowledge. Public Money & Management 26 (2), 109–116.

Collins, B., Mansell, R., 2004. Cyber trust & crime prevention: A synthesis of the state of the art science reviews. RMCS Cranfield University and London School of Economics and Political Science.

Colwill, C., 2009. Human factors in information security: The insider threat—who can you trust these days? Information Security Technical Report 14 (4), 186–196.

Comfort, L.K., Ko, K., Zagorecki, A., 2004. Coordination in Rapidly Evolving Disaster Situations—The Role of Information. Am. Behav. Sci. 48 (3), 295–313.

Committee for Disaster Medicine Studies, 2007. The Terror Attacks in Madrid, Spain, 2004— KAMEDO Report. (KAMEDO) of Sweden's National Board of Health & Welfare, September 2007. Retrieved from: http://www.socialstyrelsen.se/Lists/Artikelkatalog/Attachments/9210/2007-123-36_200712336.pdf.

Committee on National Security Systems, (2007) Program for the management and use of national reserve information assurance security equipment. [online] retrieved from http://www.cnss.gov/Assets/pdf/CNSSI-4008.pdf (last accessed 13.12.12.).

Compton, P., Jansen, R., 1990. A Philosophical Basis for Knowledge Acquisition. Knowledge Acquisition 2 (3), 241–258.

Conway, M., 2011. Privacy and Security against Cyberterrorism. Communications of the ACM 54 (2), 26–28 February 2011, ACM.

Coopey, J., Burgoyne, J., 2000. Politics & Organizational Learning. Journal of Management Studies 37 (6), 869–885.

Coroner's Office, 2011. Coroner's Inquests into the London Bombings of 7 July 2005.

Correa, D., 2011. Solutions to Detect & Analyze Online Radicalization. IIITD PhD Comprehensive Report, vol. V, No. N, November 2011. http://www.isiconference.org/

Costa Jr., P.T., McCrae, R.R., 1992. Revised NEO Personality Inventory (NEO PI–R) & NEO Five-Factor Inventory (NEO FFI) professional manual. Psychological Assessment Resources, Odessa, FL.

Coughlan, J., Macredie, R.D., 2002. Effective communication in requirements elicitation: A comparison of methodologies. Requirements Engineering 7, 47–60.

Crabtree, A., Hemmings, T., Rodden, T., Cheverst, K., Clarke, K., Dewsbury, G., et al., 2003. Designing with care: Adapting cultural probes to inform design in sensitive settings. In: Proceedings of OzCHI 2003, pp. 4–13.

Croitoru, M., Liang, X., Dupplaw, D., Lewis, P., 2008. Expressive security policy rules using Layered Conceptual Graphs. (In special issue: The 27th SGAI International Conference on Artificial Intelligence) Knowledge Based Systems, pp. 209–216.

Crowe, A., 2010. The social media manifesto: A comprehensive review of the impact of social media on emergency management. Journal of Business Continuity & Emergency Planning 5.

Cummings, A., Lewellen, T., McIntire, D., Moore, A., Trzeciak, R., 2012. Insider Threat Study: Illicit Cyber Activity Involving Fraud in the U.S. Financial Services Sector. Published by CERT, Software Engineering Institute, Carnegie Mellon.

Cummings, A., Lewellen, T., McIntire, D., Moore, A., Trzeciak, R., 2012. Insider threat study: Illicit cyber activity involving fraud in the U.S. financial services sector. (CMU/SEI-2012-SR-004). Retrieved from http://www.sei.cmu.edu/library/abstracts/reports/12sr004.cfm (Last accessed 29.08.12.).

Cyert, R.M., March, J.G., 1963. A Behavioural Theory of the Firm. Prentice Hall, Upper Saddle River, NJ.

Dadashi, N., Stedmon, A., Pridmore, T., 2012. Semi-automated CCTV surveillance: The effects of system confidence, system accuracy & task complexity on operator vigilance, reliance and workload. Appl. Ergon., Special Issue: Detecting Terrorism, in press.

Dam, M., Jacobs, B., Lundblad, A., Piessens, F., 2009. Security Monitor Inlining for Multithreaded Java. ECOOP 2009— Object-Oriented Programming. Springer Verlag. LNCS 5653, 2009, 546–569.

Das, S., 2008. High-Level Data Fusion. Artech House, 2008.

Dastbaz, M., Cesta, A., 2011. Simulations for Crisis Management – PANDORA's Box. In: Akhgar, B., Yates, S. (Eds.), Intelligence Management Advanced Information and Knowledge Processing, 2011, Part 2, pp. 111–131.

Davis, A., 2000. Tracking Tigers in Phuket. A secret Tamil guerrilla base embarrasses Bangkok. Asiaweek 29 (23), June 16, 2000.

De Cerchio, R., Riley, C., 2011. Aircraft systems cyber security. 30th IEEE/AIAA Digital Avionics Systems Conference (DASC). Seattle, WA, pp. 1C3–1–1C3–7, 16–20 October.

De Moor, A., 2004. Improving the Testbed Development Process in Collaboraties. In: 12th International Conference on Conceptual Structures. Huntsville, Alabama. Springer, New York.

de Oude, P., Pavlin, G., Groot, J., 2010. Robust Bayesian detection: A case study. In: Proceedings of the 13th International Conference on Information Fusion (Fusion 2010). Edinburgh, UK, pp. 26–29.

de Oude, P., Pavlin, G., 2009. Efficient Distributed Bayesian Reasoning via Targeted Instantiation of Variables. In: IAT 2009. pp. 323–330.

Dean, J., Ghemawat, S., 2008. MapReduce: simplified data processing on large clusters. Communications of the ACM 51 (1), 107–113. December 2005. [online]. Retrieved from http://csis.org/files/media/csis/pubs/051214_china_titan_rain.pdf.

Deelman, E., Blythe, J., Gil, Y., Kesselman, C., Mehta, G., Vahi, K., et al., 2003. Mapping abstract complex workflows onto grid environments. J. Grid Comput. 1 (1), 25–39.

Dekker, S., Hansén, D., 2004. Learning under Pressure: The Effects of Politicization on Organizational Learning in Public Bureaucracies. Journal of Public Administration Research and Theory 14 (2), 211–230.

Del Carmen, M., Alonso, G., 2003. National-security export-quality restrictions in segmented & non-segmented markets. European Journal of Political Economy 19, 377–390.

Delugach, H.S., 2009. ISO/IEC 24707—Common Logic. [online]. Retrieved from http://metadata-standards.org/24707/index.html (Last accessed 09.08.12.).

Denning, D.E., 2000. Cyberterrorism: The Logic Bomb versus the Truck Bomb. Global Dialogue Volume 2 Number 4 Autumn 2000–Terrorism: Image and Reality.

Denning, D.E., 2012. Stuxnet: What Has Changed? Future Internet 4 (3), 672–687.

Deverell, E., Hansén, D., 2009. From Post-Crisis Fantasy Documents to Actual Learning in the Heat of Crisis. Journal of Contingencies and Crisis Management 17 (3), 143–145.

Deverell, E., 2010. Crisis-induced Learning in Public Sector Organizations. Elanders, Stockholm, Sweden.

DHS, 2011. Cyber Security Experiments & Pilots. [online]. Retrieved from http://www.cyber.st.dhs.gov/experimentsandpilots/.

DiGiovanni, C., 1999. Domestic terrorism with chemical or biological agents. Am. J. Psychiatry 156, 1500–1505.

Digman, L., 1990. Strategic Management: Concepts, Decisions, Cases, second ed. Irwin, Homewood, IL.

Dignan, L., 2011. DoD: 24,000 files swiped in March from military contractor systems, for Zero Day, July 14, 2011, [online]. Retrieved from http://www.zdnet.com/blog/security/dod-24000-files-swiped-in-march-from-military-contractor-systems/9026.

Dobbs, M., 2001. Anthrax scare spreads around the world. Washington Post. 18 October. A15.

Dragoni, N., Massacci, F., Naliuka, K., Siahaan, I., 2007. Security-by-Contract: Toward a Semantics for Digital Signatures on Mobile Code. Public Key Infrastructure. Springer Verlag LNCS 4582, 297–312.

Drug Enforcement Agency, 2011. Office of Diversion Control; Drug & Chemical Evaluation Section. 3, 4-Methylenedioxypyrovalerone (MDPV).[online]. Retrieved from www.deadiversion.usdoj.gov/drugs_concern/mdpv.pdf (Last accessed 25.04.12.).

Drury, J., Cocking, C., Reicher, S., 2009. The nature of collective resilience: Survivor reactions to the 2005 London Bombings. International Journal of Mass Emergencies and Disasters 27 (1), 66–95.

Duda, R.O., Hart, P.E., Stork, D.G., 2001. Pattern Classification, second ed. Wiley.

Dyer, G., Bozorgmehr, N., Blitz, J., 2012. Cyberattack Clouds US-Iran Nuclear Talks. Financial Times, 1st June 2012, [online]. Retrieved from www.ft.com/cms/s/0/08b8b06e-ac04-11e1-923a-00144feabdc0.html#ixzz207yv40FY (Last accessed August 2012).

Earle, T.C., 2004. Thinking Aloud About Trust: A Protocol Analysis of Trust in Risk Management. Risk Anal. 24 (1), 169–183.

Eason, K., 1987. Information Technology & Organizational Change. Taylor & Francis, London.

Easterby-Smith, M., Lyles, M.A., 2003. Introduction: Watersheds of Organizational Learning & Knowledge Management. In: Easterby-Smith, Lyles (Eds.), Handbook of Organizational Learning and Knowledge Management. pp. 1–15.

Eaton, C., DeRoos, D., Deutsch, T., Lapis, G., Zikopoulos, P., 2012. Understanding Big Data: IBM. McGraw-Hill, New York.

EC, 2010. The EU Counter-Terrorism Policy: main achievements & future challenges. European Commission, Brussels.

Edgar, T., Manz, D., Carroll, T., 2011. Towards an experimental testbed facility for cyber-physical security research. In: Sheldon, F.T., Abercrombie, R., Krings, A. (Eds.), Proceedings of the Seventh Annual Workshop on Cyber Security & Information Intelligence Research (CSIIRW '11). ACM, New York, NY, Article 53, 12–14 October.

Edwards, C., 2009. Resilient Nation. DEMOS.

Ellul, J., Merton, R.K., 1967. The Technological Society. Vintage, New York.

Ericson, R., Doyle, A., 2004. Catastrophe risk, insurance, and terrorism. Economy and Society 33, 135–173.

Eriksson, J., 2004. Kampen om hotbilden. Santérus Förlag, Stockholm, Sweden.

Erlingsson, U., Schneider, F.B., 2000. IRM enforcement of Java stack inspection, Security & Privacy, 2000. S&P 2000 Proceedings. 2000 IEEE Symposium on on, pp. 246–255.

European Commission, n.d. Communicating with the outside world— Guidelines for All Staff on the Use of Social Media. [online]. Retrieved from http://ec.europa.eu/ipg/docs/guidelines_social_media_en.pdf.

Falkenrath, R., 2001. Problems of preparedness: U.S. readiness for a domestic terrorist attack. International Security 25, 47–186.

Falliere, N., Murchu, L.O., Chien, E., 2011. W32. Stuxnet Dossier, Version 1.4 (February 2011) White paper, [online]. http://www.symantec.com/security_response/whitepapers.jsp (accessed July 2012).

Farwell, J.P., Rohozinski, R., 2011. Stuxnet & the Future of Cyber War Survival. Global Politics & Strategy 53 (1), 23–40.

Fasolo, B., Ni, Z., Phillips, L.D., 2008. A Study Of The Impact Of The July Bombings On Londoners' Travel Behaviour. Non-published Research Reports. Paper 17. [online]. Retrieved from http://research.create.usc.edu/cgi/viewcontent.cgi?article=1058&context=nonpublished_reports (Last accessed 01.09.12.).

FBI, 2009. Arlington Security Guard Arrested on Federal Charges for Hacking into Hospital's Computer System. FBI Press release, 30 June. [online]. Retrieved from http://www.fbi.gov/dallas/press-releases/2009/dl063009.htm.

Fijnaut, C., 2001. Crisis and Reform in Belgium: The Dutroux Affair and the Criminal Justice System. In: Rosenthal, U., Boin, R.A., Comfort, L.K. (Eds.), Managing Crises: Threats, Dilemmas, Opportunities. Charles C Thomas, Springfield, IL.

Fiore, F., Francois, J., 2010. "Cyberterrorism prevention checklist". [online]. Retrieved from http://ptgmedia.pearsoncmg.com/images/art_fiore-francois1_doz/elementLinks/dozenlst.pdf (Last accessed December 2011).

Fischer, F., 1990. Technocracy & the politics of expertise. Sage Publications, Newbury Park, CA.

Fischer, L., 1997. A wasted life: The case of Roderick Ramsay. Security Awareness Bulletin. Department of Defense Security Institute, Richmond, VA.

Fischhoff, B., Bruine de Bruin, W., Perrin, W., Downs, J.S., 2004. Travel risks in a time of terror: Judgments and choices. Risk Anal. 24, 1299–1307.

Flin, R., 2001. Decision Making in Crisis: The Piper Alpha Disaster. In: Rosenthal, U., Boin, R.A., Comfort, L.K. (Eds.), Managing Crises: Threats, Dilemmas, Opportunities. Charles C Thomas, Springfield, IL.

Floyd, D., Prentice-Dunn, S., Rogers, R., 2000. A meta-analysis of research on protection motivation theory. Journal of Applied Social Psychology 30, 407–429.

Flynn, J., Slovic, P., Kunreuther, H., 2001. Risk media & stigma: Understanding public challenges to new technologies. Earthscan, London.

Foley, S.N., Quillinan, T.B., O'Connor, M., Mulcahy, B.P., Morrison, J.P., 2004. A Framework for Heterogeneous Middleware Security. In: Proceedings of the 13th International Heterogeneous Computing Workshop, (IPDPS 2004). Santa Fe, NM, USA. April 2004.

Former U.S. Navy Sailor, 2009. Former U.S. Navy Sailor Sentenced in Terror Case, April 3, 2009, [online]. Retrieved from http://www.adl.org/main_Terrorism/abujihaad_sentenced.htm (Last accessed 08.09.12.).

Forsythe, S., Pette, T., Kim, J.O., 2005. Modeling Consumer Behavior in On-line Environments. [online]. Retrieved from . www.ntcresearch.org: www.ntcresearch.org/pdf-rpts/AnRp02/S02-AC23-A2.pdf.

Foster, I., Kesselman, C., 1997. Globus: A Metacomputing Infrastructure Toolkit. The International Journal of Supercomputer Applications and High Performance Computing 11 (2), 115–128, Summer.

Fox, S., 2000. Communities of Practice, Foucault & Actor-Network Theory. Journal of Management Studies 37 (6), 853–867.

Framingham, M.A., 2007. Overconfidence Is Pervasive Amongst Security Professionals E-Crime Watch SurveyTM conducted by the United States Secret Service, the CERT® Coordination Center (CERT/CC) Microsoft, & CSO Magazine.

Franz, D., Jahrling, P., Friedl&er, A., McCain, C., Hoover, D., Bryne, W., 1997. Clinical recognition & managements of patients exposed to biological warfare agents. JAMA 278, 399–411.

Freeman, L.A., Jessup, L.M., 2004. The power & benefits of concept mapping: Measuring use, usefulness, ease of use, & satisfaction. International Journal of Science Education 26 (2), 151–169.

Freudenburg, W., 2003. Institutional failure & the organisational amplification of risk. In: Pidgeon, N., Kasperson, R., Slovic, P. (Eds.), The Social Amplification of Risk. Cambridge University Press, Cambridge, UK.

Friedman, V.J., Berthoin Antal, A., 2005. Negotiating reality—A theory of action approach to intercultural competence. Management Learning 6 (1), 23–30.

Friedrich, W.R., van der Poll, J.A., 2007. Towards a methodology to elicit tacit domain knowledge from users. Interdisciplinary Journal of Information, Knowledge & Management 2, 179–193.

Fukuyama, F., 1995. Trust: The social virtues & the creation of prosperity. Free Press, New York.

Fullerton, C., Ursano, R., Norwood, A., Holloway, H., 2003. Trauma, terrorism, & disaster. In: Ursano, R., Fullerton, C., Norwood, A. (Eds.), Terrorism & disaster: Individual & community mental health interventions. Cambridge University Press, Cambridge, UK, pp. 1–21.

Funabashi, Y., Kitazawa, K., 2012. Fukushima in review: A complex disaster, a disastrous response. Bulletin of the Atomic Scientists 68 (2), 9–21.

Ganter, B., Wille, R., 1998. Formal Concept Analysis: Mathematical Foundations. Springer-Verlag, Berlin.

Gantz, J.F., et al., 2008. The Diverse & Exploding Digital Universe. IDC Whitepaper.

Garrison, L., Grand, M., 2001. Cyberterrorism: An evolving concept. In: NIPC Highlights.

Gaver, B., Dunne, T., Pacenti, E., 1999. Design: Cultural probes. Interaction 6 (1), 21–29.

Geason, S., Wilson, P.R., 1988. Crime Prevention Theory & Practice. Crimes Prevention Series. Australian Institute of Criminology, Canberra.

Geeraerts, S., 2012. Digital radicalization of youth. Social Cosmos 3 (1), 25–32.

Geerts, G., McCarthy, W., 1991. Database Accounting Systems. In: Williams, B., Sproul, B.J. (Eds.), Information Technology Perspective in Accounting: an integrated approach. Chapman and Hall Publishers, London, UK, pp. 159–183.

Gelles, M.G., Brant, D.L., Geffert, B., 2008. Building a secure workforce: Guard against insider threat. [online]. Retrieved from http://www.deloitte.com/view/en_US/us/Industries/US-federal-government/764ef33b4010e110VgnVCM100000ba42f00aRCRD.htm#.

Gelles, M.G., Mahoutchian, T., Deloitte Consulting, 2012. Mitigating the Insider Threat Building a Secure Workforce. [online] retrieved from http://csrc.nist.gov/organizations/fissea/2012-conference/presentations/fissea-conference-2012_mahoutchian-and-gelles.pdf (last accessed 13.12.12.).

Gentry, C., 2009. Fully homomorphic encryption using ideal lattices. In: Proc. of the 41st annual ACM symposium on Theory of computing (STOC '09). ACM, New York, NY, pp. 169–178.

George, A.L., 1999. Knowledge for Statecraft—Lecture Given by the Winner of the Johan Skytte Prize in Political Science, Uppsala, Sweden, October 3, 1998. Scandinavian Political Studies 22 (2), 89–97.

George, A.L., 1997. From Groupthink to Contextual Analysis of Policy-making Groups. In: 't Hart, Stern, Sundelius (Eds.), Beyond Group Think, Political Group Dynamics and Foreign Policy Making. University of Michigan Press, Ann Arbor, pp. 35–53.

Get Safe Online, 2011. UK Internet Security: State of the Nation Report, November 2011 [online]. Retrieved from http://www.getsafeonline.org/media/GSOL_2011_Annual_Report.pdf (Last accessed 28.07.12.).

Gigerenzer, G., 2004. Dread risk, September 11, and fatal traffic accidents. Psychol. Sci. 15 (4), 286–287.

Gigerenzer, G., 2006. Out of the frying pan into the fire: Behavioural reactions to terrorist attacks. Risk Anal. 26 (2), 347–351.

Glass, K., Colbaugh, R., 2012. Estimating the sentiment of social media content for security informatics applications. Security Informatics 1–16.

Glass, T., Schoch-Spana, M., 2002. Bioterrorism & the People: How to Vaccinate a City against Panic. Clin. Infect. Dis. 34, 217–223.

Gleick, P.H., 2006. Water & Terrorism. Water Policy 8, 481–503.

Glik, D., 2007. Risk communication for public health emergencies. Annu. Rev. Public Health 28, 33–54.

Glik, D., Harrison, K., Davoudi, M., Riopelle, D., 2004. Public perceptions & risk communications for botulism. Biosecur. Bioterror. 2, 216–223.

Glucksmann, A., 2012 interviewed by the German weekly magazine Der Spiegel in Philosopher André Glucksmann: A Dark Vision of the Future of Europe, Der Spiegel Online, 23 August 2012, [online]. Retrieved from http://www.spiegel.de/international/europe/philosopher-andre-glucksmann-a-dark-vision-of-the-future-of-europe-a-851266.html (Last accessed 23.08.12.).

Goh, S.C., 2002. Managing effective knowledge transfer: an integrative framework & some practical implications. Journal of Knowledge Management 6 (1), 23–30.

Goldberg, I., Wagner, D., Thomas, R., Brewer, E., 1996. A Secure Environment for Untrusted Helper Applications (Confining the Wily Hacker). In: Proceedings of the Sixth USENIX UNIX Security Symposium. St. Petersburg, Russia.

Good Communication Awards, 2011. http://www.communicator.gcawards.co.uk/

Gordon, J., 2007. The Mobile Phone & the Public Sphere: Mobile Phone Usage in Three Critical Situations, Convergence. The International Journal of Research into New Media Technologies Sage Publications 13 (3), 307–319.

Gorman, S., 2009. Electricity Grid in U.S. Penetrated By Spies. The Wall Street Journal, Technology, 8th April. [online]. Retrieved from http://online.wsj.com/article/SB123914805204099085.html.

Gorman, S., Dreazen, Y., Cole, A., 2009. Insurgents Hack U.S. Drones. The Wall Street Journal, 17 December. [online]. Retrieved from http://online.wsj.com/article/SB126102247889095011.html.

Gorra, A., 2007. An analysis of the relationship between individuals' perceptions of privacy and mobile phone location data - a grounded theory study. Ph.D. thesis. Leeds Metropolitan University. Available from:http://www.leedsmet.ac.uk/inn/alic/thesis.htm.

Gostev, A., 2012. The Flame: Questions & Answers, May 2012 [online]. Retrieved from https://www.securelist.com/en/blog/208193522/The_Flame_Questions_&_Answers (Last accessed August 2012).

Gourlay, S., 2006. Conceptualizing knowledge creation: critiquing Nonaka's theory. Journal of Management Studies 43 (7), 1415–1436.

Government Office for Science (GOS), 2009. A Practical Guide to Public Risk. GSO, Washington, D.C.

Government Office for Science (GOS), 2011. The Blackett Review of High Impact Low Probability Risks. [online]. Retrieved from http://www.bis.gov.uk/assets/goscience/docs/b/12-519-blackett-review-high-impact-low-probability-risks (Last accessed 03.06.12.).

Graham, S., Wood, D., 2003. Digitizing surveillance: categorization, space and inequality. Critical Social Policy 20 (2), 227–248.

Gray, G.M., Ropeik, D.P., 2002. Dealing with the dangers of fear: The role of risk communication. Health Aff. 21 (6), 106–114.

Greater London Authority, 2006. Report of the 7 July Review Committee. [online]. Retrieved from http://www.london.gov.uk/who-runs-london/the-london-assembly/publications/safety-policing/report-7-july-review-committee.

Greenemeier, L., Choi, C., 2010. WikiLeaks breach highlights insider security threat. 1st December 2010. Sci. Am. [online]. Retrieved from http://www.scientificamerican.com/article.cfm?id=wikileaks-insider-threat.

Greengard, S., 2010. The New Face of War. Commun. ACM 53 (12), 20–22, December 2010, ACM.

Grönvall, J., 2000. Managing Crisis in the European Union: The Commission & "Mad Cow Disease". Överstyrelsen för civil beredskap, Stockholm, Sweden, Crisis Management Europe Research Program 10.

Groumpos, P.P., 2010. Fuzzy Cognitive Maps: Basic Theories & Their Application to Complex Systems. In: Dr. Glykas, M. (Ed.), Fuzzy Cognitive Maps Advances in Theory, Methodologies, Tools & Applications. Springer-Verlag, Berlin, Heidelberg, pp. 1–22.

Gruber, T., 1993. A Translation Approach to Portfolio Ontology Specification. Knowledge Acquisition: special issue - Current Issues in Knowledge Modelling 5 (2), 199–220.

Guangrong, R., 1998. The Negative Impact of the Internet & Its Solutions. The Chinese Defense Science & Technology Information Monthly 5 (121).

Guingona, T.M., 2002. Opening Remarks on the Occasion of the First ASEAN-EU Experts' Group Meeting on Maritime Security, 25-26 February 2002. Westin Philippine Plaza, Manila. [online]. Retrieved from http://www.dfa.gov.ph/moac/newsletter-main.htm (Last accessed 25.08.12.).

Gungor, V.C., Hancke, G.P., 2009. Industrial Wireless Sensor Networks: Challenges, Design Principles,

& Technical Approaches. IEEE Transactions on Industrial Electronics 56 (10), 4258–4265.

Gusfield, J., 1979. 'Buddy, Can You Paradigm?' The Crisis of Theory in the Welfare State. Pacific Sociological Review 22 (1), 3–22.

Hall, A., 2011. Media rushed to judgment in Norway attacks. 26th July 2011 ABC Sidney. [online]. Retrieved from http://www.abc.net.au/news/2011-07-25/media-rushed-judgment-in-norway-under-fire/2809786.

Halperin, D., Heydt-Benjamin, T.S., Fu, K., Kohno, T., Maisel, W.H., 2008. Security & Privacy of Implantable Medical Devices. IEEE Pervasive Computing 7 (1), 30–39.

Halpin, E.F., Fisher, S.M., 1998. The Use of the Internet for the European Parliament's Activities for the Promotion and Protection of Human Rights. European Parliament, Luxembourg.

Hammes, T.X., 2007. Fourth Generation Warfare Evolves, Fifth Emerges. Military Review, May-June 2007, 14. [online]. Retrieved from http://www.army.mil/professionalWriting/volumes/volume5/july_2007/7_07_1.html (Last accessed 15.08.12.).

Hancock, P.A., Hart, S.G., 2002. Defeating terrorism: what can human factors/ergonomics offer? Ergonomics in Design 10, 6–16.

Hari, B., Crossan, M., 2004. From Questions to Answers: Reviewing Organizational Learning Research. Management Learning 35 (4), 397–417.

Hellweg, E., 2005. Technology Responds to Hurricane Katrina. Technology Review, MIT , September 2nd, [online]. Retrieved from: http://www.technologyreview.com/computing/14780/.

Henderson, J.N., Henderson, J.C., Raskob, G.E., Boatright, D.T., 2004. Chemical (VX) terrorist threat: Public knowledge, attitudes & responses. Biosecur. Bioterror. 2 (3), 1–5.

Hendler, J., Goldbeck, J., 2008. Metcalfe's law, web 2.0 and the Semantic Web. J. Web. Sem 6 (1), 14–20.

Herbig, K.L., Wiskoff, M.F., 2002. Espionage against the United States by American citizens 1947–2001. (PERSEREC Technical Report, 02-5).

Hermann, C.F., 1963. Some Consequences of Crises Which Limit the Viability of Organizations. Adm. Sci. Q. 8 (1), 61–82.

Hermann, C.F. (Ed.), 1972a. International Crises: Insights from Behavioural Research. The Free Press, New York.

Hermann, C.F., 1972b. Some Issues in the Study of International Crisis. In: Hermann, C.F. (Ed.), (1972a) International Crises: Insights from Behavioural Research. The Free Press, New York.

Hess, M., Adams, D., 2002. Knowing & Skilling in Contemporary Public Administration. Australian Journal of Public Administration 61 (4), 68–79.

Heuer Jr., R.J., 1999. Psychology of Intelligence Analysis. Center for the Study of Intelligence. Central Intelligence Agency, Washington, D.C.

Hill, R., 2012. A Requirements Elicitation Framework for Agent-Orientated Software Engineering. Lambert, Saarbrucken.

Hiltz, S.R., Turoff, M., 1978. The Network Nation. Addison-Wesley, Reading, MA.

Hine, D., 2010. The 2009 Influenza Pandemic. An independent review of the UK response to the 2009 influenza pandemic. July 2010. [online]. Retrieved from http://www.dhsspsni.gov.uk/the2009influenzap&emic_acc.pdf (Last accessed 30.08.12.).

HM Government, 2011. A Strong Britain in an Age of Uncertainty: The National Security Strategy. [online]. Retrieved from http://www.official-documents.gov.uk/.

HMIC, 2011. The rules of engagement: A review of the August 2011 disorders. Retrieved from http://www.hmic.gov.uk/publication/rules-engagement-review-august/ (Last accessed 03.09.12.).

Hofer, C., Schendel, D., 1984/1978. Strategy Formulation: Analytical concepts. West Publishing St. Paul, MN.

Holloway, H., Norwood, A., Fullerton, C., Engel, C., Ursano, R., 1997. The Threat of Biological Weapons: Prophylaxis & Mitigation of Psychological & Social Consequences. Journal of American Medical Associations 278, 425–427.

Holmes, J.R., Yoshihara, T., 2008. Chinese Naval Strategy in the 21st Century. Routledge, New York.

Holmqvist, M., 2003. A dynamic model of intra- & interorganizational learning. Organization Studies 24 (1), 95–123.

Holsti, O.R., 1972. Crisis, escalation, war. McGill-Queens University Press, London, England.

Home Office, 2012. Coroner's Inquests into the London Bombings of 7 July 2005. Review of Progress. [online]. Retrieved from http://www.homeoffice.gov.uk/publications/counter-terrorism/inquest-7-7-progress-report?view=Binary (Last accessed 30.08.12.).

Home Office, 2011. CONTEST: The United Kingdom's Strategy for Countering Terrorism, July 2011. [online]. Retrieved from http://www.homeoffice.gov.uk/counter-terrorism/uk-counter-terrorism-strat (Last accessed September 2012).

Hooper, R.H., 2000. Is British police uniform adequate for the task? Proceedings of the Human Factors & Ergonomics Society Annual Meeting 44 (33), 6-125–6-128 July 2000.

Hopple, G.W., Kuhlman, J.A. (Eds.), 1981. Expert-Generated Data—Applications in International Affairs. Westview Press, Boulder, CO.

Hsinchun, C., Thoms, S., Tianjun, F., 2008. Cyber extremism in Web 2.0: An exploratory study of international Jihadist groups, Intelligence and Security Informatics. In: ISI 2008. IEEE International Conference. June 20, 2008. pp. 98–103.

Huber, G.P., 1991. Organizational Learning: The Contributing Process & the Literature. Organizational Science 2 (1), 88–115.

Huijsen, W.O., 1998. Controlled Language—An Introduction. In: Proceedings of the Second International Workshop on Controlled Language Applications (CLAW98). Language Technologies Institute, Carnegie Mellon University, Pittsburgh, PA, May, pp. 1–15.

Hunton, P., 2009. The growing phenomenon of crime & the internet: a cybercrime execution & analysis model. Computer Law & Security Review 25, 528–535.

Hussey, D., 1998. Strategic management: from theory to implementation, fourth ed. BH publication.

Huysman, M., 2000. An Organizational Learning Approach to the Learning Organization. European Journal of Work & Organizational Psychology 9 (2), 133–145.

Hyams, K., Murphy, F., Wessely, S., 2002. Responding to Chemical. Biological or Nuclear Terrorism: The Indirect & Long-term Health Effects May Present the Greatest Challenge. J. Health Polit. Policy Law 27, 273–291.

IAEA, 1998. The radiological accident in Goiania. IAEA, Vienna.

ICT4Peace, 2011. Peacebuilding in the Information Age— Sifting hype from reality. ICT4Peace Foundation, January. [online]. Retrieved from http://ict4peace.org/updates/peacebuilding-in-the-information-age-sifting-hype-from-reality.

InfoSecurity, 2010. SCADA systems can be located via public search engine says CERT, 03 November 2010. [online]. http://www.infosecurity-magazine.com/view/13690/scada-systems-can-be-located-via-public-search-engine-says-cert (accessed July 2012).

Initiative OpenNet, 2004. A Starting Point: Legal Implications of Internet Filtering. OpenNet Initiative. [online]. Retrieved from http://opennet.net/docs/Legal_Implications.pdf.

Institute for Strategic Dialogue, I. f., 2011. Radicalisation: The Role of the Internet. A working paper of the PPN. Institute for Strategic Dialogue. [online]. Retrieved from http://www.strategicdialogue.org/StockholmPPN2011_BackgroundPaper_FOR%20WEBSITE.pdf.

International Atomic Energy Agency (IAEA), 1998. The radiological accident in Goiania. IEAA, Vienna.

Introna, L., Wood, D., 2004. Picturing Algorithmic Surveillance: The Politics of Facial Recognition Systems. Surveillance & Society CCTV Special (eds. Norris, McCahill and Wood) 2 (2/3), 177–198 .

Investment Detection, Responses and Recovery Technology (IDRRT) Working Group, 2007. Systems security for the 21st century. Frequently Asked Questions on Incidents and Spills. [online]. Retrieved from http://www.cnss.gov/Assets/pdf/CNSS-079-07.pdf.

ISF, 2011. Information Security Governance: Raising the Game. Informational document ISF 11 ISG (Marketing) September 2011. Information Security Forum Limited.

ISF, 2012a. Providing good practice for information security. [online]. Retrieved from Information Security Forum: www.securityforum.org (Last accessed 11.07.12.).

ISF, 2012b. Information Risk Analysis Methodology. [online]. Retrieved from www.securityforum.org/IRAM (Last accessed 11.07.12.).

ISO2702, 2005. Information Technology—Security Techniques—Information Security Management Systems - Requirements. British Standards Institution, London.

Jahankhani, H., Al- Nemrat, A., 2010. Cybercrime victimisations/criminalisation & Punishment. In: 6th International Conference on Global Security, Safety & Sustainability, Communications in Computer & Information Science, CCIS, Springer Proceedings of the 6th ICGS3–10, 1–3 September 10. Springer, Braga, Portugal.

Jahankhani, H., Al-Nemrat, A., 2011. Cybercrime Profiling & Trend Analysis. In: Akhgar, B., Yates, S. (Eds.), Intelligence Management, Knowledge Driven Frameworks for Combating Terrorism & Organised Crime. Springer, pp. 181–195.

Janis, I.L., 1982. Groupthink: Psychological Studies of Policy Decisions & Fiascos, second ed. Houghton Mifflin, Boston, MA.

Jenkins, B., 1975. International Terrorism: A New Mode of Conflict. In: Carlton, D., Schaerf, C. (Eds.), International Terrorism & World Security. Croom Helm, London.

Jentleson, B.W., 2002. The Need for Praxis—Bringing Policy Relevance Back In. International Security 26 (4), 169–183.

Jia, X., 2004. Building the Sea Base for Future Expeditionary Wars. WISE Journal, August 2004, 19. [online]. Retrieved from http://www.wise-intern.org/journal/2004/WISE2004-JiaXuFinalPaper.pdf (Last accessed 22.08.12.).

Johansson, G., 1973. Visual perception of biological motion and a model for its analysis. Percept. Psychophys. 14 (2), 201–211.

Johnson, G., Scholes, K., 1993. Exploring corporate strategy, third ed. Prentice Hall, pp. 17–28; 115–129.

Jones, J., Averbeck, R., 2011. The 3 types of insider threat, 12th May 2011 CSO Security and Risk. [online]. Retrieved from http://www.csoonline.com/article/682205/the-3-types-of-insider-threat?page=1 (Last accessed August 2012).

Judea, P., 1988. Probabilistic Reasoning in Intelligent Systems: Networks of Plausible Inference. Representation & Reasoning Series, second ed. Morgan Kaufmann, San Francisco, CA.

Kagal, L., Finin, T., Joshi, A., 2003. A Policy Based Approach to Security for the Semantic Web. In: Proceedings of the 2nd International Semantic Web Conference (ISWC2003). Springer-Verlag, Florida, US, October 2003.

Kakabadse, N.K., Kakabadse, A., Kouzmin, A., 2003. Reviewing the knowledge management literature: towards a taxonomy. Journal of Knowledge Management 7 (4), 75–91.

Kallberg, J., Lowther, A.B., 2012. The return of Dr. Strangelove. The Diplomat. Aug 20, 2012. [online]. Retrieved from http://kallberg.blogs.com/jan_kallberg/2012/09/the-return-of-dr-strangelove-the-diplomat.html.

Kallberg, J., Thuraisingham, B., 2012. Towards cyber operations -The new role of academic cyber security research & education. In: 2012 IEEE International Conference on Intelligence & Security Informatics (ISI). Washington D.C., June 11–14, 2012.

Kallberg, J., 2012a. The common criteria meets realpolitik: Trust, alliances, & potential betrayal. IEEE Security & Privacy 10 (4), 50–53

Kallberg, J., 2012b. Designer satellite collisions from covert cyber war. Strategic Studies Quarterly, Spring 2012, p. 124.

Kan, P.R., 2009. Drugs & contemporary warfare. Potomac Books, Dulles, VA.

Kaplan, R., 2011. Monsoon: The Indian Ocean & the Future of American Power. Random House Trade Paperbacks, New York, 2011, p. 328.

Kasperson, R.E., Palmlund, I., 2005. Evaluating Risk Communication. In: Kasperson, J.X., Kasperson, R.E. (Eds.), The Social Contours of Risk. Volume 1: Publics, Risk Communication & the Social Amplification of Risk. Earthscan, London, pp. 51–67.

Kasperson, R., Golding, D., Tuler, S., 1992. Social distrust as a factor in siting hazardous facilities & communicating risk. Journal of Social Issues 48, 161–187.

Kavakli, E., Kalloniatis, C., Gritzalis, S., 2005. Addressing privacy: Matching user requirements to implementation techniques. In: 7th Hellenic European Research on Computer Mathematics & its Applications Conference (HERCMA 2005) Athens, Greece, September 22–24, 2005.

Keeney, M.M., Kowalski, E.F., Cappelli, D.M., Moore, A.P., Shimeall, T.J., Rogers, S.N., 2005. Insider Threat Study: Computer System Sabotage in Critical Infrastructure Sectors, Joint SEI & U.S. Secret Service Report, Section 2, 11 May 2005. [online]. http://www.cert.org/archive/pdf/insidercross051105.pdf (accessed August 2012).

Kelly, W., 1987. We have met the enemy & he is us. Simon & Schuster, New York.

Khakee, A., Barbarente, A., Borri, D., 2000. Expert & experimental knowledge in planning. J. Oper. Res. Soc. 51 (7), 776–788.

Kiesler, S., Kraut, R., Cummings, J., Boneva, B., Helgeson, V., Crawford, A., 2002. Internet Evolution & Social Impact. IT & Society 1(1) Summer, 120–134.

Kim, W., Jeong, O.-.R., Kim, C., So, J., 2009. The dark side of the Internet: Attacks, costs & responses. Special Issue of WISE 2009 - Web Information Systems Engineering, Information Systems 36(3), 675–705 May 2011.

Kirwan, B., Ainsworth, L.K., 1992. A Guide to Task Analysis. CRC Press, Boca Raton, FL.

Knight, C., 2002. Essential Elements Missing in the National Security Strategy of 2002. PDA Commentary adapted from a presentation at the Center for International Relations Boston University. October 2002.

Koraeus, M., 2007. Knowledge in Crisis—Learning from a Parallel Management Field. Master's Thesis. Uppsala University, Department of Government.

Koraeus, M., 2008. Who Knows?—The Use of Knowledge Management in Crisis. Crisis Management Europe Research Program, vol. 36. Swedish National Defense College & CRISMART, Stockholm, Sweden.

Koscher, K., Czeskis, A., Roesner, F., Patel, S., Kohno, T., Checkoway, S., McCoy, D., Kantor, B., Anderson, D., Shacham, H., Savage, S., 2010. Experimental Security Analysis of a Modern Automobile. In: Evans, D., Vigna, G. (Eds.), Proceedings of the IEEE Symposium on Security & Privacy. IEEE Computer Society, Oakland, CA, pp. 447–462, May 2010.

Kouzmin, A., Jarman, A.M.G., 2004. Policy Advice in Crisis: A Political Redefinition of Crisis Management. Int. Stud. Rev. 6, 182–189.

Kraemer, S., Carayon, P., Sanquist, T.F., 2009. Human & organisational factors in security screening & inspection systems: Conceptual framework & key research needs. Cogn. Tech. Work 11 (1), 29–41.

Krajca, P., Outrata, J., Vychodil, V., 2008. Parallel Recursive Algorithm for FCA. In: Proceedings of the Sixth International Conference on Concept Lattices & their Applications. Palacky University, Olomouc, Czech Republic, pp. 71–82.

Krekel, B., Adams, P., Bakos, G., 2012. Occupying the Information High Ground: Chinese Capabilities for Computer Network Operations & Cyber Espionage, March 2012, [online]. Retrieved from http://www.uscc.gov/RFP/2012/USCC%20Report_Chinese_CapabilitiesforComputer_NetworkOperationsandCyberEspionage.pdf (Last accessed August 2012).

Krieger, K., Rogers, M., 2012. Systematic review of existing research - Public perceptions & risk communication in times of crisis. In: PRACTICE (Ed.), Work Package 8. King's College London, London.

Krieger, K., 2012. Norms, structures, procedures & variety in risk-based governance: the case of flood management in Germany & England. Jerusalem Papers in Regulation and Governance 47.

Kriikku, P., Wilhelm, L., Schwarz, O., et al., 2011. New designer drug of abuse: 3,4-Methylenedioxypyrovalerone (MDPV). Findings from apprehended drivers in Finland. Forensic Sci. Int. 210 (1–3), pp. 195–200.

Kriwaczek, P., 2012. Babylon. Mesopotamia & the Birth of Civilization. Atlantic Books (paperback ed.), London, p. 93.

Kshetri, N., 2005. Information & communications technologies, strategic asymmetry & national security. Journal of International Management 11, 563–580.

Kumari Rajani, M., Chandio, M.S., 2004. Use of Internet & its effects on our Society. In: National Conference on Emerging Technologies 2004 1. IEEE, pp. 157–161.

Kyung, S.C., 2008. Computer Crime Victimisation & Integrated Theory: An Empirical Assessments. International Journal of Cyber Criminology (IJCC) 2 (1), 308–333.

L'Huillier, G., Alvarez, H., Rios, A., Aguilera, F., 2010. Topic-Based Social Network Analysis for Virtual Communities of Interests in the Dark Web. SIGKDD Explorations 12 (2), 66–73.

Lanfranchi, V., Ireson, N., 2009. User requirements for a collective intelligence emergency response system. In: Proceedings of the 23rd British HCI Group Annual Conference on People & Computers: Celebrating People & Technology. pp. 198–203.

Langner, R., 2011. Cracking Stuxnet, a 21st-century cyber weapon. [online]. Retrieved from http://www.ted.com/talks/ralph_langner_cracking_stuxnet_a_21st_century_cyberweapon.html (Last accessed Feb 2012).

Lannon, J., Halpin, E., 2012. Human Rights and Information Communication Technologies: Trends and Consequences of Use. IGI Global.

Laroch, M., 2009. New development in modeling Internet consumer behavior: Introduction to the special issue. Journal of Business Research 63 (9–10), 915–918.

Launders, I., Polovina, S., 2010. Semantics & Pragmatics in Enterprise Architecture through Transaction Agent Modelling. In: Proceedings of the 12th International Conference on Informatics & Semiotics in Organizations. SciTePress, Reading, UK, pp. 285–291.

Launders, I., 2011. Evaluating the Transaction Graph through a Financial Trading Case Study. In: 19th International Conference on Conceptual Structures ICCS 2011. Springer, Derby. pp. 145–158.

Launders, I., 2012. The Transaction Graph: Requirements capture in Semantic Enterprise Architectures. Lambert, Sarbrucken.

Lawson, G., D'Cruz, M., 2011. Ergonomics methods & the digital factory. In: Canetta, L., Redaelli, C., Flores, M. (Eds.), Intelligent Manufacturing System DiFac. Springer, London.

Lawson, G., Sharples, S., Cobb, S., Clarke, D., 2009. Predicting the human response to an emergency. In: Bust, P.D. (Ed.), Contemporary Ergonomics 2009. Taylor & Francis, London, pp. 525–532.

Lee, D., 2012. Flame: Attackers 'sought confidential Iran data', 4th June 2012 BBC News Technology. [online]. Retrieved fromhttp://www.bbc.co.uk/news/technology-18324234 (Last accessed August 2012).

Lemyre, L., Clement, M., Corneil, W., Craig, L., Boutette, P., Tyshenko, M., et al., 2005. A psycho-social risk assessment & management framework to enhance response to CBRN terrorism threats & attacks. Biosecur. Bioterror. 3 (4), 1–15.

Lemyre, L., Corneil, W., Johnson, C., Boutette, P., 2010. Psychosocial considerations about children & radiological events. Radiat. Prot. Dosimetry 142, 17–23.

Lenselink, L., 2011. Radicalization Online. Patterns of Social Interaction on the Al-Faloja and As-Ansar Forums. Master's Thesis. Retrieved from http://www.libsearch.com: http://igitur-archive.library.uu.nl/student-theses/2011-0824-202430/UUindex.html.

Leveson, N.G., Turner, C.S., 1993. An investigation of the Therac-25 accidents. IEEE Computer 26 (7), 18–41, July.

Levy, D., 1994. Chaos Theory and strategy: Theory, application, and managerial implications. Strategic Management Journal 15, Special Issue: Strategy: Search for New Paradigms (Summer, 1994), 167–178.

Lewin, K., 1951. Field Theory in Social Science. Harper & Row, New York.

Lewis, J.A., 2005. Computer Espionage, Titan Rain, & China. Center for Strategic and International Studies - Technology and Public Policy Program.

Liechti, O., Sumi, Y., 2002. Editorial: Awareness & the WWW. International Journal of Human-Computer Studies 56, 1–5.

Lind, W.S., 2003. Cultures in Conflict: The Four Generations of Modern War, 21st April 2003, Counter Punch. [online]. Retrieved from http://www.counterpunch.org/lind04232003.html (Last accessed 15.08.12.).

Lind, W.S., 2004. Understanding Fourth Generation War. Military Review, September-October 2004, 13. [online]. Retrieved from http://www.au.af.mil/au/awc/awcgate/milreview/lind.pdf (Last accessed 21.08.12.).

Lindkvist, L., 2005. Knowledge Communities & Knowledge Collectives: A Typology of Knowledge Work in Groups. Journal of Management Studies 42 (6), 1189–1210.

Lindsay, B., 2011. Social Media & Disasters: Current Uses, Future Options & Policy Considerations: Analyst in American National Government, 2011. Congressional Research Service.

Liu, R., Qian, X., Mao, S., Zhu, S., 2011. Research on anti-money laundering based on core decision tree algorithm. In: Proceedings from Control and Decision Conference (CCDC), 2011 Chinese, vol., no., pp. 4322–4325, 23-25 May 2011.

Lofstedt, R., Bouder, F., Wardman, J., Chakraborty, S., 2011. The changing nature of communication & regulation of risk in Europe. Journal of Risk Research 14, 409–429.

Lohse, M., 2011. Bridging the gap between users' expectations & system evaluations. In: Proceedings from 20th IEEE International Symposium on Robot & Human Interactive Communication, 31 July–3, August 2011. IEEE RO-MAN IEEE Conference Publications, Atlanta, GA pp. 485–490.

López-Rousseau, A., 2005. Avoiding the death risk of avoiding a dread risk: The Aftermath of March 11 in Spain. Psychology Science 16 (6), 426–428.

Loukas, G., Oke, G., 2010. Protection against Denial of Service Attacks: A Survey. Computer Journal 53 (7), 1020–1037, BCS.

Luhmann, N., 1979. Trust & Power. Wiley, New York.

Luiijfa, E., Alib, M., Zielstrab, A., 2011. Assessing & improving SCADA security in the Dutch drinking water sector. International Journal of Critical Infrastructure Protection 4 (3–4), 124–134, December 2011.

Lumension, 2011. What Every CEO should know about IT Security. http://www.lumension.com/eBook—What-Every-CEO-Should-Know-About-IT-Security.aspx (last checked 28.7.12).

Lunden, W.J., 1962. The theory of crime prevention. British Journal of Criminology 2, 213–228.

Lynch, R., 1997. Corporate Strategy. Financial Times, Pitman Publishing, London.

Maddux, J., Rogers, R., 1983. Protection motivation theory & self-efficacy: a revised theory of fear appeals & attitude change. J. Exp. Soc. Psychol. 19, 469–479.

Maguire, M., Bevan, N., 2002. User requirements analysis. A review of supporting methods. In: Proceedings of IFIP 17th World Computer Congress, 25–30 August. Kluwer Academic Publishers, Montreal, Canada, pp. 133–148.

Mahony, E.H., 2010. European Court Blocks Extradition Of Terror Suspects To Connecticut, 8th July, 2010 Courant.com. [online]. Retrieved from http://articles.courant.com/2010-07-08/news/hc-terror-0709-20100708_1_human-rights-court-extradition-european-court.

Maiden, N., 2008. User requirements & system requirements. IEEE Software 25 (2), 90–91.

Mainz, 2009. Right-wing Extremism Online –Targeting teenagers with stylish websites. jugendschutz.net. [online]. Retrieved from http://www.jugendschutz.net/pdf/report_cyberhate_2008.pdf.

Mak, J.N., 2011. Pirates, Renegades, & Fishermen: The Politics of 'Sustainable' Piracy in the Strait of Malacca. In: Lehr, P. (Ed.), Violence at Sea. Piracy in the Age of Global Terrorism. Routledge 2011 (paperback ed.), New York, pp. 199–224.

Malik, H., Malik, A.S., 2011. Towards Identifying the Challenges Associated with Emerging Large Scale Social Networks. Procedia Computer Science 5, 458–465.

Mandel, D.R., 2010. Radicalization: What does it mean? In: Pick, T., Speckhard, A. (Eds.), Indigenous terrorism: Understanding and addressing the root causes of radicalisation among groups with an immigrant heritage in Europe. IOS Press, Amsterdam.

Manso, M., Manso, B., 2012. The Role of Social Media in Crisis: A European Holistic Approach to the Adoption of Online & Mobile Communications in Crisis Response & Search & Rescue Efforts. Paper presented at the 17th ICCRTS (International Communications & Control Research & Technology Symposium) Fairfax, VA, June 19–21, 2012.

Marx, G., 2002. What's New About the "New Surveillance"? Classifying for Change and Continuity. Surveillance & Society 1 (1), 9–29.

Masten, A.S., Osofsky, J.D., 2010. Disasters & Their Impact on Child Development: Introduction to the Special Section. Child Dev. 81, 1029–1039.

Matrosov, A., Rodionov, E., 2012. Flame, Duqu & Stuxnet: in-depth code analysis of mssecmgr.ocx, July 20, 2012. [online]. Retrieved from http://blog.eset.com/2012/07/20/flame-in-depth-code-analysis-of-mssecmgr-ocx#.UBvIFiQdovg.email (Last accessed July 2012).

Matrosov, A., Rodionov, E., Harley, D., 2011. Win32/Duqu: It's A Date, October 26, 2011, [online]. Retrieved from http://blog.eset.com/2011/10/25/win32duqu-it%E2%80%99s-a-date (Last accessed July 2012).

Matrosov, A., Rodionov, E., Harley, D., Malcho, E., 2012. Stuxnet Under the Microscope, Revision 1.31, [online]. Retrieved from http://go.eset.com/us/resources/white-papers/Stuxnet_Under_the_Microscope.pdf (Last accessed July 2012).

McArdle, S.C., Rosoff, H., John, R.S., 2012. The Dynamics of Evolving Beliefs Concerning Emotions, & Behavioural Avoidance Following 9/11: A Longitudinal Analysis of Representative Archival Samples. Risk Anal. 32 (4), 744–761.

McCarthy, W., 1979. An Entity-Relationship View of Accounting Models. The Conference Review 14 (4), 667–686.

McCurry, J., 2011. Japan: The aftermath. Lancet 377, 1061–1062.

McKinsey, et al., 2002. McKinsey Report - Emergency Medical Service Response. [online]. Retrieved from http://www.nyc.gov/html/fdny/pdf/mck_report/ems_response.pdf.

McNabb, D.E., 2007. Knowledge Management in the Public Sector—A Blueprint for Innovation in Government. M.E. Sharpe, Armonk, NY & London, England.

McNeese, M.C., Zaff, B.S., Citera, M., Brown, C.E., Whitaker, R., 1995. AKADAM: Eliciting user knowledge to support participatory ergonomics. International Journal of Industrial Ergonomics 15, 345–363.

Memon, N., Larsen, H.L., 2006. Investigative Data Mining Toolkit: A Software Prototype for Visualizing, Analyzing and Destabilizing Terrorist Networks. In: Visualising Network Information. Meeting Proceedings RTO-MP-IST-063, Paper 14. RTO, Neuilly-sur-Seine, France, pp. 14-1–14-24.

Meredith, L., Eisenmann, D., Rhodes, H., Ryan, G., Long, A., 2007. Trust influences response to public health messages during a bioterrorist event. Journal of Health Risk Communication 12, 217–232.

Miller, E.L., Long, D.E.E., Freeman, W.E., Reed, B.C., 2002. Strong security for network-attached storage. In: Proceedings of the 1st USENIX conference on File & storage technologies (FAST'02). USENIX Association, Berkeley, CA.

Milne, S., Sheeran, P., Orbell, S., 2000. Prediction & intervention in health-related behaviour: a meta-analytic review of protection motivation theory. Journal of Applied Social Psychology 30, 106–143.

Ministry of Defence (UK), 2012. Design unveiled of Royal Navy's future warship. Equipment & Logistics News, 20 August 2012, [online]. Retrieved from http://www.mod.uk/DefenceInternet/DefenceNews/Equipment & Logistics/DesignUnveiledOfRoyalNavysFutureWarships.htm (Last accessed 28.08.12.).

Mintzberg, H., 1976. An Emerging Strategy of "Direct" Research. Adm. Sci. Q. 24 (4), 582–589.

Mirsham, E., 1980. Cost/benefit analysis. Allen and Unwin, London.

Mitchell, R.K., Agle, B., Wood, D.J., 1997. Toward a theory of stakeholder identification & salience: Defining the principle of who & what really counts. The Acad. Manage. Rev. 22 (4), 853–886.

Moltz, J.C., 2011. The politics of space security, second ed. Stanford University Press, Stanford, CA.

Moore, A., Cappelli, D., Trzeciak, R., 2008. The "Big Picture" of Insider IT Sabotage Across U.S. Critical Infrastructures. (CMU/SEI-2008-TR-009). The Software Engineering Institute, Carnegie Mellon University.

Moore, A., Hanley, M., Mundle, D., 2012. A Pattern for increased Monitoring for Intellectual Property Theft by Departing Insiders. Published by CERT, Software Engineering Institute, Carnegie Mellon University.

Morgan, M., Fischhoff, B., Bostrom, A., Atman, C., 2001. Risk communication: The Mental Models Approach. Cambridge University Press, New York.

Morgan, T., 2002. Business Rules & Information Systems. Addison, Reading, PA.

Moscaritolo, A., 2010. Stuxnet should serve as wake-up call, say experts. SC Magazine News, 28th September, 2010, [online]. Retrieved from http://www.scmagazine.com/stuxnet-should-serve-as-wake-up-call-say-experts/article/179858/ (Last accessed July 2012).

Moskal, M.D., Sudit, M., Sambhoos, K., 2011. The Role of Information Fusion in Providing Analytical Rigor for Intelligence Analysis. In: Proceedings of the 14th International Conference on Information Fusion. Chicago, USA.

Munro, R., 2010. Haiti Emergency Response: The Power of Crowdsourcing & SMS. Mission 4636, Relief 2.0 in Haiti Conference, February 26th, 2010.

Munster, R., 2004. The War on Terrorism: When the Exception Becomes the Rule. Int. J. Semiot. Law 17, 141–153.

Myers, D.G., 2001. Do we fear the right things? American Psychological Society 14, (10).

Nalchigar, S., Nasserzadeh, S.M.R., Akhgar, B., 2011. Simulating strategic information systems planning process using fuzzy cognitive map. Business Information Systems 8 (3), 286–306.

Naraine, R., 2010. Shodan search exposes insecure SCADA systems, November 2010. [online]. Retrieved from http://www.zdnet.com/blog/security/shodan-search-exposes-insecure-scada-systems/7611 (Last accessed August 2012).

The 9-11 Commission. National Commission of Terrorist Attacks upon the United States, 2004. The 9-11 Commission Report. [online]. Retrieved from http://www.9-11commission.gov/report/911Report.pdf (Last accessed 30.08.12.).

National Infrastructure Advisory Council (U.S.), Noonan, T., Archuleta, E., 2008. In: (Ed.), The national infrastructure advisory council's final report & recommendations on the insider threat to critical infrastructures (Trans.). DHS/NIAC, Washington, D.C.

National Security Strategy, May 2010. Office of the President of the United States. The White House. [online]. Retrieved from http://www.whitehouse.gov/sites/default/files/rss_viewer/national_security_strategy.pdf.

National Strategic Studies: 1998 Strategic Assessment. US National Defense University. [online]. Retrieved from http://www.ndu.edu/inss/Strategic%20Assessments/sa98/sa98ch11.html (Last accessed 20.08.12.).

Neutze, J., 2012. Cybersecurity in Germany - Toward a Risk-based Approach. American Institute for Contemporary German Studies. John Hopkins University [online] retrieved from http://www.aicgs.org/publication/cybersecurity-in-germany-toward-a-risk-based-approach/ (last accessed 13.12.12.).

Newman, A.R., 2002. Confidence, pedigree, & security classification for improved data fusion, Information Fusion, 2002. In: Proceedings of the Fifth International Conference, vol. 2(2), pp. 1408–1415.

Nickols, F., 2000. Strategy: Definitions and Meaning. Retrieved from http://home.att.net/~nickols/strategy_definition.htm IDC Publication.

Nonaka, I., Takeuchi, H., 1995. The Knowledge-Creating Company: How Japanese Companies Create the Dynamics of Innovation. Oxford University Press, Oxford, England.

Nonaka, I., Toyama, R., 2003. The knowledge-creating theory revisited: knowledge creation as a synthesizing process. Knowledge Management Research & Practice 1 (1), 2–10.

Nonaka, I., von Krogh, G., Voelpel, S., 2006. Organizational knowledge creation theory: Evolutionary paths & future advances. Organization Studies 27 (8), 1179–1208.

North Yorkshire Police Authority, 2011. Police app for national award. July 1st, 2011 [online]. Retrieved from http://www.northyorkshire.police.uk/index.aspx?articleid=7028.

Nuseibeh, B., Easterbrook, S., 2000. Requirements engineering: A roadmap. In: Proceedings of International Conference of Software Engineering (ICSE-2000) 4-11 July 2000. ACM, Limerick, Ireland, pp. 37–46.

O'Rourke, R., 2012. Navy Irregular Warfare & Counterterrorism Operations: Background & Issues for Congress, CRS Report for Congress August 10, 2012, p. 3. [online]. Retrieved from http://www.fas.org/sgp/crs/natsec/RS22373.pdf (Last accessed 22.08.12.).

Office of the Inspector General, 2003, August 14. A review of the FBI's performance in deterring, detecting, & investigating the espionage activities of Robert Philip Hanssen. [online]. Retrieved from http://www.fas.org/irp/agency/doj/oig/hanssen.html.

O'Harrow Jr., R., 2012. Cyber search engine Shodan exposes industrial control systems to new risks, June 2012. [online]. Retrieved from. http://smartgridsecurity.blogspot.co.uk/2012/06/shodan-again-search-engine-you-need-to.html (Last accessed August 2012).

Oman, P., Schweitzer, E.O., Frincke, D., 2000. Concerns About Intrusions into Remotely Accessible Substation Controllers & SCADA Systems, 2000. [online]. Retrieved from http://citeseerx.ist.psu.edu/viewdoc/summary? (accessed August 2012).

Oppel, R., Bowley, G., 2012. In Afghanistan, hitting pause on local police training. New York Times. [online]. Retrieved from http://mobile.nytimes.com/2012/09/03/world/asia/in-afghanistan-hitting-pause-on-local-police-training.xml.

O'Reilly, T., 2005. Web 2.0: compact definition.[Online]. http://radar.oreilly.com/archives/2005/10/web_20_compact_definition.html (accessed 02.12.10.).

O'Reilly, T., Battelle, J., 2009. Web Squared: Web 2.0 Five Years on.[online] retrieved from http://assets.en.oreilly.com/1/event/28/web2009_websquared-whitepaper.pdf (last accessed 13.12.12.).

Ott, R., Gutiérrez, M., Thalmann, D., Vexo, F., 2006. Advanced virtual reality technologies for surveillance and security applications. In: Proceedings of the 2006 ACM international conference on Virtual reality continuum and its applications, pp. 163–170.

Page, L., Petrie, K., Wessely, S., 2006. Psychosocial responses to environmental incidents: A review & proposed typology. J. Psychosom. Res. 60, 413–422.

Palen, L., Liu, S.B., 2007. Citizen Communications in Crisis: Anticipating a Future of ICT-Supported Public Participation. In: CHI 2007 Proceedings Emergency Action, San Jose, CA, April 28t–May 3.

Palen, L., Vieweg, S., Sutton, J., Liu, S.B., Hughes, A., 2007. Crisis Informatics: Studying Crisis in a Networked World, In: Third International Conference on e-Social Science, Ann Arbor, MI, October 7–9.

Paleri, P., 2008. National Security: Imperatives And Challenges. Tata McGraw-Hill, New Delhi p. 521.

Pantucci, R., Rubin, H., 2011. A Typology of Lone Wolves: Preliminary Analysis of Lone Islamist Terrorists. ICSR, London.

Papageorgiou, E.I., Stylios, C.D., Groumpos, P.P., 2004. Active Hebbian learning algorithm to train fuzzy cognitive maps. International Journal of Approximate Reasoning 37, 219–249.

Papageorgiou, E.I., Salmeron, J., 2012. Learning Fuzzy Grey Cognitive Maps using Nonlinear Hebbian-based approach. International Journal of Approximate Reasoning 53, 54–65.

Papazoglou, M.P., Traverso, P., Dustdar, S., Leymann, F., 2007. Service oriented computing: State of the art and research challenges. IEEE Computer 40 (11), 38–45.

Parkin, S.E., Moorsel, A., Coles, R., 2009. An Information Security Ontology Incorporating Human-Behaviour Implications. ACM. In: Proceeding SIN '09 Proceedings of the 2nd international conference on Security of information and networks. pp. 46–55.

Pavlin, G., de Oude, P., Maris, M., Nunnink, J., Hood, T., 2010. A multi-agent systems approach to distributed bayesian information fusion. Information Fusion 11 (3), 267–282.

Pavlin, G., Kamermans, M., Scafes, M., 2010. Dynamic Process Integration Framework: Toward Efficient Information Processing in Complex Distributed Systems. Informatica 34 (4), 477–490.

PCI, 2006. Welcome to the PCI Security Standards Council. Retrieved December 07, 2012, from PCI Security Standards Council www.pcisecuritystandards.org.

Pearce, J.M., Rogers, M.B., 2011. CIE Toolkit WP8: Risk & crisis communication requirements following a chemical incident or emergency. Unpublished report to the European Commission.

Pearce, J., Rubin, G., Selke, P., Amlot, R., Mowbray, F., Rogers, M., 2012a. Communicating with the public following radiological terrorism. Results from a series of focus groups & national surveys in Britain & Germany. Journal of Health Communication in press.

Pearce, J.M., Rubin, G.J., Amlôt, R., Wessely, S., Rogers, M.B., 2012b. Communicating public health advice following a chemical spill: Results from national surveys in the UK and Poland. Disaster Medicine and Public Health Preparedness in press.

Pearse, D., 2011. Al-Qaida hoped to blow up oil tankers, Bin Laden documents reveal. The Guardian, 20th May 2011. [online]. Retrieved from. http://www.

guardian.co.uk/world/2011/may/20/al-qaida-oil-tankers-bin-laden (Last accessed 24.08.12.).

Pehrson, C.J., 2006. String of Pearls: Meeting the Challenge of China's Rising Power Across the Asian Littoral. Strategic Studies Institute, Carlisle, PA, July 2006. [online]. Retrieved from http://www.strategicstudiesinstitute.army.mil/pdffiles/PUB721.pdf (Last accessed 28.08.12.).

Peirce, C.S., 2012. Manuscript 514. Transcribed by Michel Balat with commentary by J. F. Sowa. [online]. Retrieved from http://www.jfsowa.com/peirce: http://www.jfsowa.com/peirce/ms514.htm (Last accessed 15.11.09.).

Penders, A., Pavlin, G., Kamermans, M., 2010. A Collaborative Approach to Construction of Complex Service Oriented Systems. In: Proceedings of the Fourth International Symposium on Intelligent & Distributed Computing. Springer, New York.

Perla, P.P., Nofi, A.A., Markowitz, M.C., 2006. Wargaming Fourth-Generation Warfare (U), CAN, September 2006. [online]. Retrieved from http://www.cna.org/research/2006/wargaming-fourth-generation-warfare (Last accessed 21.08.12.).

Petterson, J., 1988. Perception versus Reality of Radiological Impact: The Goiania Model. Nuclear News November, 84–90.

Pfitzmann, A., Hansen, M., 2005. Anonymity, unlinkability, unobservability, pseudonymity & identify management—a consolidated proposal for terminology. [online], version v0.25, Dec 6, 2005. Retrieved from. http://dud.inf.tu-dresden.de/Anon_Terminology.shtml (Last accessed 11.09.12.).

Pfleeger, C.P., 2008. Reflections on the Insider Threat, Insider Attack and Cyber Security. Advances in Information Security 39, 5–16.

Pfleeger, S.L., Stolfo, S.J., 2009. Addressing the insider threat. Security & Privacy, IEEE 7 (6), 10–13.

Piazza, P., 2006. Technofile: Anti-social networking sites. Security Management 1–5.

Pidgeon, N., Hood, C., Jones, D., Turner, B., Gibson, R., 1992. Risk perception. In: R.S. (Ed.), Risk analysis, perception & management: Report of the Royal Society Study Group. The Royal Society, London, pp. 89–134.

Pijawka, K., Mushkatel, A., 1991–1992. Public opposition to the siting of the high-level nuclear waste repository: The importance of trust. Policy Studies Review 10, 180–194.

Plaskoff, J., 2003. Intersubjectivity and community building: Learning to learn organizationally. In: Easterby-Smith, M., Lyles, M. (Eds.), The Blackwell Handbook of Organizational Learning and Knowledge Management. Blackwell Publishing, pp. 161–184.

PNLD, Staniforth, A., 2009. Blackstone's Counter Terrorism Handbook. Oxford University Press, Oxford.

Polanyi, M., 1966. The Tacit Dimension. Routledge & Kegan Paul, London.

Policy-making. 1997. University of Michigan Press, Ann Arbor, MI, pp. 55–93.

Pollitt, M.M., 2003. Cyberterrorism—Fact or Fancy? FBI Laboratory. [online]. Retrieved from http://www.cs.georgetown.edu/~denning/infosec/pollitt.html (Last accessed 01.09.12.).

Polovina, S., 1993. The Suitability of Conceptual graphs in Strategic Management Accountancy. PhD thesis. Department of Computer Studies, Loughborough University, Loughborough.

Polovina, S., 2007. An Introduction to Conceptual Graphs. Conceptual Structures: Knowledge Architectures for Smart Applications. In: Proceedings of the 15th International Conference on Conceptual Structures. Springer, Sheffield, pp. 1–14.

Polovina, S., Hill, R., 2009. A Transaction Pattern for Structuring Unstructured Corporate Information in Enterprise Applications. International Journal of Intelligent Information Technologies 5 (2), 34–48.

Poortinga, W., Pidgeon, N., 2003. Exploring the dimensionality of trust in risk regulation. Risk Anal. 23, 961–972.

Poortinga, W., 2012. Community resilience & health: The role of bonding, bridging, & linking aspects of social capital. Health Place 18, 286–295.

Porter, M.E., 1985. Competitive Advantage. The Free Press, New York.

Powel, J.H., Swart, J., 2005. This is what the fuss is about: a systemic modelling for organizational knowing. Journal of Knowledge Management 9 (2), 45–58.

Prahalad, C.K., 1975. The strategic process in a Multinational Corporation. Harvard Business School 54 (4), 54.

Preece, J., Rogers, Y., Sharp, H., 2007. Interaction design: Beyond human-computer interaction, second ed. John Wiley & Sons, Hoboken, NJ.

Preston, T., 2001. The President & His Inner Circle—Leadership Style & the Advisory Process in Foreign Affairs. Columbia University Press, New York.

Preston, T., 't Hart, P., 1999. Understanding & Evaluating Bureaucratic Politics: The Nexus Between Political Leaders & Advisory. Political Psychology 20 (1), 49–98.

Priss, U., 2005. Theoretical Perspectives—Formal Concept Analysis in Information Science. In: Cronin, B. (Ed.), Annual Review of Information Science & Technology, vol. 40. ARIST.

Priss, U., 2006. Formal Concept Analysis in Information Science. In: Cronin, B. (Ed.), Annual Review of Information Science & Technology, vol. 40, 2006, pp. 521–543.

Probst, C.W., Hunker, J., 2011. Insider & insider threats: An overview of definitions & mitigation techniques. Journal of Wireless Mobile

Networks, Ubiquitous Computing, & Dependable Applications 2, 4–27.

Proulx, G., Fahy, R.F., 2003. Evacuation of the World Trade Centre: What Went Right? In: Proceedings of the CIB-CTBUH International Conference on Tall Buildings, October 20–23, 2003. Malaysia CIB Publication No. 290, 27–34.

Prutsalis, M., et al., 2010. The Sahana Software Foundation response to the 2010 Haiti Earthquake:A New Standard for Free & Open Source Disaster Data Management Systems. In: Proceedings of the 7th International ISCRAM Conference, USA, May 2010. Psychological Science, vol. 16 (6), pp. 426–428.

Putnam, R., 1993. Making democracy work. Civic traditions in modern Italy. Princeton University Press, Princeton.

Qin, J., Zhoub, Y., Reidc, E., Lai, G., Chenc, H., 2007. Analyzing terror campaigns on the internet: Technical sophistication, content richness, & Web interactivity. International Journal of Human-Computer Studies 65, 71–84.

Quillinan, T.B., et al., 2010. Negotiation & Monitoring of Service Level Agreements. Grids & Service Oriented Architecture for Service Level Agreements, Core Grid Series. Springer, New York.

Radim Belohlavek, J.D., 2007. Fast factorization by similarity in formal concept analysis of data with fuzzy attributes. Journal of Computer & System Sciences 73, 1012–1022.

Rainie, L., Wellman, B., 2012. Networked: The New Social Operating System. MIT Press, Cambridge, MA.

Rainer, R.T., 1989. A review and Synthesis of the theory of strategy. LOM Publication.

Randazzo, M.R., et al., 2004. Insider Threat Study: Illicit Cyber Activity in the Banking and Finance Sector. Carnegie Mellon Software Engineering Institute.[online] Retrieved from http://www.cert.org/insider_threat/study.html (Last accessed 13.12.12.).

Rashman, L., Hartley, J., 2002. Leading & Learning? Knowledge transfer in the beacon council scheme. Public Administration 80 (3), 523–542.

Rasmussen, M., 2001. Reflexive security: NATO & International Risk Society. Millennium 30, 285–309.

Reed, R.H., 1975. On deterrence: A Broaden Perspective. Air University Review. May–June, pp. 2–17

Reed, T.C., 2004. At the Abyss: An Insider's History of the Cold War. Presidio Press, New York.

Regester, M., Larkin, J., 2005. Risk issues & crisis management: a casebook of best practice, third ed. Kogan Page, London.

Rein, K., Hieb, M.R., Schade, U., 2009. Battle Management Language (BML) as an Enabler. In: Proceedings of IEEE Communications. Communications, 2009. ICC '09. IEEE International Conference on Communications. Wachtberg Germany 14–18 June.

Richards, A., 2011. Countering the psychological impact of terrorism. In: Silke, A. (Ed.), The Psychology of Counter-Terrorism. Routledge, London, pp. 186–199.

Richards, J., 2010. The Art and Science of Intelligence Analysis. Oxford University Press, Oxford, UK.

Richelson, J.T., 2012. Documenting the FBI: Declassified Documents Provide New Detail on Confronting the Terrorist Threat—from al-Qaeda to Skinheads. New Archive Posting Opens Window into Broad Range of Recent FBI Operations. National Security Archive Electronic Briefing Book No. 386. Posted July 19, 2012 [online]. Retrieved from http://www.gwu.edu/~nsarchiv/NSAEBB/NSAEBB386/.

Riggs Fuller, S., Ramon, J.A., 1997. Challenging the Mind Guards. In: 't Hart, Stern, Sundelius (Eds.), In Beyond Groupthink—Political Group Dynamics and Foreign.

Riggs, J., 1987. Production Systems. John Wiley & Sons, New York.

Roberts, D., 1973. The Existential Graphs of Charles S. Peirce, Mouton & Co, Clearwater, FL.

Robertson, J., 2012. Facebook Widens 'Bug Bounty' Program to Combat Internal Breaches. Bloomberg News Jul 26, 2012, [online]. Retrieved from http://www.bloomberg.com/news/2012-07-26/facebook-widens-bug-bounty-program-to-combat-internal-breaches.html (Last accessed August 2012).

Robinson, J.A., 1972. Crisis: An Appraisal of Concepts & Theories. In: Hermann, C.F. (Ed.), International Crises: Insights from Behavioral Research. Free Press, New York.

Rogers, C., 2000. Military Revolutions & Revolutions in Military Affairs: A Historian's Perspective. In: Gongora, T., van Riekhoff, H. (Eds.), Toward a Revolution in Military Affairs. Defense & Security at the Dawn of the Twenty-first Century. Greenwood Press, Westport, CT, p. 22.

Rogers, M.B., Amlôt, R., Rubin, G.J., Wessely, S., Krieger, K., 2007. Mediating the Social & Psychological Impacts of Terrorist Attacks. The International Review of Psychiatry 19 (3), 279–288.

Rogers, M.B., Pearce, J.M., Amlôt, R., Rubin, G.J., Wessely, S., Krieger, K., 2010. Communicating with the public about CBRN Terrorism. Magazine for National Safety & Security & Crisis Management. The Directorate General for Public Safety & Security of the Netherlands Ministry of the Interior.

Rogers, M., Amlot, R., Rubin, G., Wessely, S., Krieger, K., 2007. Mediating the social & psychological impacts of terrorist attacks: The role of risk perception & risk communication. Int. Rev. Psychiatry 19, 279–288.

Rogers, R., 1975. A protection motivation theory of fear appeals & attitude change. J. Psychol. 91, 93–114.
Romm, J.J., 1993. Defining national security: the nonmilitary aspects. Pew Project on America's Task in a Changed World (Pew Project Series). Council on Foreign Relations, p. 122. A Strong Britain in an Age of Uncertainty: The National Security Strategy Presented to Parliament by the Prime Minister by Command of Her Majesty October 2010 http://www.official-documents.gov.uk/.
Rose, R., 1993. Lesson-drawing in Public Policy: A Guide to Learning Across Time & Space. Chatham House Publishers, Chatham, NJ.
Rosneau, J.N., 2003. Distant Proximities: Dynamics beyond Globalization. Princeton University Press.
Rosenhead, J., 1989. Robustness Analysis; keeping your options open, in rational analysis for a problematic world. Wiley, New York.
Rosenthal, U., 't Hart, P., 1991. Experts & Decision Makers in Crisis Situations. Knowledge: Creation, Diffusion, Utilization 12 (4), 350–372.
Rosenthal, U., Michael, T.C., 't Hart, P. (Eds.), 1989. Coping With Crisis: The management of disasters, riots & terrorism. Charles C Thomas, Springfield, IL.
Rosenthal, U., Boin, R.A., Comfort, L.K., 2001b. The Changing World of Crises & Crisis Management. pp. 5–27.
Rosenthal, U., Arjen Boin, R., Bos, C.J., 2001. Shifting Identities: The Reconstructive Mode of the Bijlmer Plane Crash. pp. 200–215.
Rosenthal, U., Arjen Boin, R., Comfort, L.K. (Eds.), 2001a. Managing Crises—Threats, Dilemmas, Opportunities. Charles C Thomas, Springfield, IL.
Rosseau, D., Sitkin, S., Burt, R., Camerer, C., 1998. Not so different after all: A cross-discipline view of trust. Acad. Manage. Rev. 23, 393–404.
Royal Canadian Mounted police, 2011. Youth Online & at Risk: Radicalisation Facilitated by the Internet. In: R. C. Program. Canadian Electronic Library (Firm), Ottawa.
Rubin, G.J., Potts, H.W.W., Michie, S., 2010. The impact of communications about swine flu (influenza A H1N1v) on public responses to the outbreak: results from 36 national telephone surveys in the UK. Health Technol. Assess. 14 (34), 183–266.
Rubin, G.J., Page, L., Morgan, O., Pinder, R.J., Riley, P., Hatch, S., 2007. Public information needs after the poisoning of Alexander Litvinenko with polonium-210 in London: Cross sectional telephone survey & qualitative analysis. BMJ 335, 1143–2246.
Rubin, G., Amlot, R., Page, L., Wessely, S., 2009. Public Perceptions, Anxiety & Behavioural Change in Relation to the Swine Flu Outbreak: A Cross-Sectional Telephone Survey. BMJ 339, 2651.
Rubin, G.J., Amlôt, R., Rogers, M.B., Hall, I., Leach, S., Simpson, J., et al., 2010. Perceptions & reactions with regard to pneumonic plague. Emerg. Infect. Dis. 16, 120–122.
Ruppert, B., 2009. Protecting Against Insider Attacks, Reading Room SANs. [online]. Retrieved from www.sans.org/reading_room/whitepapers/incident/protecting-insider-attacks_33168.
Russell, S.J., Norvig, P., 2010. Artificial Intelligence - A Modern Approach (3. internat. ed.). Pearson Education 2010, pp. I–XVIII, 1–1132.
Ryle, G., 1949. The Concept of Mind. University of Chicago Press, Chicago, IL.
Rynes, S.L., Bartunek, J.M., Daft, R., 2001. Across the great divide: Knowledge creation & transfer between practitioners & academics. Acad. Manage. J. 44 (2), 340–355.
Sandhu, R.S., Coyne, E.J., Feinstein, H., Youman, C.E., 1996. Role-Based Access Control Models. IEEE Computer 29 (2), 38–47.
Saalbach, K., 2012. Cyber war Methods & Practice Version 4.0 – 25 Mar 2012. [online]. Retrieved from http://ebookbrowse.com/saalbach-cyberwar-methods-&-practice-pdf-d47450965 (Last accessed July 2012).
Saathoff, G., DeFrancisco, G., Benedek, D., Everett, A., Holstege, C., Johnson, S., et al., 2011. The Amerithrax case: Report of the Expert Behavioral Analysis Panel. Research Strategies Network, Vienna, VA.
Saaty, T., 1980. The Analytic Hierarchy Process. McGraw Hill, NY.
SABSA, 2011. Integrated with TOGAF based on the objective of developing security architecture. Retrieved from http://www.sabsa-institute.org/the-sabsa-method/what-is-sabsa.aspx (Last accessed 11.06.12.).
Sackett, P.R., 2002. The Structure of Counterproductive Work Behaviors: Dimensionality & Relationships with Facets of Job Performance. International Journal of Selection & Assessment 10, 5–11.
Sageman, M., 2008. Leaderless Jihad. Terror Networks in the 21st Century. University of Pennsylvania Press, Philadelphia, Pennsylvania.
Sahdra, B., Thagard, P., 2003. Procedural knowledge in molecular biology. Philosophical Psychology 16 (4), 477–498.
Saikayasit, R., Stedmon, A.W., Lawson, G., Fussey, P., 2012. User requirements for security & counter-terrorism initiatives. In: Vink, P. (Ed.), Advances in Social & Organisational Factors. CRC Press, Boca Raton, FL, pp. 256–265.
Sakellari, G., 2010. The Cognitive Packet Network: A Survey. Computer Journal 53 (3), 268–279.
Salem, M.B., Hershkop, S., Stolfo, S.J., 2008. A Survey of Insider Attack Detection Research. Insider Attack & Cyber Security - Beyond the Hacker, Series: Advances in Information Security, vol. 39, 2008, XII, 223, Chapter 5, p. 69
Salgado, J.F., 2002. The Big Five Personality Dimensions & Counterproductive Behaviors. International Journal of Selection & Assessment 10, 117–125.

Salmeron, J.L., 2010. Modelling grey uncertainty with Fuzzy Grey Cognitive Maps. Expert Systems Appl. 37, 7581–7588.

Sarraf, Q., Ellis, G., 2006. Business Rules in retail: the tesco.com story. Business Rules Community. [online]. Retrieved from http://www.brcommunity.com/p-n014.php (Last accessed 12.06.12.).

SBVR, 2011. Semantics of Business Vocabulary & Business Rules. [online]. Retrieved from www.businessrulesgroup.org/sbvr.shtml (Last accessed 21.04.11.).

Schade, U., Hieb, M.R., 2006. Development of Formal Grammars to Support Coalition Command Control: A Battle Management Language for Orders, Requests, & Reports. In: Proceedings from The International Command and Control Research and Technology Symposium (ICCRTS) 11th ICCRTS - Coalition Command and Control in the Networked Era–Cambridge, UK. 26–28 September.

Scherer, R., 2009. Why did it take so long to catch spy for China? Christian Science Monitor, 18 July 2009 [online]. Retrieved from http://www.csmonitor.com/USA/Justice/2009/0718/p05s01-usju.html.

Schineier, B., 2012. Liars & Outliers: Enabling the Trust that Society Needs to Thrive. John Wiley & Sons, New York.

Schoch-Spana, M., 2000. Implications of Pandemic Influenza for Bioterrorism Response. Clin. Infect. Dis. 31 (6), 1409–1413.

Schouten, R., Saathoff, G., 2010. Biosurety in the Post-9/11 Era. In: Microbial Forensics. Elsevier, Amsterdam,pp. 221–238.

Schreft, L., 2007. Risks of Identity Theft: Can the Market Protect the Payment System? Economy Review. Fourth Quarter 2007. [online]. Retrieved from www.kc.frb.org/Publicat/econrev/PDF/4q07schreft.pdf.

Scott, D., 2006. India's 'Grand Strategy' for the Indian Ocean: Mahanian Visions. Asia-Pacific Review 13 (2), (2006), 97–129; Holmes, James/Winner, Andrew C./Yoshihara, Toshi: Indian Naval Strategy in the Twenty-first Century. London: Routledge 2009.

Semafone, 2012. Secure Call Centre Voice Transactions. Retrieved December 07, 2012, from Semafone: www.semafone.com.

Senate Select Committee on Intelligence, 1994. An assessment on Aldrich H. Ames espionage case & its implications for U.S. intelligence. [online]. Retrieved from http://www.fas.org/irp/congress/1994_rpt/ssci_ames.htm.

Senge, P.M., 1990. The Fifth Discipline: The Art & Practice of the Learning Organization. Century Business, London, England.

Shariq, S.Z., 1999. How does knowledge transform as it is transferred? Speculations on the possibility of a cognitive theory of knowledge escapes. Journal of Knowledge Management 3 (4), 243–251.

Shaw, E., Ruby, K.G., Post, J.M., 1998. The insider threat to information systems: The psychology of the dangerous insider. Security Awareness Bulletin 2, 1–10.

Sheng, Q.Z., Benatallah, B., Dumas, M., Oi-Yan Mak, E., 2002. Selfserv: A platform for rapid composition of web services in a peer-to-peer environment. In: VLDB. pp. 1051–1054.

Sheppard, B., Rubin, G., Wardman, J., Wessely, S., 2006. Terrorism & dispelling the myth of a panic prone public. J. Public Health Policy 27, 216–245.

Shiva Kumar, V., 2003. Cybercrime—Prevention and Detection. [online]. Retrieved from http://www.cidap.gov.in/documents/Cyber%20Crime.pdf.

Siegrist, M., Keller, C., Kiers, H., 2005. A new look at the psychometric paradigm of perception of hazards. Risk Analysis 25, 2110–2223.

Sinclair, M.A., 2005. Participative assessment. In: Wilson, J.R., Corlett, E.N. (Eds.), Evaluation of Human Work: A Practical Ergonomics Methodology. third ed. CRC Press, Boca Raton, FL, pp. 83–112.

Sjoberg, L., 1996. A discussion of the limitations of the psychometric & cultural theory approaches to risk perception. Radiat. Prot. Dosimetry 68, 219–225.

Slovic, P., 1987. Perception of risk. Science 236, 280–285.

Slovic, P., Finucane, M.L., Peters, E., MacGregor, D.G., 2004. Risk as analysis & risk as feelings: Some thoughts about affect, reason, risk & rationality. Risk Anal. 24 (2), 311–322.

Slovic, P., Fischhoff, B., Lichtenstein, S., 1980. Facts & fears: Understanding perceived risk. In: Schwing, R.C., Albers Jr., W.A. (Eds.), Societal risk assessment: How safe is safe enough? Plenum, New York.

Slovic, P., Fischhoff, B., Lichtenstein, S., 1981. Perceived risk: Psychological factors & social implications. Proceedings of the Royal Society London A376, 17–34.

Slovic, P., Fischhoff, B., Lichtenstein, S., 1985. Characterizing perceived risk. In: Kates, R.W., Hohenemser, C., Kasperson, J. (Eds.), Perilous progress: Managing the hazards of technology. Westview, Boulder, CO, pp. 91–125.

Slovic, P., Fischhoff, B., Lichtenstein, S., 1986. The psychometric study of risk perception. In: Covello, V.T., Menko, J., Mumpower, J. (Eds.), Risk evaluation & management. Plenum, New York.

Smets, P., Kennes, R., 1994. The transferable belief model. Artificial Intelligence 66 (2), 191–234 April 1994.

Snowden, D., 2002. Complex acts of knowing: paradox & descriptive self-awareness. Journal of Knowledge Management 6 (2), 100–111.

Solingen, E., 1993. Between markets & the state—Scientists in comparative perspective. Comparative Politics 26 (1), 31–51.

Song, H.H., et al., 2009. Scalable proximity estimation & link prediction in online social networks. In: Internet Measurement Conference 2009, 322–335.

Sood, A.K., Enbody, R., 2012. Targeted Cyber Attacks–A Superset of Advanced Persistent Threats. IEEE Security & Privacy. [online]. Retrieved from. http://doi.ieeecomputersociety.org/10.1109/MSP.2012.90 (Last accessed August 2012).

Sowa, J., Zachman, J., 1992. Extending & Formalising the Framework for Information System Architecture. IBM Systems Journal 31 (3), 590–616.

Sowa, J., 1984. Conceptual structures: Information Processing in Mind & Machine. Addison-Wesley, London.

Stacey, R.D., 1993. Strategic Management and Organisational Dynamics. Pitman Publishing, London.

Stach, W., Kurgan, L., Pedrycz, W., Reformat, M., 2005. Genetic learning of fuzzy cognitive maps. Fuzzy Sets & Systems 153, 371–401.

Stamos, A., 2010. "Aurora" Response Recommendations, February 17th, 2010 [online]. Retrieved from https://www.isecpartners.com/blog/2010/3/1/aurora-response-paper-released.html (Last accessed August 2012).

Staniforth, A., 2011. Blackstone S Counter-terrorism Handbook. Oxford University Press.

Stanton, N.A., Salmon, P.M., Walker, G.H., Baber, C., Jenkins, D.P., 2005. Human Factors Methods: A Practical Guide for Engineering & Design. Ashgate Publishing Limited, Aldershot, pp. 21–44.

Stern, E.K., Sundelius, B., 1997. Understanding Small Group Decisions in Foreign Policy: Process Diagnosis & Research Procedure. In: 't Hart, Stern, Sundelius. (Eds.), Beyond Groupthink: Political Group Dynamics and Foreign Policy-Making. pp. 123–150. University of Michigan Press.

Stern, E.K., 1997a. Crisis & Learning: A Conceptual Balance Sheet. Journal of Contingencies & Crisis Management 5 (2), 69–86.

Stern, E.K., 1997b. Probing the Plausibility of Newgroup Syndrome: Kennedy & the Bay of Pigs. pp. 153–189.

Stern, E.K., 2000. Crisis Decision-making: A Cognitive Institutional Approach. Överstyrelsen för civil beredskap, Stockholm, Sweden, Crisis Management Europe Research Program 6.

Stern, E.K., 2003. Crisis Studies & Foreign Policy Analysis: Insights, Synergies, & Challenges. Int. Stud. Rev. 5 (2), 183–191.

Stevens, T., Neumann, R., 2009. Countering Online Radicalisation A Strategy for Action. ICSR, London.

Stone, F., 2007. The "Worried Well" response to CBRN events: Analysis & solutions. USAF Counterproliferation Center Working Paper Series. Strategic & International Studies - Technology & Public Policy Program. December 2005.

Substance Abuse & Mental Health Services Administration, Center for Behavioral Health Statistics & Quality, 2010. The DAWN report: Highlights of the 2010 Drug Abuse Warning Network (DAWN) findings on drug-related emergency department visits. Rockville, MD. http://www.samhsa.gov/data/2k12/DAWN096/SR096EDHighlights2010.pdf

Sutcliffe, A., 1998. Scenario-based requirements analysis. Requirements Eng. 3, 48–65.

Sutton, J., Palen, L., Shklovski, I., 2008. Backchannels on the Front Lines: Emergent Uses of Social Media in the 2007 Southern California Wildfires. In: Proceedings of the 5th International ISCRAM Conference-Washington, DC, USA, May 2008.

Suryanarayan, V., 2004. Fishing in Choppy Waters. The Hindu, 25 February 2004.

Symantec, 2011. W32.Duqu The precursor to the next Stuxnet. Version 1.4 November 23, 2011 [online]. Retrieved from http://www.symantec.com/security_response/whitepapers.jsp (Last accessed July 2012).

't Hart, P., Boin, R.A., 2001. Between Crisis & Normalcy: The Long Shadow of Post-crisis Politics. pp. 28–46.

't Hart, P., 1997. From Analysis to Reform of Policy-making Groups. pp. 311–336.

't Hart, P., Stern, E.K., Sundelius, B., Eds. 1997. Beyond Groupthink—Political Group Dynamics & Foreign Policy-making. University of Michigan Press, Ann Arbor, MI.

Tabatabaei, F., Akhgar, B., Nasserzadeh, S.M.R., Yates, S., 2012. Simulating online consumer satisfaction using Fuzzy Cognitive Mapping. ITNG2012. IEEE.

Tanaka, Y., 1988. Psychological dimensions of risk assessment: Risk perception & risk. Progress in Nuclear Energy 32 (3/4), 243–253.

Taylor, C., 2011. Twitter Users React To Massive Quake, Tsunami In Japan. Mashable, March 11th, 2011 [online]. Retrieved from http://mashable.com/2011/03/11/japan-tsunami/.

Taylor, D.G., Strutton, D., 2009. e-marketing come of age? Modeling historical influences on post-adaption era internet consumer behaviors. Business Research.

Ten, C.W., Manimaran, G., Liu, C.C., 2010. Cybersecurity for Critical Infrastructures: Attack & Defense Modeling. IEEE Trans. Syst. Man. Cybern. 40 (4), 853–865, July 2010.

The French White Paper on Defence and National Security. 2008. Odile Jacob Publishing Corporation, New York.

Thompson, J., 1993. Strategic management, Awareness and change, second ed. Chapman and Hall, London.

Thompson, R., 2011. Radicalization & the Use of Social Media. Journal of Strategic Security 4 (4), 167–190.

Thornburgh, N., 2005. The Invasion of the Chinese Cyberspies (and the Man Who Tried to Stop Them) Aug. 29, 2005 [online]. Retrieved from http://www.time.com/time/magazine/article/0,9171,1098961,00.html (Last accessed August 2012).

Thorup, M., 2009. Enemy of Humanity: The Anti-Piracy Discourse in Present-Day Anti-Terrorism. Terrorism & Political Violence 21 (3), 401–411, July–September 2009.
Thuraisingham, B., 2005. Security standards for the semantic web. Computer Standards & Interfaces 27, 257–268.
TOGAF, 2009. The Open Group version 9. Van Haren Publishing.
Tonella, P., 2004. Formal Concept Analysis in Software Engineering. In: 26th International Conference on Software Engineering (ICSE'04). IEEE, Edinburgh, UK.
Triggle, N., 2011. Mobile phone brain cancer link rejected. BBC News, Health, 21st October 2011. Retrieved from http://www.bbc.co.uk/news/health-15387297 (Last accessed 01.09.12.).
Trustwave, 2012. Global security report. Spiderlab [online] retrieved from https://www.trustwave.com/global-security-report?utm_expid=5000628-0&utm_referrer=http%3A%2F%2Fwww.google.co.uk%2Furl%3Fsa%3Dt%26rct%3Dj%26q%3Dtrustwave%25202012%2520global%2520security%2520report%26source%3Dweb%26cd%3D1%26sqi%3D2%26ved%3D0CCwQFjAA%26url%3Dhttps%253A%252F%252Fwww.trustwave.com%252Fglobal-security-report%26ei%3DLtbJUOXoN4nM0AWu5IHgDQ%26usg%3DAFQjCNH8SVNhVVn7Pn19P2qXKMuxuhIPSw (last accessed 13.12.12.).
Tseng, Y.C., Wang, Y.C., Cheng, K.Y., 2005. An integrated mobile surveillance and wireless sensor (iMouse) system and its detection delay analysis. In: International Workshop on Modeling Analysis and Simulation of Wireless and Mobile Systems - MSWiM. pp. 178–181, 2005. http://65.54.113.26/Conference/1677/mswim-international-workshop-on-modeling-analysis-and-simulation-of-wireless-and-mobile-systems.
Tsfati, Y., Weimann, G., 2002. www.terrorism.com: Terror on the Internet. Studies in Conflict & Terrorism 25, 317–332.
Tsoukas, H., 2003. Do We Really Understand Tacit Knowledge? pp. 410–427.
Turk, R.J., 2005. Cyber Incidents Involving Control Systems. US-CERT Control Systems Security Center, Idaho Falls, Idaho 83415, INL/EXT-05-00671, October.
Tyler, T., Degoey, P., 1996. Trust in organisational authorities. In: Kraemer, R., Tyler, T. (Eds.), Trust in Organisations. Frontiers of Theory & Practice. Sage, London.
Ulph, S., Heffelfinger, C., 2004. Anti-Ship Warfare and Molotov Cocktails at the Siege of Acre 1190'. In: Sirrs, J., Mahan, A., Hefflefinger, C. (Eds.), Unmasking Terror: A global Review of Terrorists Activities. A compendium of articles from Jamestown's Terrorism Monitor, September 11 2003 - September 11 2004. The Jamestown Foundation, Washington, pp. 86–90.
United States Space Command, 1997. Vision for 2020. [online] retrieved from http://www.fas.org.spp/military/docops/usspac/visbook.pdf.
U.S Airforce, 1998. Information Operations. Airforce Doctrine Document 2–5. [online] retrieved from http://www.fas.org/man/dod-101/usaf/docs/afdd/afdd2-5.pdf (last accessed 13.12.12.).
U.S. Army War College Guide to Strategy. 2001. Cerami, J.R., Holcomb Jr., J.F. (Eds.). DIANE Publishing.
U.S. House of Representatives, 2006. Select Bipartisan Committee to Investigate the U.S. v. Bradley Manning, Charge Sheet (2010). [online]. Retrieved from http://www.scribd.com/collections/3376478/US-v-Bradley-Manning.
UKCO, 2008. The National Security - Strategy of the United Kingdom Security in an interdependent world. UK Cabinet Office, London.
Upton, D., 1994. The Management of Manufacturing Flexibility. Calif. Manage. Rev. 36 (2).
Ursano, R., Fullerton, C., Norwood, A., 2004. Bioterrorism: Psychological & Public Health Interventions. Cambridge University Press, Cambridge, UK.
U.S. Department of Defense, 1983. Trusted Computer System Criteria (Orange Book). US National Computer Security Center. (CSC-STD-001-83).
van Aart, C.J., Wielinga, B., Schreiber, G., 2004. Organizational building blocks for design of distributed intelligent system. International Journal of Human-Computer Studies 61 (5), 567–599.
van de Rlet, R., Junk, A., Gudes, E., 1997. Security in Cyberspace: A knowledge-base approach. Data Knowl. Eng. (24)69–96.
Van de ven, A.H., Johnson, P.E., 2006. Knowledge for theory & practice. Acad. Manage. Rev. 31 (4), 802–821.
Van Dijk, J., 1991. The Network Society. SAGE Houten.
Vera, D., Crossan, M., 2003. Organizational Learning & Knowledge Management: Toward an Integrative Framework. pp. 122–141.
VeriSign, 2012. iDefence Cyber Threats & Trends report 2012. Verisign iDefense Security Intelligence Services White Paper.
Verizon, 2009. 2009 Data Breach Investigations. [online]. Retrieved from www.verizonbusiness.com/resources/security/reports/rp_2009-data-breach-investigations-supplemental-report_en_xg.pdf (Last accessed 30.08.12.).
Vertzberger, Y.Y.I., 1990. The World in Their Minds—Information Processing, Cognition, & Perception in Foreign Policy Decision making. Stanford University Press, Stanford, CA.
Vonderford, M.L., 2004. Breaking new ground in WMD risk communication: The pre-event messages development project. Biosecur. Bioterror. 2 (3), 193–194.

Walt, S.M., 2005. The Relationship Between Theory & Policy in International Relations. Annual Review of Political Science 8, 23–48.

Waltz, E., 2003. Knowledge management in the intelligence enterprise. Artech House Inc, Norwood, MA.

Waltz, K.N., 1990. Nuclear myths & political realities. American Political Science Review 84 (3).

Wang, T.Y., Wu, C.H., Hsieh, C.C., 2009. A virus prevention model based on static analysis an data mining methods. In: Proceedings from IEEE 8th international conference on computer & information technology workshops. Washington, D.C., USA.

Wang, J.L., Loui, M.C., 2009. Privacy and ethical issues in location-based tracking systems. In: Proceedings of the IEEE International Symposium on Technology and Society. Phoenix, Ariz, May 18–20, 2009.

Waterman, S., Jones, A., Saathoff, G., 2008. Cyber incursions: Erosions of security & social trust? Critical Incident Analysis Group (CIAG) University of Virginia. Charlottesville, VA.

Wei, Z., Lu, L., Yanchun, Z., 2008. Using fuzzy cognitive time maps for modeling & evaluating trust dynamics in the virtual enterprises. Expert Systems Appl. 35, 1583–1592.

Weidemann, P.M., Clauberg, M., Schütz, H., 2003. Understanding amplification of complex risk issues: The risk story model applied to the EMF case. In: Pidgeon, N., Kasperson, R.E., Slovic, P. (Eds.), The social amplification of risk. Cambridge University Press, Cambridge, UK, pp. 286–301.

Weimann, G., 2004. How Modern Terrorism Use the Internet. Special Report, US Institute of Peace, 2004, [online]. Retrieved from http://www.usip.org/pubs/specialreports/sr116.pdf.

Weisaeth, L., Tonnessen, A., 2003. Responses of individuals & groups to consequences of technological disasters & radiation exposure. In: Ursano, R., Fullerton, C., Norwood, A. (Eds.), Terrorism & disaster: Individual & community mental health interventions. Cambridge University Press, Cambridge, UK, pp. 236–258.

Weiss, G.W., 1996. The Farewell Dossier: Duping the Soviets. CSI Publications, Studies in Intelligence. [online]. Retrieved from https://www.cia.gov/library/center-for-the-study-of-intelligence/csi-publications/csi-studies/studies/96unclass/farewell.htm.

WenJie, W., Yufei, Y., Archer, N., 2006. A contextual framework for combining identity theft. Security & Privacy, IEEE 4 (2), 30–38.

Wessely, S., Fischhoff, B., Krasnov, V., 2003. Guidelines for communicating the risk of chemical, biological or nuclear terrorism: How to inform the public, improve resilience & not generate panic. NATO/Russia Committee on Psychological & Social Consequences of Chemical, Biological & Nuclear Terrorism.

Wexler, M.N., 2001. The who, what & why of knowledge mapping. Journal of Knowledge Management 5 (3), 249–263.

White, B.S., Walker, M., Humphrey, M., Grimshaw, A.S., Legion, F.S., 2001. A Secure and Scalable File System Supporting Cross-Domain High-Performance Applications. In: SC2001: High Performance Networking and Computing November 10–16.

Whitney, L., 2012. Flame Virus can Hijack PCs by Spoofing Windows Updates, June 2012, [online]. Retrieved from http://news.cnet.com/8301-10805_3-57447277-75/flame-virus-can-hijack-pcs-by-spoofing-windows-update/ (Last accessed August 2012).

Whoriskey, P., Jenkins, C., 2001. Mailbox trip meets with trepidation. Washington Post, 24th October.

Wickham, P., 2000. Corporate Strategy Casebook. Prentice Hall, Upper Saddle, NJ.

Wiig, K.M., 2002. Knowledge management in public administration. Journal of Knowledge Management 6 (3), 224–239.

Wilber, D.Q., Sheridan, M.B., 2009. State Dept. Retiree, Wife Accused of Spying for Cuba for Decades. The Washington Post , June 6, 2009, [online]. retrieved from http://www.washingtonpost.com/wp-dyn/content/article/2009/06/05/AR2009060502359.html?hpid=topnews.

Wille, R., 2005. Formal Concept Analysis as Mathematical Theory of Concepts & Concept Hierarchies. In: Bernhard Ganter, G.S., Radukf Wille, J.C. (Eds.), Formal Concept Analysis foundations & applications. Springer, New York.

Wille, R.B., Stumme, G., Ganter, R., 2005. Formal Concept Analysis as Mathematical Theory of Concepts. Formal Concept Analysis: Foundations & Applications. Springer-Verlag, Berlin.

Willman, D., 2011. The mirage man: Bruce Ivins, the anthrax attacks, & America's rush to war. Bantam Books, New York.

Wilson, J.R., 1995. Ergonomics & participation. In: Wilson, J.R., Corlett, E.N. (Eds.), Evaluation of Human Work: A Practical Ergonomics Methodology, second & Revised ed. Taylor & Francis, London.

Winstock, A.R., Mitcheson, L.R., Deluca, P., 2011. Mephedrone, new kid for the chop? Addiction 106 (1), 154–161.

Wired, 2011. Hackers Targeted U.S. Government Satellites. [online]. Retrieved from http://www.wired.com/threatlevel/2011/10/hackers-attack-satellites/ (Last accessed 18.09.12.).

Wiseman, C., 1988a. Strategic Information Systems. Irwin, Ontario, Canada.

Wiseman, C., 1988b. Attack and Counterattack: The New Game in Information Technology. Planning Review 16 (5), 6–12(Sep/Oct 1988).

Wolf, N., 2012. The New Totalitarianism of Surveillance Technologies. The Guardian. 15th August. [online] retrieved from http://www.guardian.co.uk/commentisfree/uk-edition (last accessed 13.12.12.).

Wood, D.M., Greene, S.L., Dargan, P.I., 2011. Clinical pattern of toxicity associated with the novel synthetic cathinone mephedrone. Emergency Medical Journal 28, 280–282.
Woodcock, J., (unpublished 2009). Leveraging Social Media To Engage The Public In Homeland Security, Thesis for the Master of Arts in Security Studies. Pacific Lutheran University, California, September 2009. [online]. Retrieved from: http://www.dtic.mil/cgi-bin/GetTRDoc?AD=ADA509065&Location=U2&doc=GetTRDoc.pdf.
Woods, D., Matters, T., 2006. Enterprise SOA: Designing IT for Business Innovation. O'Reilly Media.
Wray, R., Jupka, P., 2004. What does the public want to know in the event of a terrorist attack using plague? Biosecur. Bioterror. 2 (3), 208–218.
Wynne, B., 1992. Misunderstood misunderstanding: Social identities & public uptake of science. Public Underst. Sci. 1, 281–304.
Xiang, Y., 2002. Probabilistic Reasoning in Multiagent Systems: A Graphical Models Approach. Cambridge University Press, Cambridge, UK.
Xirogiannis, G., Glykas, M., 2007. Intelligent modeling of e-business maturity. Expert Systems Appl. 32, 687–702.
Yan, J., 2010. Disaster Risk Assessment: Disaster Risk Mapping, Global Risk Identification Programme. In: Training Workshop on Drought Risk Assessment for the Agricultural Sector, Slovenia, September 20–24, 2010.
Yar, M., 2006. Cybercrime & society. Sage Publications, Thousand Oaks, CA.
Yunos, Z., Hafidz, S., 2011. Cyber Terrorism and Terrorist use of ICT and Cyberspace. In: SEARCCT, Ramli (Eds.), SEACCTs Selection of Articles, vol. 2. Southeast Asia Regional Centre for Counter-Terrorism (SEARCCT), Kuala Lumpur, Malaysia.
Zachman, J.A., 1987. A framework for information systems architecture. IBM Systems Journal 26 (3), 276–292.
Zachman, J.A., 2012. Zachman Framework. Retrieved December 07, 2012, from Zachman International www.zachman.com.
Zellner, C., Fornahl, D., 2002. Scientific knowledge & implications for its diffusion. Journal of Knowledge Management 6 (2), 190–198.
Zhang, Q., Man, D., Yang, W., 2009. Using HMM for Intent Recognition in Cyber Security Situation Awareness. In: Proceedings of the Second International Symposium on Knowledge Acquisition & Modeling. IEEE Computer Society, Wuhan, China, vol. 2, pp. 231–236, November 30–December 1, 2009.
Zhou, Y., Reid, E., Qin, J., Chen, H., Lai, G., 2005. U.S. Domestic Extremist Groups on the Web: Link & Content Analysis. Intelligent Systems 20 (5), 44–51.

INDEX

Note: Page number followed by *b* indicate boxes, *f* indicate figures, and *t* indicate tables.